Lecture Notes in Mechanical Engineering

Lecture Notes in Mechanical Engineering (LNME) publishes the latest developments in Mechanical Engineering - quickly, informally and with high quality. Original research reported in proceedings and post-proceedings represents the core of LNME. Volumes published in LNME embrace all aspects, subfields and new challenges of mechanical engineering. Topics in the series include:

- Engineering Design
- Machinery and Machine Elements
- Mechanical Structures and Stress Analysis
- Automotive Engineering
- Engine Technology
- Aerospace Technology and Astronautics
- Nanotechnology and Microengineering
- Control, Robotics, Mechatronics
- MEMS
- Theoretical and Applied Mechanics
- Dynamical Systems, Control
- Fluid Mechanics
- Engineering Thermodynamics, Heat and Mass Transfer
- Manufacturing
- Precision Engineering, Instrumentation, Measurement
- Materials Engineering
- Tribology and Surface Technology

To submit a proposal or request further information, please contact the Springer Editor in your country:

China: Li Shen at li.shen@springer.com
India: Dr. Akash Chakraborty at akash.chakraborty@springernature.com
Rest of Asia, Australia, New Zealand: Swati Meherishi at swati.meherishi@springer.com
All other countries: Dr. Leontina Di Cecco at Leontina.dicecco@springer.com

To submit a proposal for a monograph, please check our Springer Tracts in Mechanical Engineering at http://www.springer.com/series/11693 or contact Leontina.dicecco@springer.com

Indexed by SCOPUS. The books of the series are submitted for indexing to Web of Science.

More information about this series at http://www.springer.com/series/11236

Vidosav D. Majstorovic · Numan Durakbasa
Editors

Proceedings of the 12th International Conference on Measurement and Quality Control - Cyber Physical Issue

IMEKO TC 14 2019

Springer

Editors
Vidosav D. Majstorovic
University of Belgrade
Belgrade, Serbia

Numan Durakbasa
TU Wien
Vienna, Austria

ISSN 2195-4356 ISSN 2195-4364 (electronic)
Lecture Notes in Mechanical Engineering
ISBN 978-3-030-18176-5 ISBN 978-3-030-18177-2 (eBook)
https://doi.org/10.1007/978-3-030-18177-2

This Springer imprint is published by the registered company Springer Nature Switzerland AG
The registered company address is: Gewerbestrasse 11, 6330 Cham, Switzerland

Preface

The 12th International Symposium on Measurement and Quality Control - Cyber Physical Issue - (IMEKO TC 14 2019), was held in Mechanical Engineering Faculty, Belgrade, Serbia, from 4 to 7 June 2019. It is organized by the Faculty of Mechanical Engineering, University of Belgrade, Belgrade, and the Faculty of Mechanical and Industrial Engineering, TU Vienna, Vienna. We expect that a total of over 100 participants will attend this year's symposium—academics, practitioners and scientists from 15 countries, who will contribute 26 papers on the plenary, special, workshop and ordinary sessions. The event follows the symposium programme which includes keynote addresses (opening/closing session), breakout sessions and workshop discussions, a gala dinner and a closing session on the final day.

The previous International Symposium on Measurement and Quality Control, according IMEKO TC 14 Measurement of Geometrical Quantities umbrella of were:

Name	Year	Place	Country
TC14 LMPMI Symposium 2014	2014	Tsukuba	Japan
TC14 ISMQC 2013	2013	Cracow and Kielce	Poland
TC14 LMPMI Symposium 2011	2011	Braunschweig	Germany
TC14 ISMQC 2010	2010	Osaka	Japan
TC14 LMPMI Laser Symposium 2008	2008	Singapore	Singapore
TC14 ISMQC 2007	2007	Chennai/Madras	India
TC14 LMPMI Symposium 2005	2005	Merida	Mexico
TC14 ISMQC 2004	2004	Erlangen	Germany
TC14 LMPMI Symposium 2002	2002	Novosibirsk	Russia
TC14 ISMQC 2001	2001	Cairo	Egypt
TC14 LMPMI Symposium 1999	1999	Florianopolis/SC	Brazil
TC14 ISMQC Symposium 1999	1999	Tokyo	Japan
TC14 ISMQC 1998	1998	Vienna	Austria

(continued)

(continued)

Name	Year	Place	Country
TC14 LMPMI Symposium 1996	1996	Lyngby	Denmark
TC14 ISMQC 1995, Zaragoza	1995	Zaragoza	Spain
TC14 LMPMI Symposium 1994	1994	Heidelberg	Germany
TC14 ISMQC 1992	1992	Tampere	Finland
TC14 LMPMI Workshop 1990	1990	Balatonfüred	Hungary
TC14 ISMQC 1989, Aachen	1989	Aachen	Germany
TC14 ISMQC 1989, Beijing	1989	Beijing	China
TC14 LMPMI 1986	1986	Budapest	Hungary
TC14 ISMQC 1984	1984	Tokyo	Japan

The main goal of the symposium is to bring together experts from academia and industries, and it represents an excellent occasion for exchange of knowledge, ideas, experiences, research results and information in the field of manufacturing metrology and quality control, based on advanced approaches, like Industry 4.0 model and cyber-physical systems.

We acknowledge the outstanding contributions of the all keynote and opening/closing sessions speakers to this year's symposium.

This year's symposium will have various special, workshop and ordinary sessions.

The scope of the symposium encompasses the following topics:

- 3D measurement of GPS characteristics
- Measurement of gears and threads
- Freeform measurements
- Measurement of roughness
- Large-volume metrology
- Micro- and nano-metrology
- Laser metrology for precision measurements
- Optical measurement techniques
- Machine vision
- Industrial computed tomography
- Multi-sensor techniques
- Advanced sensors
- Intelligent measurement systems
- Calibration and testing methods for measuring and manufacturing devices
- Evaluation of measurement uncertainty
- Dimensional management in industry
- Product quality assurance methods
- Pre-, in- and post-process measurement
- Measurement validation and assurance methods
- Education and training in metrology

- Entrepreneurship in metrology
- I4.0 and metrology/quality management issue
- Internet of things and metrology
- Cyber-physical metrology
- Big data analytics

Officers of Symposium 2019 are:

Symposium Chair: Prof. Vidosav MAJSTOROVIC, University of Belgrade, Serbia.

Symposium Co-chair: Prof. Numan DURAKBASA, TU Vienna, Austria.

International Programme Committee Members: Nicholas Brown, Australia; Eric G. Thwaite, Australia; Jorge Bauer, Argentina; Bernhard Zagar, Austria; Brigitte Weiss, Austria; NerminaZaimović-Uzunović, Bosnia and Herzegovina; Wim Dewulf, Belgium; Wellington Barros, Brazil; Gustavo Daniel Donatelli, Brazil; Armando Albertazzi Gonçalvez Jr., Brazil; Mauricio Nogueira Frota, Brasil; Maurício de Campos Porath, Brazil; J.C. De Oliviera-Valente, Brazil; Vitor Nardelli, Brazil; Guoxiong Zhang, China; Boxiong Wang, China; Vedran Mudronja, Croatia; Miroslav Skopal, Czech Republic; Vit Zeleny, Czech Republic; Hans Norgaard Hansen, Denmark; Leonardo De Chiffre, Denmark; Sarwat Z.A. Zahwi, Egypt; Thomas Grégoire Mathia, France; Harald Bosse, Germany; Markus Bartscher, Germany; Frank Haertig, Germany; Eberhard Manske, Germany; Reiner Tutsch, Germany; Albert Weckenmann, Germany; Jürgen Leopold, Germany; Heinrich Schwenke, Germany; Jürgen Fleischer, Germany; Joerg Seewig, Germany; Wolfgang Osten, Germany; Gert Goch, Germany; Eberhard Manske, Germany; Gerhard Linss, Germany; Martin Molitor, Germany; Horst Bodschwinna, Germany; Klaus-Dieter Sommer, Germany; Ralf Woll, Germany; Robert Schmitt, Germany; Gerd Jäger, Germany; Tilo Pfeifer, Germany; AgotaDregelyi-Kiss, Hungary; Ferenc Alpek, Hungary; László Monostori, Hungary; BalakrishnanRamamoorthy, India; Vladimir Portman, Israel; Alessandro Balsamo, Italy; Enrico Savio, Italy; Ryoshu Furutani, Japan; Toshiyuki Takatsuji, Japan; Yasuhiro Takaya, Japan; Masaji Sawabe, Japan; Vytautas Giniotis, Lithuania; Pieter H.J. Schellekens, Netherland; José Sanchez Vizcaino, Mexico; Stanislav Adamczak, Poland; Zbigniew Humienny, Poland; Ryszard Jablonski, Poland; Wojciech Plowucha, Poland; KrystofStepien, Poland; Jerzy Sladek, Poland; Malgorzata Kujawinska, Poland; Józef Gawlik, Poland; Eugeniusz Ratajczyk, Poland; Marcel Sabin Popa, Romania; Sorin Popescu, Romania; Yuri V. Chugui, Russia; Sinisa Delcev, Serbia; Igor Budak, Serbia; Nemanja. Majstorovic, Serbia; Srdjan Zivkovic, Serbia; Karol Karovic, Slovakia; Igor Brezina, Slovakia; Ales Krsek, Slovakia; Bojan Acko, Slovenia; EugenTrapet, Spain; Fernando Torres-Leza, Spain; Wolfgang Knapp, Swiss; Claus P. Keferstein, Swiss; Kuang-Chao Fan, Taiwan; M.E. Yurci, Turkey; Alistair Forbes, UK; K.T.V. Grattan, UK; Liam Blunt, UK; Richard Leach, UK; Derek G. Chetwynd, UK; Jay Raja, USA.

Chair of the Organizing Committee: Dr.-Ing. Slavenko M. Stojadinovic, Assistant Professor, Faculty of Mechanical Engineering, Department of Production Engineering, University of Belgrade, Belgrade, Serbia.

IMEKO TC 14 Symposium 2019 can be regarded as very successful due to its special dimensions: (i) it presented new research directions in the field of advanced manufacturing metrology and quality management, cyber-physical metrology and Industry 4.0, and (ii) it offered practical applications and solutions for various problems in the manufacturing metrology.

IMEKO TC 14 Symposium planning, preparation and realization required the engagement of a large number of personnel and a number of organizations. We express our gratitude to all of them, especially to:

- IMEKO TC 14 Officers: Prof. Dr. Yasuhiro Takaya, Chairperson; Prof. Dr. Rainer Tutsch, Vice Chairperson; and Prof. Dr. Tilo Pfeifer, Honorary Chairperson.
- All authors, and especially the authors that prepared keynote papers, thus contributing to the high scientific and professional level of the symposium
- All members of International Programme Committee for the review of the papers and chairing the symposium sessions
- Springer and Mr. Pierpaolo Riva for publishing symposium proceedings within the edition Lecture Notes in Mechanical Engineering
- Ministry of Education, Science and Technological Development of the Republic of Serbia for the support in the symposium.

We wish to express my special gratitude to Miss Julija Kostić as a member of Organizing Committee for the invested efforts in preparing of symposium proceedings in the best way.

March 2019

Vidosav D. Majstorovic
Numan Durakbasa

Contents

Advanced Manufacturing Metrology in Context of Industry 4.0 Model

Vidosav D. Majstorovic[1(✉)], Numan Durakbasa[2], Yasuhiro Takaya[3], and Slavenko Stojadinovic[1]

[1] Faculty of Mechanical Engineering, University of Belgrade, Belgrade, Serbia
vidosav.majstorovic@sbb.rs
[2] TU Viena, Viena, Austria
[3] Mechanical Engineering Faculty, University of Osaka, Osaka, Japan

Abstract. The Cyber-Physical Manufacturing Metrology Systems (CPM^{2}Ss) are based on integration of the Cyber Physical Systems (CPSs) and connection between Internet of Things (IoTs) and Cloud technology (CT). These are high-level methodologies for development of new generation manufacturing metrology systems, which are more intelligent, flexible and self-adaptable. CPM^{2}Ss generates Big Data, horizontally by integration (network of machines/CMMs, processes and sensors) and vertically by control (usually defined over five levels) which should be analytically processed and managed by the Cyber-Physical Manufacturing Metrology (CPM^2). In this paper was given, a detailed analysis of the current framework of development the Industry 4.0 model, with a special place and role of CPM^2 in current model. A brief overview of the concept CPM^2 research, particularly in Serbia is given as well.

Keywords: Industry 4.0 · Metrology · Manufacturing

1 Introduction

Today rapidly changing industrialized world, where globalization, product customization and automation are playing an imposing role in the development of the manufacturing industry, where is on the top of the Industry 4.0. This model connecting with it advanced technologies and techniques that will change the products, processes and supply chains involved in every aspect of industry.

This technology ushers in even greater connectivity that will allow manufacturers to maintain their competitive edge in a rapidly changing world, and respond flexibly and quickly to customers' requirements [1]. Manufacturing metrology is integral part of advanced manufacturing system based on Industry 4.0.

By PTB's approach [2] in digital era, general speaking, following new key points have been determined: (a) *the digital transformation of metrological services*. The digital upgrading of the quality infrastructure by developing reference architectures, validated statistical procedures for predictive maintenance, an infrastructure for digital calibration certificates and, last but not least, by setting up a "metrology cloud" for the harmonization and development of conformity assessment and market surveillance; (b) *the analysis of large quantities of data*. The developing metrological analytical

V. D. Majstorovic and N. Durakbasa (Eds.): IMEKOTC14 2019, LNME, pp. 1–11, 2019.
https://doi.org/10.1007/978-3-030-18177-2_1

methods for large quantities of data and in assessing machine learning methods for big data; (c) *the communication systems for digitalization.* This focus concerns the securing and metrological validation of reliable, secured and efficient communication in complex scenarios; (d) *simulations and virtual measuring instruments.* By developing analytical methods and procedures for interconnected and virtualized measuring systems, the simulation of complex measuring systems for the planning and analysis of experiments, procedures and measurement standards for automated process control and virtual measurement processes for the automatic assessment of measured data is actively supported.

Industry 4.0 in manufacturing sector, in context metrology, there are three areas where it will support [3]: (a) smart supply chains (*cyber-physical metrological traceability*) – greater coordination and real time flow of information across supply chains and relationships allows better tracking of assets and inventory and integrated business planning and manufacturing. This unlocks new ownership and collaboration models across supply chains; (b) smart manufacturing (*cyber-physical metrology in manufacturing*) – the use of data analytics and new manufacturing techniques and technologies (such as autonomous robots, multi-purpose manufacturing lines and augmented reality) helps to improve yield and speed up manufacturing. This allows new business models to be pursued such as mass customization, and (c) smart products (*intelligent metrology*) – rapid innovation and a faster time to market is enabled by data collected from products along with user feedback, whether direct or collected via social sentiment on the internet. This data also allows remote diagnostics, predictive maintenance and also *metrological characteristics.*

2 Industry 4.0 Framework and Basic Pillars from Metrology Approach

Original definition of Industry 4.0 is: "*Industry 4.0 is a German-government-sponsored vision for advanced manufacturing. The underlying concept of Industry 4.0 is to connect embedded systems and smart manufacturing facilities to generate a digital convergence between industry, business and internal functions and processes.*

Industry 4.0 refers to a fourth industrial revolution (following water/steam power, mass manufacturing and automation through IT and robotics) and introduces the concept of "cyber-physical systems" to differentiate this new evolutionary phase from the electronic automation that has gone before" [4].

This definition contains several key words, and the most important is – advanced manufacturing including advanced metrology. This means that advanced manufacturing/metrology is the basis for the fourth industrial revolution, with industrial manufacturing being integrated into digital technologies on the Internet. Industry 4.0 is the original German term. In the same context, the following terms are used worldwide: a smart factory (metrology), and intelligent manufacturing (metrology).

Based on references [5–10] we can concluded that main trends in manufacturing metrology are conceded with same trends in manufacturing and Industry 4.0 of course, Fig. 1.

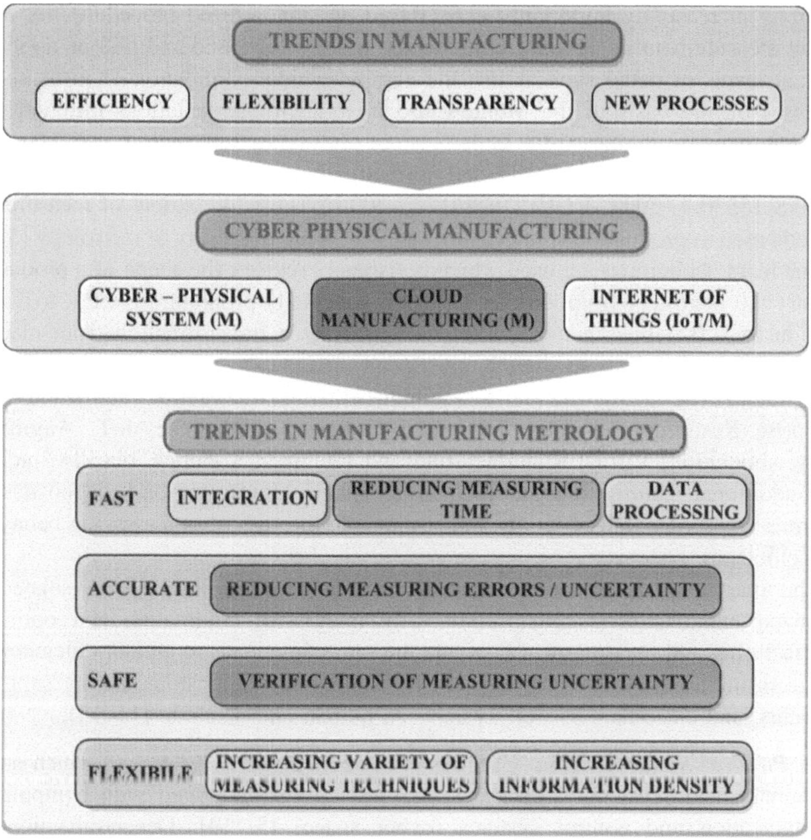

Fig. 1. Manufacturing trends and their impact on challenges and trends in manufacturing metrology (adopted according [5]).

The challenges and trends in manufacturing metrology can be described with the key words: faster, safer, more accurately and more flexibly [5].

Fast or speed means the development and application of metrological procedures by which information about product quality can be obtained in a shorter time. In context Industry 4.0 this characteristic means used of cloud technology for automated transmission of data, first, and second - direct installation of coordinate measuring machine (CMM) in production as a cyber-physical system. Also we have integration into the automatic material and information flow based on using robot, again as CPS in this context. The accuracy of measurement technology is also increasing in conjunction with stricter quality requirements [5]. From the point of industrial application today we have the next rank of metrological accuracy: transmissions - electron microscope (atomic diameter), electron microscopy (one nanometer), surface measuring (between 0,01 and 0,1 μm), and (CMM between 0,1 and 100 μm). Based on Industry 4.0 approach in this characteristic is support as CPS and Internet of Things (IoT). Measurement uncertainty and taking it into consideration in the conformity assessment are becoming increasingly important today. Based on standardized procedures for determining measurement uncertainty will become more established and will be applied at different levels of detail such as [5]: the computer-aided simulation of measurement processes on the basis of the Monte-Carlo method, implementations different measurement methods, in most cases in the form of prototypes and used in particular in the calibration of individual standards, and normative publications are now also available (ISO/TS 15530-4, 2008; VDI/VDE 2617-7, 2006). The wide variety of measurement methods used in production is increasing and with it the flexibility of metrology [5]. On the one hand, techniques are used which holistically register the shape of a product. In Industry 4.0 flexibility from various aspects is especially pronounced (CPS, IoT, etc.).

Industry 4.0 defines a methodology to generate a transformation from machine dominant manufacturing to digital manufacturing [11], by Cyber Physical Systems (CPS), cloud system, Big data and data mining, Machine to Machine (M2M) interfaces, Enterprise Resource Planning (ERP) and Business intelligence, IoT, Augmented reality, simulation, Virtual Manufacturing and intelligent robotics, but also includes some additional features such as; facilitating system monitoring and diagnostics, the system is environmentally friendly and sustainable through resource saving behaviors, more efficiency systems.

The nine pillars and forty two elements of Industry 4.0 will transform isolated and optimized manufacturing cells into a fully integrated, automated, and optimized manufacturing and measurement flow and the same time leads to greater efficiency and change in traditional manufacturing and measurement relationships among suppliers, producers, and customers as well as between human and machine [11–14].

Cyber Physical Systems (CPS). CPS has been defined as the systems in which natural and human made systems (physical space) are tightly integrated with computation, communication and control systems (cyber space) [5, 14]. Decentralization and autonomous behaviour of the measurement process are key characteristics of CPS. The continuous interchanging of data is carried out by linking CPSs intelligently with the help of cloud systems in real time, and digital shadow (digital twins) of measurement is defined as the representation of physical object in virtual world [11, 14]. For control, the 5C structure uses cloud computing to communicate with the machines (machine with machine or human with machine) [1, 14]. In Industry 4.0 model we have the increased connectivity and use of standard communications protocols, which the need to protect critical elements of industrial and manufacturing systems from cyber security threats increases dramatically [14]. A typical representative of CPS in manufacturing metrology today is CMM.

The Industrial Internet of Things (IIoT). The Internet of Things (IoT) is an advance paradigm that is rapidly gaining ground in the scenario of modern wireless telecommunications. The basic idea of this concept is the pervasive presence around us of a variety of things or objects – such as Radio – Frequency IDentification (RFID) tags, sensors, actuators, mobile phones, etc. – which, through unique addressing schemes, are able to interact with each other and cooperate with their neighbors to reach common goals [10]. By IoT a worldwide network of interconnected and uniform addressed objects that communicate via standard protocols [14]. In Industry 4.0 model usually we used term – IIoT, of course for cyber-physical manufacturing metrology model (CPM^3

[15]) for: internetworking of all cyber-physical systems, collect and exchange data in real time and horizontal and also vertical integration. Digitalization means that we have all elements of manufacturing metrology a digital representation.

The Cloud Computing. It is a general term that refers to delivering computational services through visualized and scalable resources over the Internet [16]. Based on recommendations from the National Institute of Standards and Technology (NIST), an ideal cloud should have five characteristics: on-demand self-service, broad network access, resource pooling, rapid elasticity, and measured service. This cloud model is composed of four deployment models—public, private, community, and hybrid—and three delivery models—“software as a service,” “platform as a service,” and “infrastructure as a service” [17]. In Industry 4.0 model [14], organization needs increased data sharing across the companies and supply chains, achieving the reaction times in milliseconds or even faster regarding measurement activities.

Big Data and Analytics (BDA). Big data typically stems from various channels, including sensors, devices, video/audio, networks, log files, transactional applications, the web, and social media feeds [18]. The collection and comprehensive evaluation of data from many different sources manufacturing equipment and systems as well as enterprise and customer-management systems will become standard to support real-time decision making [14]. Therefore, for organizations and manufacturers with an abundance of operational and shop-floor data including measurement data, advanced analytics techniques are critical for uncovering hidden patterns, unknown correlations, market trends, customer preferences, and other useful business information [19].

System Integration: Horizontal and Vertical System Integration. Integration and self-optimization are the two major mechanisms used in industrial organization by Industry 4.0 model [1].

The paradigm of Industry 4.0 (including metrology activities) is essentially outlined by three dimensions of integration: (a) horizontal integration across the entire value creation network, (b) vertical integration and networked manufacturing systems (c) end-to-end engineering across the entire product life cycle [7]. The full digital integration and automation of manufacturing/metrology processes in the vertical and horizontal dimension implies as well an automation of communication and cooperation especially along standardized processes [8]. On shop floor level, measurement activities, by Industry 4.0 are: monitor quality problem of thousands parts, optimize critical processes and sequences, easy to find bottlenecks, rapid and fast visual analysis for immediate action and data accessible at all times.

Simulation. In this case used more extensively in plant operations to leverage real-time data to mirror the physical world in a virtual model, which can include machines/CMM, products, and humans, thereby driving down machine setup times and increasing quality [3]. 3D simulations can be created for virtual commissioning and for simulation of cycle times, energy consumption or ergonomic aspects of a manufacturing facility [14]. Virtual CMM analysis based on simulation of measuring path optimization and collision avoidance.

Autonomous Robots. In Industry 4.0 concept robots are becoming more intelligent, autonomous, flexible, and cooperative, interact with one another and work safely side by side with humans and learn from them [4]. An autonomous robot is used to perform autonomous manufacturing/metrology method more precisely and also work in the places where human workers are restricted to work [14]. Collaborative Robots is a big trend in Smart factory and Industry 4.0 model, including measuring activities.

Additive Manufacturing. With Industry 4.0, additive-manufacturing methods will be widely used to produce small batches of customized products that offer construction advantages, such as complex, lightweight designs [14]. High-performance, decentralized additive manufacturing systems will reduce transport distances and stock on hand [4]. The manufacturing should be faster and cheaper with the use of additive manufacturing technologies like fused deposition method (FDM), selective laser melting (SLM), and selective laser sintering (SLS) [20].

Augmented Reality. Augmented-reality-based systems support a variety of services, such as selecting parts in a warehouse and sending repair instructions over mobile devices [14]. Industry can use of augmented reality to provide workers with real - time information to improve decision making and work procedures by augmented reality glass. Workers may receive repair instructions on how to replace a particular part as they are looking at the actual system needing repair [4].

As you see, the discovery of new technologies has made industry development from the early adoption of mechanical systems, to today's highly automated manufacturing/measuring/assembly lines, in order to be responsive and adaptive to current dynamic market requirements and demands [14]. Challenges like embedment, predictability, flexibility and robustness to unexpected conditions for Industry 4.0 in practice [1].

3 Cyber-Physical Manufacturing Metrology Model (CPM^3) – Some Research Results [19, 21–23]

The framework of CPM^3 model is presented in Fig. 2, and includes two basic components: physical and virtual. The overall model consists of the following sub - modules: (a) module for definition and recognition of geometrical features (GF) from CAD/GD&T model of the measurement part, where we used them for definition of metrological feature (MF) (b) module for building of intelligent inspection process planning (IIPP), that contains methods for prismatic parts presented, and method for freeform surfaces application, (c) CMM – generation of control data list for CMM that is transferred to CMM using cloud technology, and (d) module for analysis of results and generation of the final measuring reports. Cloud services within the organization provide the necessary information for integration of knowledge and data from various phases in product design and manufacturing/metrology into inspection planning, and make available information about inspection results to all interested parties in product lifecycle.

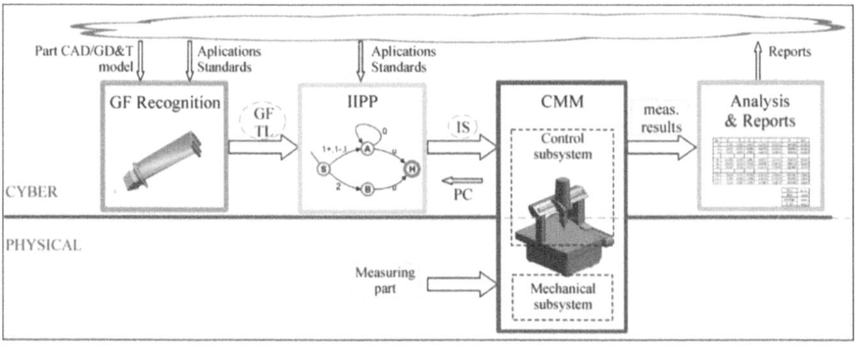

Fig. 2. Cyber-physical manufacturing metrology model (CPM^3) framework [23]

GF recognition module recognizes geometrical features from 3D model of measurement part in a neutral CAD format, such as STEP or IGES. GFs of interest depend on the type of measurement part and the applied standards. If GF recognition module does not have the application for recognition of the geometrical features for the considered measuring part in its database, the application for GF recognition needs to be provided along with the part model. The role of IIPP module is to generate inspection sequence (IS) for probe configurations (PC) that CMM supports. It has geometrical features and tolerances (TL) at input. During inspection sequence generation, it is necessary to extract metrological features (MFs) from geometrical features. The metrological, as well as the geometrical features, depend on the type of measuring part and the applied standards. If necessary, the application for metrological features extraction should be provided to IIPP module.

Currently available commercial software for inspection planning on CMM, do not have open interface for modifications and upgrades made by user.

This in particular refers to the impossibility of the inspection time reduction by measuring path optimization and generating the collision-free path. On the other hand, the relationship between geometry (feature) and tolerances (metrological feature) that is necessary for inspection planning process does not exist in a form of CAD output record. The only alternative is STEP-NC standard but software developers still have not managed to implement all forms of tolerance. This approach implies the open interface for developing CPM^3.

This section presents a feature-based model for probe path planning for sculptured surfaces using an example of turbine blades. The probe path planning model represents a part of CPM3 for free form surfaces. In this model the geometrical information for feature description is taken from CAD model of the part in IGES format. Figure 3 presents the complete inspection plan for turbine blades/free forms surfaces. The plan consists of input CAD data, recognition of metrological and geometrical primitives, definition of inspection sequences, distribution of measuring points, collision avoidance principle, and measurement and analysis of the results.

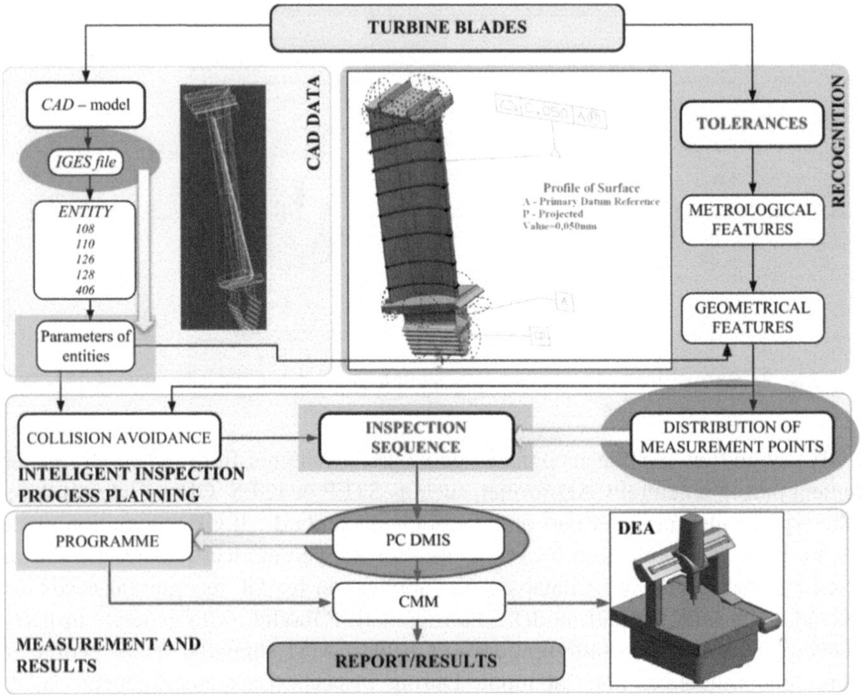

Fig. 3. Inspection planning of turbine blades on CMM [19]

The extraction of geometrical feature parameters from IGES file is based on the recognition of its structure. An IGES file is composed of the following five sections: start section, global section, directory entry section, parameter data section, and terminate section.

All geometric entities are given in the directory entry section and parameter data section.

The extraction of parameters is carried out using IGES type numbers that correspond to specific geometric entities (geometrical features). Metrological features are recognized using the tolerance of turbine blades (profile of surface) and the geometrical features parameters. They (MF) define the link between tolerance and geometrical features.

Realization of the CPM3 model in practice is being investigated for the DEA CMM, which is based on the IoT Cloud technology, Fig. 4. Connection of the physical and the virtual world is provided using an IoT device - Raspberry Pi3, which, via industrial router, has access to the Internet and realizes direct communication with the CMM control unit, using Ethernet or Wi-Fi connection.

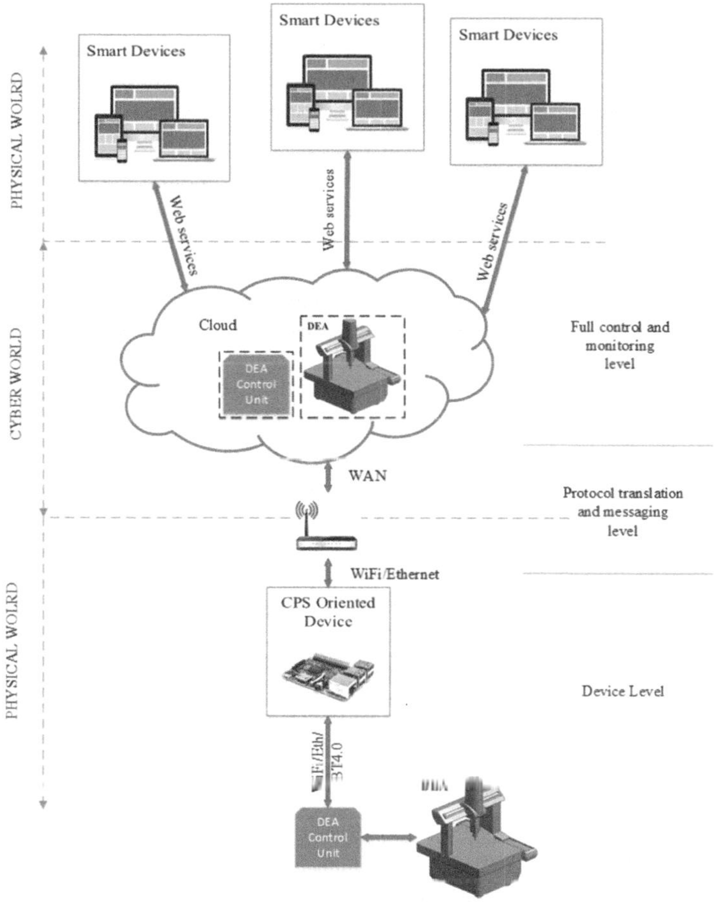

Fig. 4. Configuration of the CPM3 model for DEA CMM [23]

In this concept, CMM DEA is accessed remotely, from anywhere in the world via an Internet connection. Besides the division of physical and virtual part, the solution contains the following parts: (a) unit which provides management and monitoring of the entire system, (b) unit which defines protocols for communication Cloud - CMM and vice versa, and (c) unit which represents a physical implementation of the model, in other words, IoT device and CMM and their communication.

The starting point for the implementation of the CPM^3 model is approaching the CU (Control Unit) architecture and source code for which the drivers generate CMM execution codes, using the high level programming skills and knowledge of the IoT Cloud technology. In the next step, system creates virtual clones of CU and CMM DEA in the Cloud. Virtual twins behave as physical devices and CMM programming can be done in the cloud, without using the physical CMM.

Using the smart devices (cell phones, tablets or notebook) user accesses the Cloud by web-browser, with a unique username and password. In this research we have opted to employ Raspberry Pi3 since it is open to easy programming of communication protocols. Using our CPM^3 method new control codes are generated and directly loaded on the CMM. For GD&T definition PMI (Product Manufacturing Information) can be utilized since it contains additional information about free-form surfaces. Cloud keeps the previously used control codes (GD&T for the previously measured parts, reports) in databases, and user can access and download it, at any time, to own device. Also, it is possible to monitor the work of CMM in real time. This is an important option, in the case of an error, since the user can stop the CMM remotely. After each measurement, report is generated and automatically sent to the Cloud where it is stored. The system sends the report to a user or a predefined group of users, via email or SMS service: start/end of measurement, various alarms, environment sensor readings and other important information.

4 Conclusions

Cyber-physical manufacturing metrology play very important role in Industry 4.0 model for manufacturing. Already today, manufacturers of metrology equipment, in particular CMM, produce them as CPS, which can be used in Industry 4.0 model. On the other hand, researchers have already presented their results from the field research of CPM2. Some of them have already been presented in this paper. In the coming period, rapid diffusion of these models (Industry 4.0 and CPM2) into industrial applications should be expected, as soon as possible.

References

1. 4th Industrial Revolution: A Primer for Manufacturers, Oracle (2018). https://cloud.oracle.com/opc/saas/indmftg/reports/the-fourth-industrial-revolution-report.pdf. Accessed Feb 2019
2. Metrology for the Digitalization of the Economy and Society, Report of the German Council of Science and Humanities (2017). https://www.bipm.org/cc/PARTNERS/Allowed/2017_October/2017-Metrology-for-the-Digitalisation-of-Economy-and-Society.pdf. Accessed Feb 2019
3. Täuscher, K.: Business Models in the Digital Economy: An Empirical Study of Digital Marketplaces, Fraunhofer MOEZ. Fraunhofer Center for International Management and Knowledge Economy. Städtisches Kaufhaus Leipzig, Neumarkt, 9–19, 04109, Leipzig (2018). https://www.imw.fraunhofer.de/content/dam/moez/de/documents/Working_Paper/Working_Paper_Digital_Marketplaces_final.pdf. Accessed Feb 2019
4. Recommendations for implementing the strategic initiative INDUSTRIE 4.0: Securing the future of German manufacturing industry Final report of the Industrie 4.0 Working Group, April 2013. https://www.din.de/blob/76902/e8cac883f42bf28536e7e8165993f1fd/recommendations-for-implementing-industry-4-0-data.pdf. Accessed Feb 2019
5. Berthold, J., Imkamp, D.: Looking at the future of manufacturing metrology: roadmap document of the German VDI/VDE society for measurement and automatic control. J. Sens. Sens. Syst. **2**, 1–7 (2013). https://doi.org/10.5194/jsss-2-1-2013

6. Weinisch, J.: Trends in Manufacturing Metrology Fast, Accurate, Safe and Flexible, Carl Zeiss Industrial Metrology GmbH (2017). http://file.vogel.com.cn/2014/1111/1412591564.pdf. Accessed Feb 2019
7. Hossain, K., et al.: Vision for Metrology in the 2020s, National Physical Laboratory (2013). http://www.npl.co.uk/upload/pdf/metrology-2020vision.pdf. Accessed Feb 2019
8. Nee, J.: 3D Automated Inspection in Industry 4.0. International Industrial Automation Technology Forum Automotive Industry, Hong Kong (2017). https://bud.hkatc.org.hk/upload/files/1/file/58d1f317d06da.pdf. Accessed Feb 2019
9. Schmitt, R., Fürtjes, T.: Real-Time Machine-Vision-Systems for an Automated Quality Monitoring in Context of Industry 4.0. Laboratory for Machine Tools and Production Engineering (WZL), RWTH Aachen University (2015). https://businessdocbox.com/Biotech_and_Biomedical/66581811-Real-time-machine-vision-systems-for-an-automated-quality-monitoring-in-context-of-industry-4-0.html. Accessed Feb 2019
10. Imkamp, D., et al.: Challenges and trends in manufacturing measurement technology – the "Industrie 4.0" concept. J. Sens. Sens. Syst. **5**, 325–335 (2016). https://doi.org/10.5194/jsss-5-325-2016
11. Oztemell, E., Gursev, S.: Literature review of Industry 4.0 and related technologies. J. Intell. Manuf. **29**, 1–56 (2018). https://doi.org/10.1007/s10845-018-1433-8
12. Zezulka, F., et al.: Industry 4.0 – an introduction in the phenomenon. IFAC-Pap. OnLine **49** (25), 008–012 (2016). https://doi.org/10.1016/j.ifacol.2016.12.002
13. Majstorovic, V., et al.: Cyber-physical manufacturing systems – manufacturing metrology aspects. Proc. Manuf. Syst. **10**(1), 9–14 (2015). ISSN 2067-9238
14. Saurabh, V., Prashant, A., Santosh, B.: Industry 4.0 – a glimpse. Proc. Manuf. **20**, 233–238 (2018)
15. Xu, X.: From cloud computing to cloud manufacturing. Rob. Comput. Integr. Manuf. **28**(1), 75–86 (2012). https://doi.org/10.1016/j.rcim.2012.07.002
16. Mell, P., Grance, T.: The NIST definition of cloud computing. Gaithersburg: National Institute of Standards and Technology. Special Publication 800-145 (2011)
17. Wamba, F., et al.: How "big data" can make big impact: findings from a systematic review and a longitudinal case study. Int. J. Prod. Econ. **165**, 234–246 (2015). https://doi.org/10.1016/j.ijpe.2016.08.018
18. Recommendations for implementing the strategic initiative INDUSTRIE 4.0, Securing the future of German manufacturing industry Final report of the Industrie 4.0 Working Group, April 2013. https://www.din.de/blob/76902/e8cac883f42bf28536e7e8165993f1fd/recommendations-for-implementing-industry-4-0-data.pdf. Accessed Feb 2019
19. Majstorovic, V., et al.: Cyber-physical manufacturing metrology model (CPM^3) – big data analytics issue. Proc. CIRP **72**, 503–508 (2018). https://doi.org/10.1016/j.procir.2018.03
20. Landherr, M., et al.: The application centre industrie industry-driven manufacturing, research and development. Proc. CIRP **57**, 26–31 (2016). https://doi.org/10.1016/j.procir.2016.11.006. 49th CIRP Conference on Manufacturing Systems (CIRP-CMS 2016)
21. Durakbasa, N., et al.: Advanced metrology and intelligent quality automation for industry 4.0-based precision manufacturing systems. Solid State Phenom. **261**, 432–439 (2017). https://doi.org/10.4028/www.scientific.net/SSP.261.432. ISSN: 1662-9779. Accessed: 22 May 2017
22. Majstorović, V.D., et al.: Cyber-physical manufacturing in context of industry 4.0 program. In: Proceedings of 3rd International Conference on the Industry 4.0 Model for Advanced Manufacturing, pp. 227–238. Springer, Heidelberg (2018). https://doi.org/10.1007/978-3-3-319-89563-5
23. Majstorović, V., et al.: Cyber-physical manufacturing metrology model (CPM^3) for sculptured surfaces – turbine blade application. Proc. CIRP **63**, 658–663 (2017). https://doi.org/10.1016/j.procir.2017.03.093

System Development for Microsphere Measurement Based on Whispering Gallery Mode Resonance

Masaki Michihata[1(✉)], Yumeki Kobayashi[2], Bohuai Chu[1], Kiyoshi Takamasu[2], and Satoru Takahashi[1]

[1] Research Center for Advanced Science and Technology, The University of Tokyo, 4-6-1 Komaba, Meguro, Tokyo 153-8904, Japan
michihata@nanolab.t.u-tokyo.ac.jp

[2] Department of Precision Engineering, The University of Tokyo, 7-3-1 Hongo, Bunkyo, Tokyo 113-8656, Japan

Abstract. A sphere is frequently used as a reference for calibration of 3-dimensional measurement instruments because of its isotropic shape. Assuring an accuracy of the reference sphere is responsible for measurement uncertainty. Therefore, sphericity and diameter of the reference sphere need to be guaranteed with high accuracy. For micro-scale 3-dimentional metrology, size of the reference sphere is also micro-scale from several millimeters to a few tens of micrometers. These spheres have to be measured with accuracy of better than 10 nm. We have proposed the new measurement method of a microsphere based on whispering gallery mode resonances. One of the key points to the proposed method is to measure the whispering gallery mode resonant wavelength accurately. In this paper, the measurement instruments of whispering gallery mode resonances in the microsphere was developed.

Keywords: Diameter measurement · Micro-sphere · Whispering gallery mode · Resonances · Tapered optical fiber · Evanescent wave

1 Introduction

A sphere is often used for an artifact of three-dimensional metrology such as calibration artifact [1], ball-bar, reference for laser tracker [2, 3], probe tip of coordinate measuring machine [4]. For these purposes, typically two parameters are important for the reference sphere; sphericity (form) and diameter (size). Form evaluation was proposed by many researchers using light interference [5–7]. Size measurement was also studied using mechanical and optical ways [8, 9].

Recently, as demands for small feature products, the elements and structures are miniaturized. As a result, it is developed that the novel measuring instruments adapted to measure products smaller than 10 mm. Assuring a measurement uncertainty of these novel measuring instruments, the small-scaled microsphere with guaranteed size and shape is necessarily required. Although several methods are developed to evaluate the microsphere, measuring the microsphere is still challenging issue [10–13]. We have

V. D. Majstorovic and N. Durakbasa (Eds.): IMEKOTC14 2019, LNME, pp. 12–24, 2019.
https://doi.org/10.1007/978-3-030-18177-2_2

proposed the new method using whispering gallery mode (WGM) resonances [14]. In this method, the microsphere is treated as a spherical resonator similar to ring resonator. Because the resonant wavelength is highly sensitive to the morphology of the resonator, the information on the sphere geometry can be estimated from the multiple resonant wavelengths. In order to evaluate the microsphere in an accuracy of better than 10 nm, the accuracy to measure the resonant wavelengths have to be of the order of a few picometer. In this paper, we discuss the development of the measuring instruments that excite the WGM resonance stably and measure the resonant wavelengths with high accuracy.

2 Measurement Principle of Microsphere

2.1 WGM Based Diameter Measurement Method

Whispering gallery mode resonance. In WGM, the light propagates along with a equatorial line of a sphere as shown in Fig. 1(a). The light resonates if the phase of light oscillation is matched after one round of propagation. As like other resonators, WGM also has modes of resonances. Angular mode is a kind of longitudinal mode as shown in Fig. 1(a). Azimuthal and radial mode are recognized as transverse mode. As shown in Fig. 1(b), azimuthal mode has an extend of light field in polar direction, and radial mode has in radial direction. It is noted that the resonant wavelengths are same for different azimuthal mode number in the perfectly spherical shape, implying that the distortion of the microsphere shape can be evaluated using the resonant wavelengths of different azimuthal mode numbers.

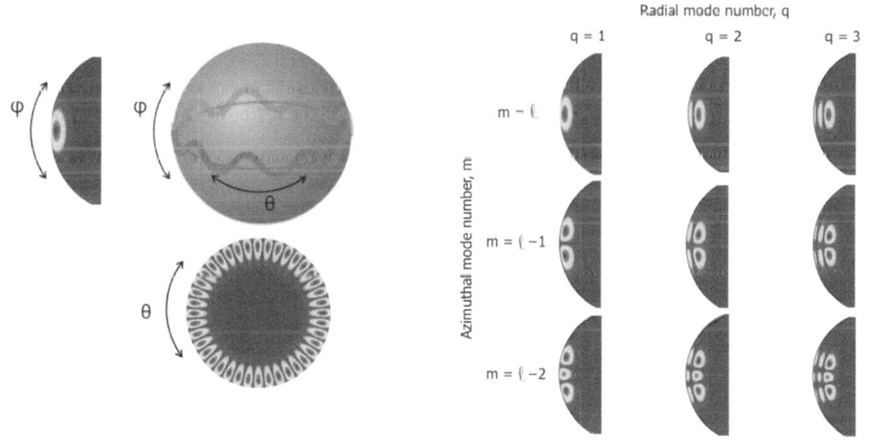

(a) Schematics. (b) Transverse mode.

Fig. 1. Whispering gallery mode.

Relation of WGM resonant wavelength and microsphere diameter. Conceptually, the optical path length of the WGM is nearly equivalent to the circumference of the sphere, which is measured by multiplying the resonant wavelength with the angular mode numbers. Therefore, the diameter can be derived from the circumference of the microsphere under the assumption that the measured target is highly spherical. Exact relation between the resonant wavelength with the corresponding mode number and the diameter of the microsphere is theoretically obtained such as follows.

$$\alpha\left(1+\frac{\rho_{in}}{j_\ell(\rho_{in})}\frac{\partial j_\ell(\rho_{in})}{\partial\rho_{in}}\right)=1+\frac{\rho_{out}}{h_\ell^{(1)}(\rho_{out})}\frac{\partial h_\ell^{(1)}(\rho_{out})}{\partial\rho_{out}} \tag{1}$$

where, ρ is size parameter that $\pi d n_{\text{in, out}}/\lambda_0$, d is the diameter of microsphere, n is the refractive index and λ_0 is the resonant wavelength, the subscript indicates inside and outside of sphere, j_ℓ and $h_\ell^{(1)}$ are the spherical Bessel function and the first ordered spherical Hankel function, respectively. ℓ is the angular mode number and α is polarization dependent factor.

In the experiment, the resonant wavelength is only measurand. Figure 2 shows typical WGM spectrum where the electric distributions of corresponding mode numbers of WGMs are shown together. Each mode has different resonant wavelengths. We identify the mode number for the measured resonant wavelength by means of analysis of the WGM spectrum. The morphological parameter such as form and size are determined with the resonant wavelength and the mode numbers.

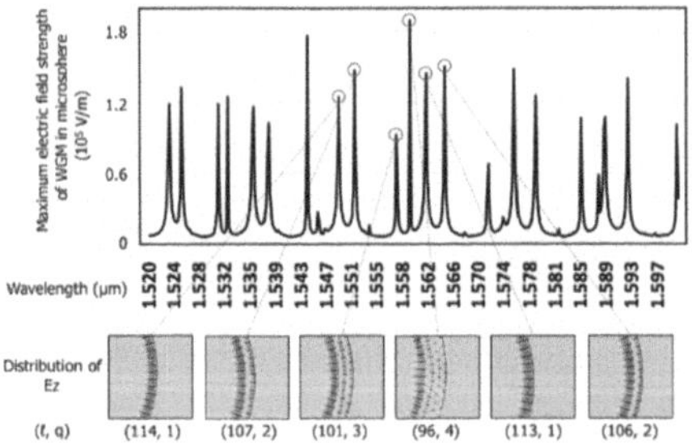

Fig. 2. Resonant wave lengths depending on the mode number of WGM. ℓ and q indicate angular mode number and radial mode number, respectively.

2.2 Requirement for WGM Resonant Wavelength Measurement

The measurement accuracy of the resonant wavelength directly affects the measurement accuracy of the microsphere. Figure 3 illustrates example of the estimation error of the diameter due to the measurement error of the resonant wavelength when measuring the glass microsphere as small as 190 μm in diameter. As shown, the picometer scale measurement accuracy for resonant wavelength is required to ensure the measurement accuracy better than 10 nm. The measurement accuracy of the resonant wavelength indirectly influences the uncertainty estimation of the mode numbers as well. The estimation error of mode number causes the measurement error of the order of 100 nm. Therefore, the measurement system to stable excitation of the WGMs and to accurate measurement of the resonant wavelength of the order of better than 10 pm are demanded.

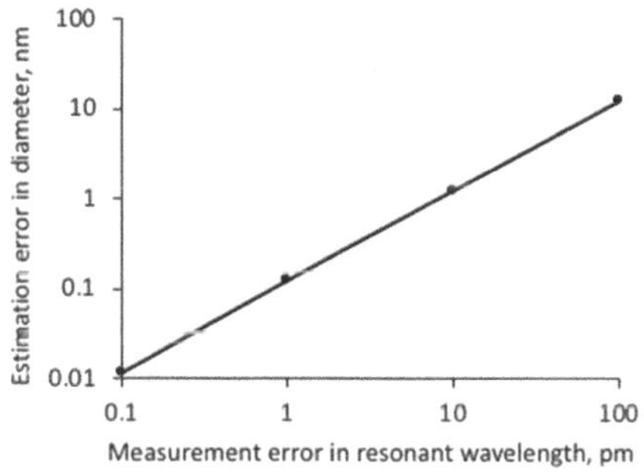

Fig. 3. Relation of estimating diameter error due to measurement error of resonance wavelength.

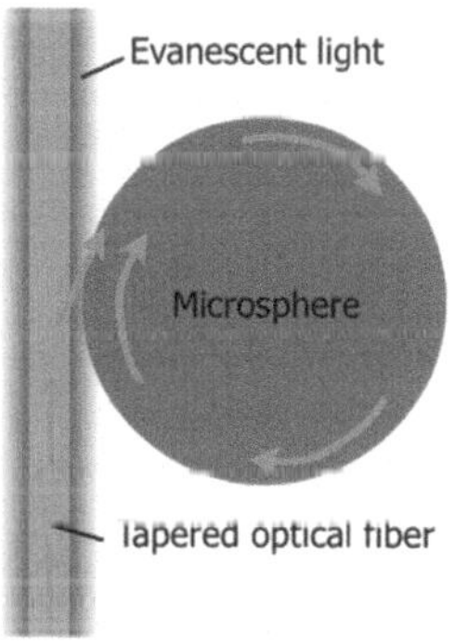

Fig. 4. Structure of light coupling into microsphere using tapered optical fiber

3 Strategy to Measure Accurate Resonant Wavelength

3.1 Structure of Exciting WGM Resonance and Measuring the WGM Resonant Wavelength

In order to excite WGMs and measure the resonant wavelength, the light is appropriately introduced into the microsphere. Usually, the evanescent coupling using prism, half polished optical fiber and tapered optical fiber is utilized due to high coupling efficiency. Among them, the coupling system using the tapered optical fiber is reported to shows higher coupling efficiency and achieve high signal-to-noise ratio of measuring the resonant wavelength. The tapered optical fiber has a diameter as small as 1 μm or less, which is fabricated by stretching the single mode optical fiber. Therefore, the optical fiber itself serves as a core of the optical fiber, the surrounding air is a clad, so the evanescent light field is generated the outside of the tapered optical fiber. The light is coupled into the microsphere when the sphere is in the evanescent light field (Fig. 4). When the wavelength of the incident light is coincident with the resonant wavelength, the light is strongly coupled into the sphere, then the transmitted light intensity of the tapered optical fiber is much decreased, otherwise, the most of the incident light is transmitted the tapered optical fiber without the coupling to the microsphere.

3.2 Stabilization System of the Gap Between the Sphere and the Fiber

The gap between the sphere and the optical fiber is highly important both to excite WGMs and to measure the resonant wavelength stably. In fact, if the gap was improper, it is not able to achieve highly accurate measurement. Therefore, in this study, the optical system to stabilize the distance between the sphere and the optical fiber, which is called "gap", was developed. We use two tunable lasers having different wavelength ranges. One is for measuring the sphere and another is for maintaining the gap. Figure 5 summarize the gap stabilizing system. As later shown, the gap affects the coupling strength that is observed via depth of the dip in the spectrum of the transmitted light intensity. Since strong coupling makes the dip deeper, the gap can be controlled by changing the displacement of the piezo-stage where the sphere is attached. At that stabilized condition, the resonant wavelength was measured by finding the dip position of the spectrum of the transmitted light intensity of another laser. A wavelength scanning range of the laser for stabilization is only around ±0.1 nm because a single dip is observed, on the other hands, the wavelength scanning range of another laser was ±20 nm to seek the resonant wavelengths over the broad wavelength range. This is why the two lasers were used. Besides, to assure the wide evanescent light field, the wavelength range for stabilization was chosen to be longer.

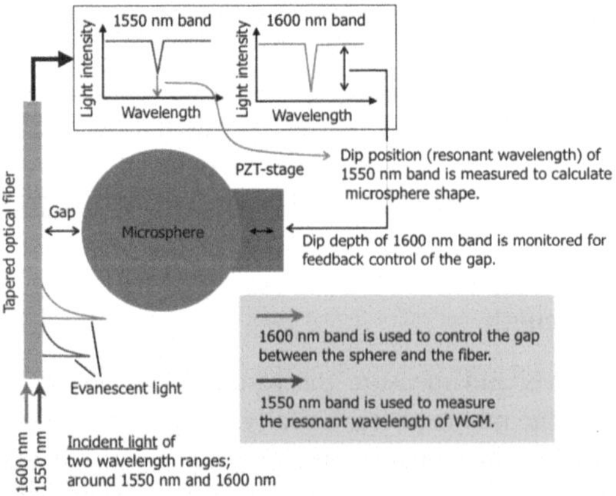

Fig. 5. Structure of optical system for gap stabilizing.

4 Simulation

4.1 Simulation Model

Properties of optical coupling are numerically studied by using the finite element method (FEM) commercial software COMSOL multi-physics. Figure 6 shows the simulation model. We constructed the 2-dimensional model because of limitation of computer memory. Most of parameter used in the simulation were found in the Fig. 6. The diameter of the tapered optical fiber was set to 0.5 μm, which is thin enough to be single mode propagation. In this simulation, the wavelength dispersion of refractive index was ignorable. Maximum mesh size was set to smaller than λ/10.

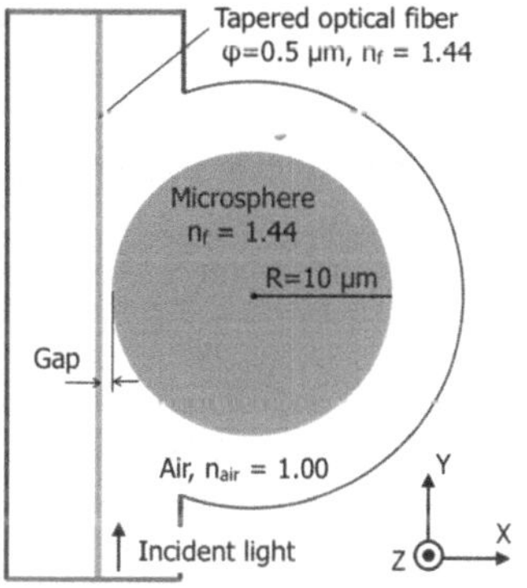

Fig. 6. Simulation model of light coupling.

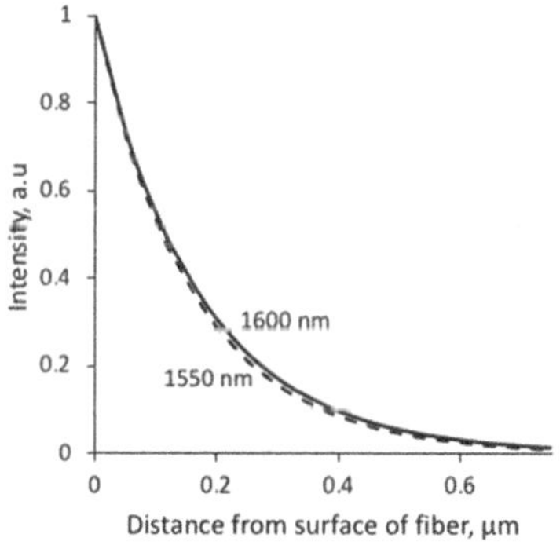

Fig. 7. Evanescent light field generated surrounding the tapered optical fiber.

4.2 Coupling Property

Evanescent light field generated surrounding the tapered optical fiber. Firstly, evanescent light field generated by the tapered optical fiber was examined. The results are shown in Fig. 7, which indicates that the gap between the sphere and the fiber should be narrower than 1 μm. Regarding with the wavelength, the effective area of the evanescent light was same level for 1550 and 1600 nm. The stronger optical interaction can be expected in narrower gap.

Coupling properties. WGM spectrum was simulated between 1520 and 1535 nm with an interval of 0.05 nm. The gaps were set to 1.0, 0.5 and 0.3 μm. When the gap is 1.0 μm, the shallow dips were seen and the number of the dips were smaller. The dips are deepened with narrowing the gap, which is because the optical coupling is stronger. In order to measure the resonant wavelength with high signal to noise ratio, the deeper dips advantageous. At next, we focused on one dip with different gaps. The gap was changed from 1.0 μm to 0.5 μm. Again, when the gap is narrow, the dip is deep. However, it seemed that position of the dip was shifted. By fitting Lorentz curve to the dip profile, the resonant wavelength was obtained. The result is shown in Fig. 10. As seen, when the sphere was approaching the optical fiber, the resonant wavelength was shifted. This is because the optical fiber disturbs the WGM resonance when they are too close, which is also confirmed that the Q value was also decreased with narrowing the gap (Fig. 10). Thus the disturbance by the fiber might change the WGM conditions. Therefore, to measure the resonant wavelength accurately of the order of picometer scale, optical fiber should be as far as possible. Consequently, it is stated that there is an appropriate gap between the optical fiber and the sphere.

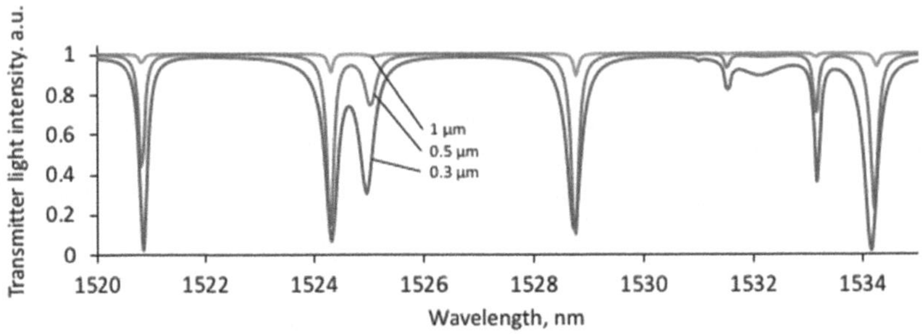

Fig. 8. Spectrum of WGM resonance with different distance between the microsphere and the tapered optical fiber.

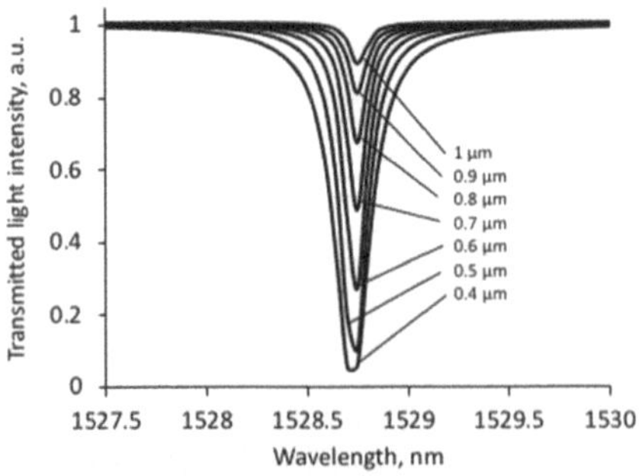

Fig. 9. Coupling signal chancing with decreasing the distance between the microsphere and the tapered optical fiber.

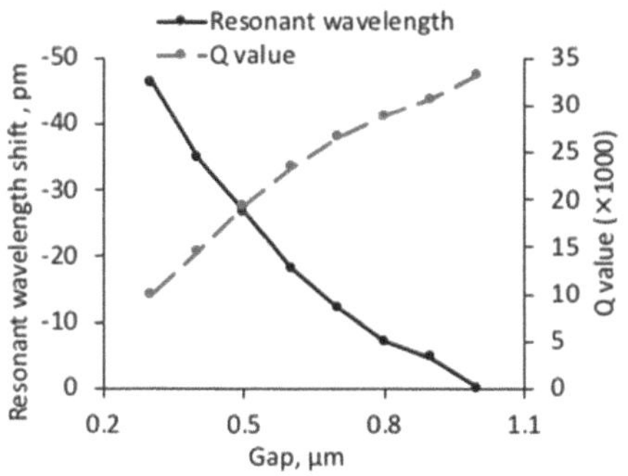

Fig. 10. Q value and resonant wavelength with distance between the microsphere and the tapered optical fiber

5 Experimental Study

5.1 Experimental Setup

In this study, we proposed the scheme to stabilized the gap between the sphere and the fiber during microsphere measurements. In this experiment, we develop the system and the performances of the measurement system was experimentally verified.

Figure 11 shows whole measurement system. The microsphere to be measured is fixed to the XYZ piezo-stage. The tapered optical fiber of a diameter of 0.9 μm is hold with holder on the XYZ micrometer-stage. There are two laser sources. The tunable laser 1 is for measuring the morphology of the microsphere, having a wavelength range of 1550 ± 25 nm with a linewidth of 100 kHz. The transmitted light of the tunable laser 1 is measured with photodiode 1 and wavelength meter. The guaranteed accuracy of the wavelength meter was 1 pm. The measured data was feed to the PC 1 to analyze WGM spectrum. The tunable laser 2 is to stabilize the gap between the sphere and the fiber, having a wavelength range of 1620 ± 60 nm with a linewidth of 500 kHz. The transmitted light of the laser 2 was measured with the photodiode 2, whose signal is used to feedback control with XYZ piezo-stage via PC 2. X-axis of the stage was actually controlled. Thus, in the light coupling region, the two evanescent lights of different wavelength are generated; one is for measuring the sphere and another is for maintaining the certain gap (Fig. 11(b)). As the measurement target, a glass microsphere was used. The tip of the single mode optical fiber was melted by the CO_2 laser to make a sphere with approximately 200 μm in diameter. The stem part was used to hold such like Fig. 11(c).

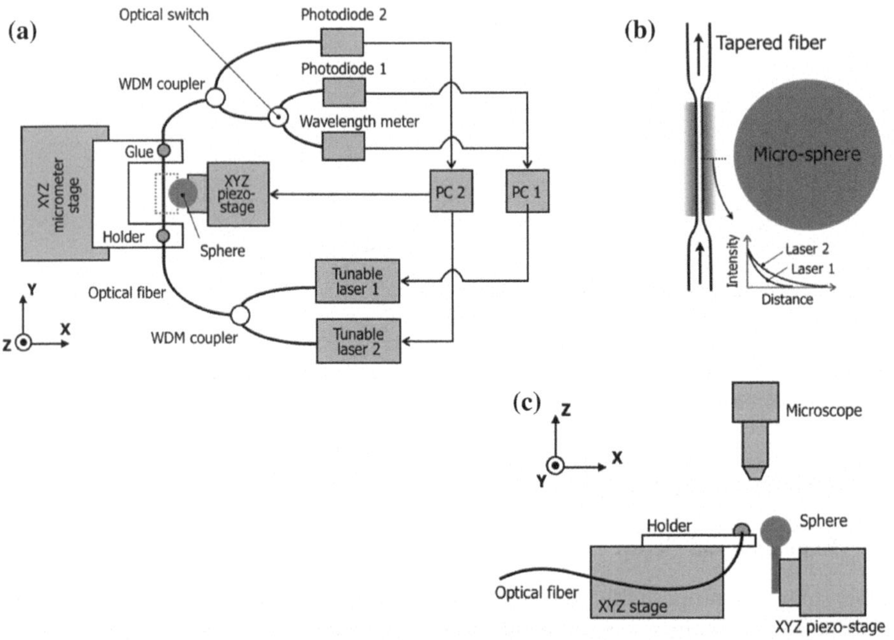

Fig. 11. Schematics of developed experimental setup. (a) Whole measurement system. (b) Magnified view of red dotted line in (a). (c) Side view.

5.2 Result of Fundamental Optical Coupling Properties

Firstly, the WGM spectrum was measured using the laser 2. The microsphere was approached step by step to the fiber using the piezo-stage with an interval of 250 nm. The measured spectrums when the piezo-stage displacement of 1.50 μm, 1.75 μm and 2.25 μm are shown in Fig. 12. At 1.50 μm, the depth of the dips are small. At 2.25 μm, the dip depths becomes deeper, and also there are so many dips appeared. This result shows good agreement with the simulated result shown in Fig. 8.

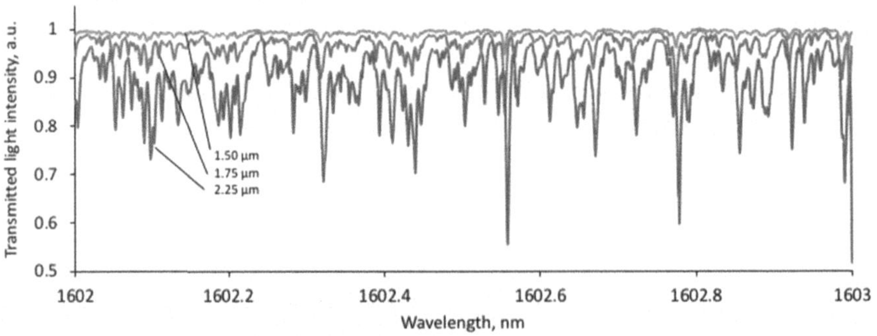

Fig. 12. Experimentally measured WGM spectrum.

In order to analyze the properties of the resonant wavelength measurement, the single dip, which is the strong dip at around 1602.55 nm shown in Fig. 12, was focused on. Lorentz curve fitting was employed to specify the dip profile. Because wavelength dependent optical responses are often expressed with the Lorentz model, the dip of the transmitted light intensity also fitted by Lorentz distribution as following.

$$I = a - \frac{b}{(\lambda - c)^2 + d^2} \tag{2}$$

Where, I is the light intensity, λ is the wavelength, c is the center wavelength of the dip, that is the resonant wavelength, b/d^2 indicates the depth of the dip, d is the parameter related to the width of the dip, and a means the base intensity of the spectrum. By fitting this equation to the measured signal such as shown in Fig. 13, the fitting parameter could be obtained. Figure 14 shows the coupling strength and the resonant wavelength shift with changing the gap. At narrower gap between the sphere and the fiber (large displacement of piezo-stage), dips are deeper, and of important the resonant wavelength was shifted of the order of picometer range. These tendencies were well agreed with the simulated results. Since the dip depth was increasing with decreasing the gap, it was confirmed that the dip depth is useable information to maintain the gap with feedback control with the piezo stage.

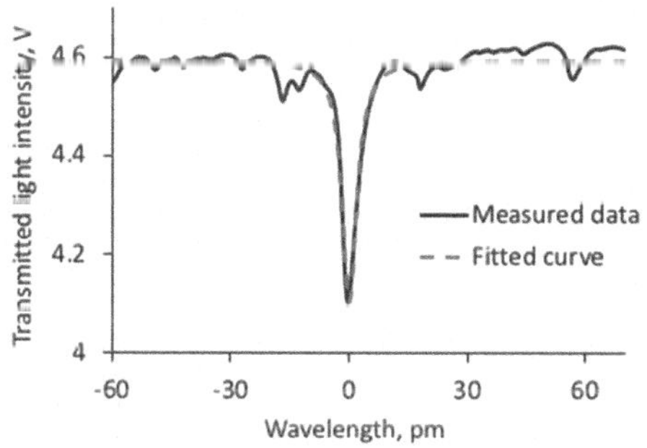

Fig. 13. Lorentz curve fitting.

5.3 Result of Gap Stabilization

At last, the gap between the sphere and the fiber was controlled to be stable. Figure 15 shows the results. Without feedback control, the dip depth was largely changed with a few V level due to the thermal and mechanical drift. As a result, the resonant wavelength on laser 1 was fluctuated ±1.2 pm for 13 min. Then the feedback was set at 0.3 V of the dip depth for laser 2. With feedback control, the dip depth was highly stable and the resonant wavelength shift on laser 1 was as small as ±0.5 pm. Furthermore, it is notable that coupling system with feedback control was rather stable from the mechanical impact at 6 min. Consequently, by using the transmitted light

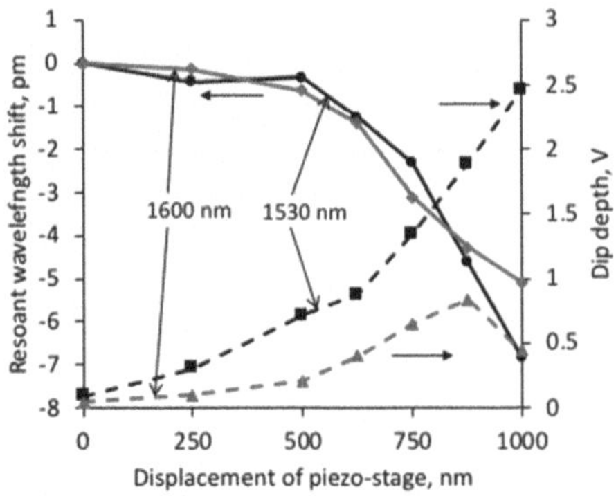

Fig. 14. Resonant wavelength shift depending on the gap between the sphere and the fiber.

signal, we could finely control the gap to stabilize the measurement of the WGM resonant wavelength. By using the system, the gap of the sphere and the fiber is easily adjustable to appropriate value for microsphere measurement.

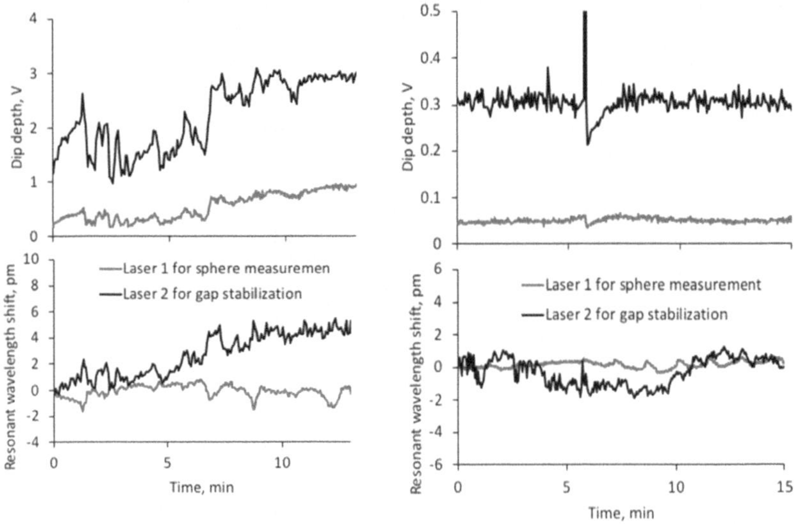

(a) without feedback. (b) with feedback.

Fig. 15. Experimental result of gap stabilization.

6 Conclusions

We proposed the novel measurement method for microsphere based on whispering gallery mode resonances. To achieve the high accuracy measurement performance, it is necessary to measure the WGM resonant wavelengths of the order of picometer scale. In this paper we developed the measurement instruments for stably exciting the WGM and accurately measuring the resonant wavelength using two lasers having different wavelength range. Experimental results showed that stability of resonant wavelength measurement with the developed system was as stable as ±0.5 pm, which is 2.4 times better than conventional system. Using this system, it is expected that accuracy of microsphere measurement can be expected better than 1 nm.

Acknowledgement. This work was financially supported by the Foundation for Interaction in Science & Technology, Mizuho Foundation for the Promotion of Sciences, A-STEP from JST, and MEXT/JSPS KAKENHI (No. 18K18803, 15H05505).

References

1. Keferstein, C.P., Marxer, M., Gotti, R., Thalmann, R., Jordi, T., Andrasd, M., Becker, J.: Universal high precision reference spheres for multisensor coordinate measuring machines. CIRP Ann. – Manuf. Technol. **61**(1), 487–490 (2012)
2. Schwenke, H., Schmitt, R., Jatzkowski, P., Warman, C.: On-the-fly calibration of linear and rotary axes of machine tools and CMMs using a tracking interferometer. CIRP Ann. – Manuf. Technol. **58**(1), 477–480 (2009)
3. Takatsuji, T., Goto, M., Osawa, S., Yin, R., Kurosawa, T.: Whole-viewing-angle cat's-eye retroreflector as a target of laser trackers. Measur. Sci. Technol. **10**, 87–90 (1999)
4. Weckenmann, A., Estler, T., Peggs, G., McMurtry, D.: Probing systems in dimensional metrology. CIRP Ann. – Manuf. Technol. **53**(2), 657–684 (2004)
5. Wu, X., Li, Y., Wei, H., Yang, H., Yang, G., Zhang, J.: Interferometric diameter determination of a silicon sphere using a traceable single laser frequency synthesizer. Measur. Sci. Technol. **24**, 115202 (2013)
6. Ramirez, C., Strojnik, M.: Estimation of the degree of asphericity of a glass sphere using a vectorial shearing interferometer. Opt. Commun. **284**, 1517–1525 (2011)
7. Pfund, J., Lindlein, N., Schwider, J., Burow, R., Blumel, T., Elssner, K.-E.: Absolute sphericity measurement: a comparative study of the use of interferometry and a Shack-Hartmann sensor. Opt. Lett. **23**(10), 742–744 (1998)
8. Kotte, G.J., Haitjema, H.: Ball diameter measuring instrument in a gauge block interferometer. In: Proceedings of SPIE 3477, Recent Developments in Optical Gauge Block Metrology (1998)
9. Fujii, K., Tanaka, M., Nezu, Y., Nakayama, K., Masui, R., Zosi, G.: Interferometric measurements of the diameters of a single-crystal silicon sphere. Rev. Sci. Instrum. **63**, 5320–5325 (1992)
10. Medicus, K.M., Jansen, M.: Diameter measurement of small spheres on a white light interferometer including uncertainty analysis. In: Proceedings of the Euspen International Conference, vol. 75 (2010)
11. Fan, K.-C., Wang, N., Wang, Z.-W., Zhang, H.: Development of a roundness measurement system for microspheres. Measur. Sci. Technol. **25**, 064009 (2014)

12. Küng, A., Meli, F., Thalmann, R.: Ultraprecision micro-CMM using a low-force 3D touch probe. Measur. Sci. Technol. **18**, 319 (2007)
13. Chen, L.-C.: Automatic 3D surface reconstruction and sphericity measurement of micro spherical balls of miniaturized coordinate measuring probes. Measur. Sci. Technol. **18**, 1748–1755 (2007)
14. Michihata, M., Hayashi, T., Adachi, A., Takaya, Y.: Measurement of stylus-probe sphere diameter for micro-CMM based on spectral fingerprint of whispering gallery mode. CIRP Ann. – Manuf. Technol. **63**(1), 469–472 (2014)

Verification of the CMM Measuring Path Based on the Modified Hammersly's Algorithm

Slavenko Stojadinovic(✉), Sasa Zivanovic, and Nikola Slavkovic

Faculty of Mechanical Engineering, Production Engineering Department, University of Belgrade, Kraljice Marije 16, 11120 Belgrade, Serbia
sstojadinovic@mas.bg.ac.rs

Abstract. This paper presents a CAI verification of the measuring path for inspection of prismatic measuring parts (PMP) which consists of the basic geometric features. The aim of the verification is to visualize collision check between the measuring sensor and the workpiece. The simulation of the measuring path was realized on the configured virtual CMM in the CAD environment. The generated measuring path for inspection planning at the CMM consists of three sets of points. The first set of points, measuring points, is generated based on the modified Hammersly's algorithm for distribution of measuring points by the features from which the workpiece consists. The other two sets, nodal points, are collision fee points, which are generated on the basis of the developed model that analyze accessibility to features at the level of one feature and based on the principle of collision avoidance when pass from one feature to another (requirement of tolerance). On the basis of these three sets of points, the total measuring path without collision is generated. The measuring path generated in this way is verified by simulations on a configured virtual CMM through several examples of standard forms of tolerance.

Keywords: Hammersly's algorithm · Prismatic parts · CMM · CAI · Simulation

1 Introduction

Coordinate measuring machines (CMM) are essential for quality assurance and production control in modern manufacturing [1]. CMMs are recognized as a flexible element of production metrology, and they are applicable for a wide range of metrological tasks. As such, CMMs have maintained their presence in almost all industries.

At the same time with their use, comes to developing of the methods of automatic planning on them [2–7], as well as the verification and simulations of these methods. This research directs the path to the development of the 5th generation of measuring machines - towards intelligent measuring machines [8–11].

This paper presents a CAI verification of the measuring path for inspection of prismatic parts which consists of the basic geometric features. The aim of the verification is to visualize collision check between the measuring sensor and the workpiece and fixture. The simulation of the measuring path was realized on the configured virtual CMM in the CAD/CAM environment.

V. D. Majstorovic and N. Durakbasa (Eds.): IMEKOTC14 2019, LNME, pp. 25–38, 2019.
https://doi.org/10.1007/978-3-030-18177-2_3

In paper [12] is proposed haptic virtual coordinate measuring machine for inspection path planning, which simulates a CMM's operation and its measuring process in a virtual environment with haptic perception. The inspection measuring path of a workpiece is generated by pointing a probe at the 3D CAD model of a part using a haptic device. An example of application of virtual CMM and virtual measuring arm is determination of uncertainty. Currently used methods for uncertainty assessment require knowledge and measuring experience therefore it is necessary to validated methods that will also be easy to implement and will not require broad metrological knowledge from the personnel [13, 14]. The paper [15] presented advanced virtual coordinate-measuring machines in an integrated virtual environment which provide virtual measuring and evaluate the uncertainty, all without the need of using a physical machine.

The simulation of the measuring path, in this paper, was realized on the configured virtual CMM configured in the CAD/CAM environment. The verification of measuring path, based on modified Hammersly's algorithm for basic geometrical features of workpiece, is first visualized in MatLab and than through simulations in CMM module of PTC Creo. The aim of the verification is to visualize collision check between the measuring sensor and the workpiece and fixture. Output from the simulation on virtual CMM generated measuring path is saved in CL file (DMIS program).

2 Outline of the Concept for CMM Simulation

Verification of the measuring path for inspection planning of workpiece is important in order to visualize collision check between the measuring sensor, workpiece and fixture. A proposed concept for verification of the measuring path through simulation, in this paper, is realized on the configured virtual CMM in the CAD/CAM environment, Fig. 1. This method starts from the modeling of the workpiece in *CAD system*. The generation of the measuring path for *Inspection planning* of modeled workpiece consists of three sets of points. The first set of points, measuring points, is generated based on the modified Hammersly's algorithm for distribution of measuring points by the features from which the workpiece consists.

The other two sets, nodal points, are collision fee points, which are generated on the basis of the developed model that analyze accessibility to features at the level of one feature and based on the principle of collision avoidance when pass from one feature to another (requirement of tolerance) presented in [9]. Those points necessary for inspection, generated for the features of the workpiece, are now sorted in MatLab environment in purpose to generate point-to-point measuring path in unique sequence. The *MatLab visualization* of such sequence is done and then it is saved in file with appropriate format for loading in CAD/CAM system. Next step, *Path simulation on Virtual CMM* in CAD/CAM system, besides modeling of the workpiece and loading of generated point-to-point measuring path, required modeling, and assembling of the CMM components using appropriate kinematics links in *Virtual CMM*, setting of coordinate systems and measuring probe, analyzing possibility of fixture application, setting of measuring parameters, and generation of measuring path. Output from this step is CL file (DMIS program).

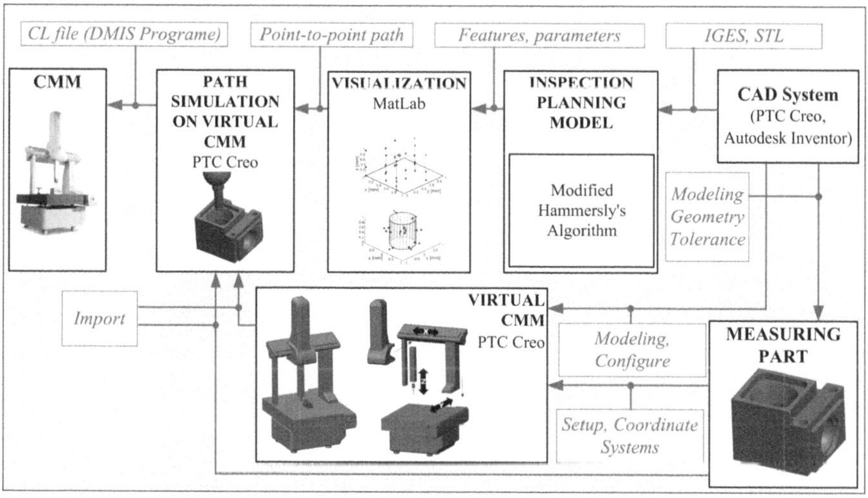

Fig. 1. The concept for CMM simulation

3 Modified Hammersly's Algorithm

Modified Hammersly's Algorithm define point-to-point measuring path for prismatic parts by three set of points [8]:

- measuring points (contact points),
- nodal points, and,
- collision fee points.

3.1 Inspection Features

A model for Inspection Feature Construction (IFC) based on the basic geometric features and their parameters, is a part of model presented in [9]. The parameters of features plane, cylinder and truncated cone are given in Fig. 2.

The parameters of features define uniquely each of them. Defining parameters of the features was performed by fully describing their geometry, as well as whether the feature is full or empty. Defining of a full and empty feature is important and done based on the feature fullness vector that will be explained in detail below, which provides information on whether the inspection of a given feature is performed inside or outside. The defined feature parameters are a basis for the development of algorithms such as the algorithm for measuring points' distribution, collision avoidance and path planning [8], where ontologically defined relationships between features and prescribed tolerances are also used [16].

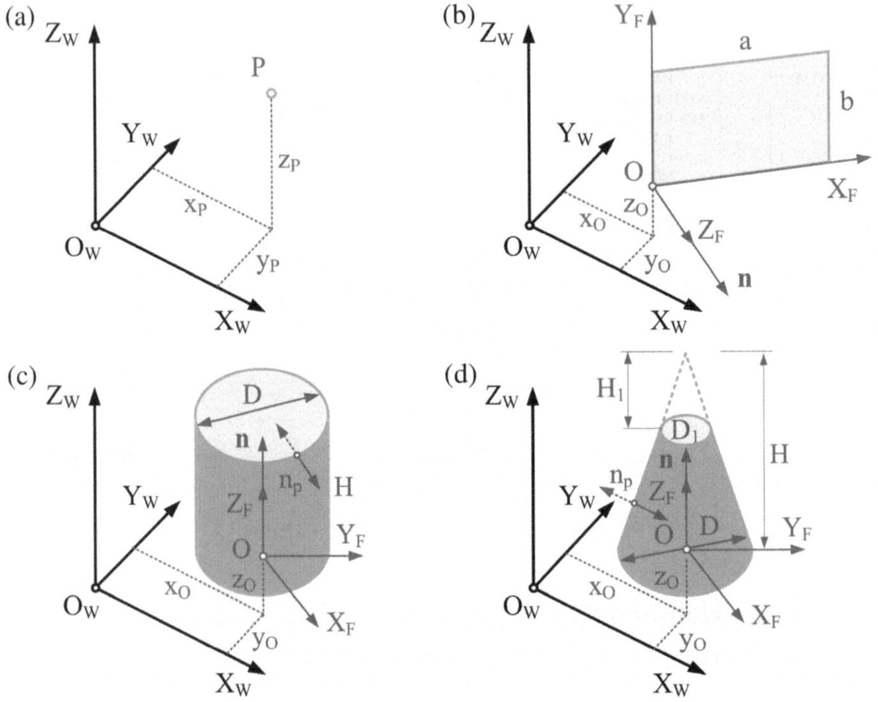

Fig. 2. Metrological features and their parameters: (a) point, (b) plane, (c) cylinder and (d) truncated cone [16]

The initial path obtained based on information (parameters) about features is the subject of optimization by ant colony application given in [17].

Widely the geometric feature term has been defined in analytical geometry, and applied later in engineering modelling. In coordinate metrology, the set of geometric features presents a basis for defining IFC from the aspects of geometry and tolerances. In the inspection model based on IFC, geometric features present the lowest level of tolerance definition. Each geometric feature is uniquely determined by the local coordinate system O_F, X_F, Y_F, Z_F and a set of corresponding parameters. These parameters could belong to the following types:

- diameter (D, D_1),
- height (H, H_1),
- width (a),
- length (b),
- normal vector of a feature (**n**), and
- fullness vector of a feature ($\mathbf{n}_p$).

3.2 Measuring Points

Distribution of measuring points is based on Hemmersly's sequences [18] for the calculation of coordinates for two axes of a feature:

$$s_i = \frac{i}{N} \text{ and } t_i = \sum_{j=0}^{k-1}\left(\left[\frac{i}{2^j}\right]\mathrm{Mod}2\right)\cdot 2^{-(j+1)} \tag{1}$$

where, $k = \log_2 N$ and N is the desired number of measuring points, $i = 0, 1, 2, \ldots, (N-1)$. By modifying the Hemmersly's sequences, the distribution of measuring points is defined for the metrological feature from the Fig. 2 that are involved in creation of PMP tolerances as follows.

To define the distribution of measuring points for each feature, the Descartes coordinates system O_F, X_F, Y_F, Z_F and polar cylindrical coordinates system O'_F, X'_F, Y'_F, Z'_F are needed. The coordinates in Descartes coordinates system are denoted by $P_i(s_i, t_i, w_i)$, and in polar cylindrical coordinates system by $P_i(s'_i, t'_i, w'_i)$. The distributions of measuring points for different features (defined by unit parameters) based on modified Hammersly's algorithm are presented in Fig. 3. The equations for calculation of measuring point coordinates are:

- Plane:

$$\begin{aligned} s_i &= \frac{i}{N}\cdot a \\ t_i &= \left(\sum_{j=0}^{k-1}\left(\left[\frac{i}{2^j}\right]\mathrm{Mod}2\right)\cdot 2^{-(j+1)}\right)\cdot b \\ w_i &= 0 \end{aligned} \tag{2}$$

where a[mm] is the plane limit for X-axis; b[mm] is the plane limit for Y-axis.

- Cylinder:

$$\begin{aligned} s_i &= R\cos\left(-\frac{\pi}{2}-\frac{2\pi}{N}\cdot i\right) \\ t_i &= R\sin\left(-\frac{\pi}{2}-\frac{2\pi}{N}\cdot i\right) \\ w_i &= \left(\sum_{j=0}^{k-1}\left(\left[\frac{i}{2^j}\right]\mathrm{Mod}2\right)\cdot 2^{-(j+1)}\right)\cdot h \end{aligned} \tag{3}$$

where R[mm] is the radius of a cylinder and h[mm] is the height of a cylinder.

– Truncated cone, in polar-cylindrical coordinates:

$$s_i' = R_1 + \left(1 - \sum_{j=0}^{k-1}\left(\left[\frac{i}{2^j}\right]\mathrm{Mod}2\right)\cdot 2^{-(j+1)}\right)^{\frac{1}{2}} \cdot (R - R_1)$$
$$t_i' = \frac{i}{N}\cdot 360°;\ w_i' = (R - s_i')\cdot\frac{h_1}{R_1} \tag{4}$$

where R_1 [mm] is the radius of the smaller cone base, and h_1 [mm] is the height of a truncated cone. In Descartes coordinates, the formulations are $s_i = s_i' \cos(t'_i)$; $t_i = s_i' \sin(t'_i)$; $w_i = w_i'$.

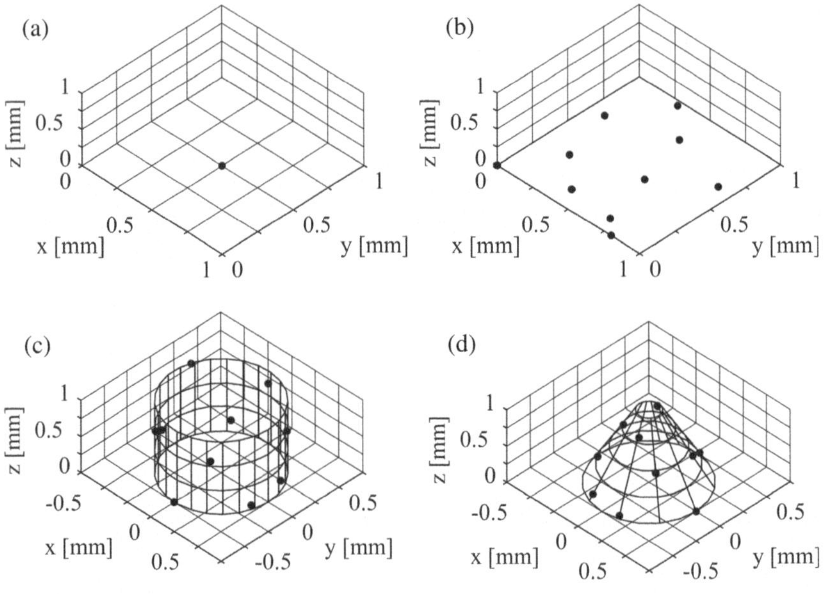

Fig. 3. Distribution of measuring points by modified Hammersly's equations of features defined with unit parameters: (a) point, (b) plane, (c) cylinder and (d) truncated cone

3.3 Nodal Points

The coordinates of measuring points are determined using distribution of measuring points as presented in the above section and are denoted as $P_i(x_i, y_i, z_i), P_{i+1}(x_{i+1}, y_{i+1}, z_{i+1}), \ldots, P_{N-1}(x_{N-1}, y_{N-1}, z_{N-1})$ where $i = 0, 1, 2, \ldots, (N-1)$, with regard to coordinate system O_F, X_F, Y_F, Z_F.

To perform an inspection, it is necessary to conduct the Probe Accessibility Analysis (PAA), as it is presented in [8]. This analysis involves the determination of two sets of points $P_{i1}(x_{i1}, y_{i1}, z_{i1})$ and $P_{i2}(x_{i2}, y_{i2}, z_{i2})$ where $i = 0, 1, 2, \ldots, (N-1)$, as well as the definition of a fullness vector of a feature $\mathbf{n}_p$. The direction of a fullness vector is defined to coincide with the direction of X-axis of a feature, and the orientation with regard to the X-axis could be positive or negative. According to this, the fullness vector for the full feature is $\mathbf{n}_p = [-1 \quad 0 \quad 0]$, and for the empty feature is $\mathbf{n}_p = [1 \quad 0 \quad 0]$. The set $P_{i1}(x_{i1}, y_{i1}, z_{i1})$ presents points for the transition from rapid to slow feed of CMM.

The distance between points $P_{i1}(x_{i1}, y_{i1}, z_{i1})$ and $P_i(x_i, y_i, z_i)$ is presented by d_1 that is a slow feed path, and the distance between points $P_{i2}(x_{i2}, y_{i2}, z_{i2})$ and $P_{i1}(x_{i1}, y_{i1}, z_{i1})$ is d_2- a rapid feed path. This approach enables the execution of PAA, in order to avoid the collision between a feature of PMP and a measuring probe. In the inspection of PMP, there are three different cases for the definition of point sets $P_{i1}(x_{i1}, y_{i1}, z_{i1})$ and $P_{i2}(x_{i2}, y_{i2}, z_{i2})$, as presented in Fig. 4.

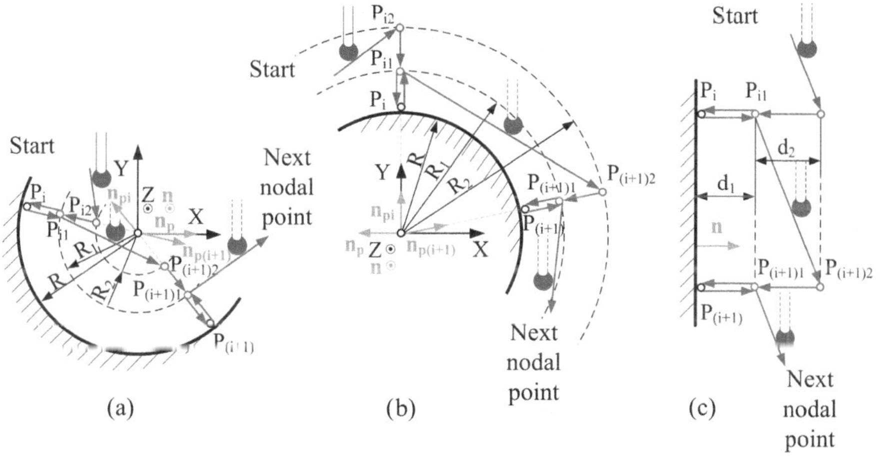

Fig. 4. Nodal points (a) empty cylindrical surface, (b) fullness cylindrical surface and (c) plane [9]

3.4 Collision Fee Points

Based on STL model for the presentation of PMP geometry, the tolerances of PMP, the coordinates of the last point $P_{(N_{F1})1}$ of a feature F1 (precedent feature) and the coordinates of the first point $P_{(N_{F2})1}$ of a feature F2 (subsequent feature), the simplified principle of Automated Collision-Free Generation (ACFG) is presented in [9].

4 Verification of the CMM Measuring Path

As it is said, verification of the measuring path for inspection planning of workpiece is importance in order to visualize collision check between the measuring sensor, workpiece and fixture. A proposed concept for verification of the measuring path is realized thought visualization of the measuring path in MatLab environment and measuring path simulation on Virtual CMM in CAD/CAM system. This concept of verification is validated through several examples of generated measuring paths for selected features and tolerance of arbitrarily designed PMP given in Fig. 5. The plane A, cylinder C, and truncated cone B are selected as features, mark with red, with belonging tolerance, mark with green, for validation of proposed concept. The parameters that define selected features are shown on Fig. 6.

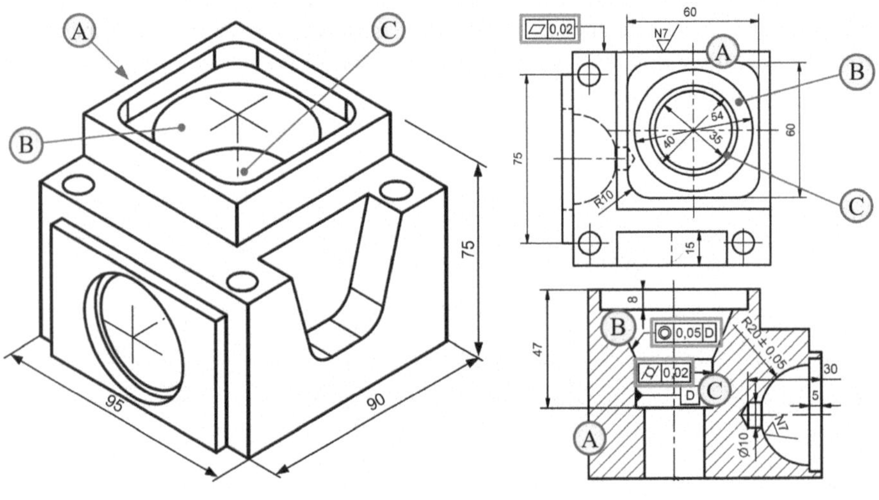

Fig. 5. Workpiece with selected features

4.1 Visualization of the Measuring Path in MatLab Environment

As it is said, as the examples for the verification of measuring paths on virtual CMM, features of plane, cylinder and truncated cone is selected from workpiece given in Fig. 5. For each of mentioned features the appropriate parameters are defined, Fig. 6, e.g. for plane that parameters are a and b, as also the homogenous transformation matrix ${}^{W}_{F}T$ [19] that defined position and orientation of coordinate frames of the feature, O_F, X_F, Y_F, Z_F, in accordance to coordinate frame of workpiece O_W, X_W, Y_W, Z_W, Fig. 6.

For extracted parameter of features from measuring workpiece, Fig. 5, defined equation of modified Hammersly's algorithm, and derived equation for nodal points from Fig. 4, in MatLab environment the point-to-point measuring path is generated as initial sequence. The examples of generated point-to-point measuring paths for plane, cylinder and truncated cone are given in Fig. 6(a), (b) and (c). Such generated points are sorted in unique sequences – measuring path for each of feature, Fig. 6(d), (e) and (f). This unique sequence je first visualized in MatLab and then saved in file with appropriate format for loading in CAD/CAM system.

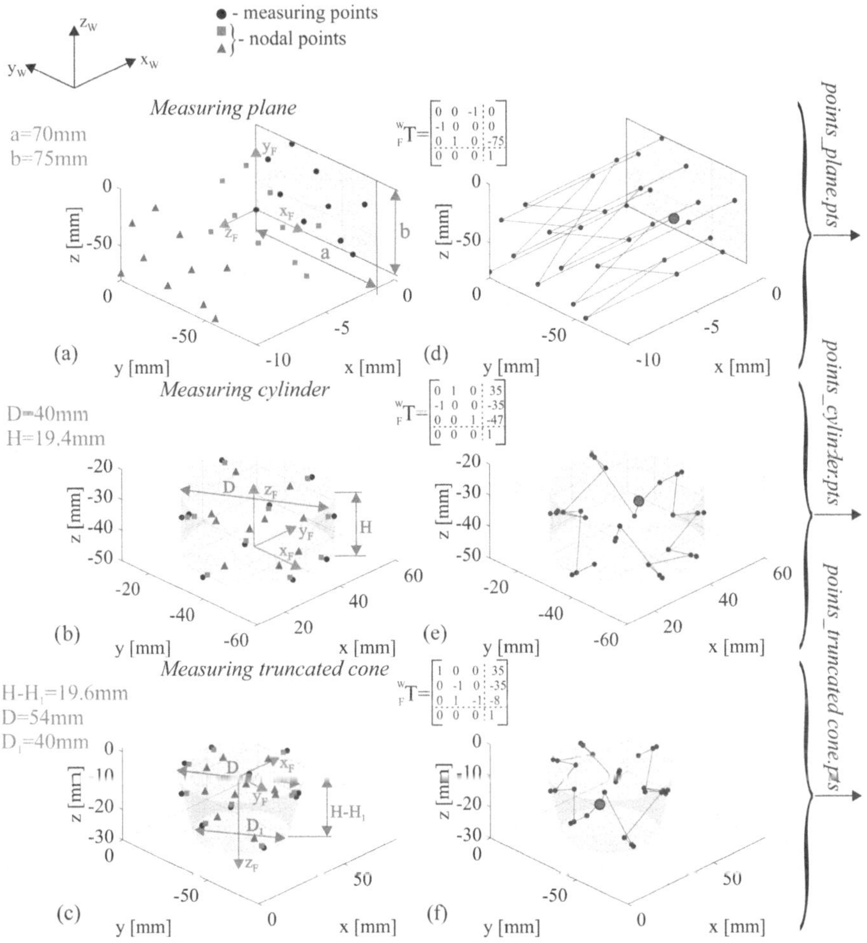

Fig. 6. Visualization of measuring paths in MatLab environment for (a) plane, (b) cylinder and (c) truncated cone

4.2 Path Simulation on Virtual CMM in CAD/CAM System

Path simulation on Virtual CMM is based on generated point-to-point measuring path as unique sequence which is saved in file with appropriate format for loading in CAD/CAM system. Path simulation that uses the virtual CMM is the safest and the most cost-effective way for program verification. For programming of such measuring path CMM module of CAD/CAM system PTC Creo 2.0 was used [20].

For generation of measuring path in CMM module of CAD/CAM system PTC Creo, it is necessary to realize a few activities: (i) loading of reference CAD workpiece for inspection; (ii) create of a CMM; (iii) defining of measuring operation and selecting of task' coordinate frame; (iv) setting of measuring probe; (v) selection of the step in accordance with of feature that have to be measured (plane, cone, cylinder,…);

(vi) setting of measuring parameters for selected step; (vii) selection of measuring points, given as results from MatLab code for visualization, on the feature that have to be measured; (viii) generation of measuring path; (ix) simulation of measuring path including probe and (x) measuring path simulation on configured Virtual CMM, using Machine Play option.

Based on previous research in configuring virtual machine tools [21, 22] and robots [23–25], virtual CMM is configured accordance of procedure described below. The configured virtual CMM are used for the verification of the programmed measuring path in a CAD/CAM environment based on the generated measuring probe path. This simulation of measuring path is important in order to [25]: (i) configure the off-line CMM programming environment, (ii) verify the program before measuring, and (iii) detect the possible collisions of probe with: component, fixture, CMM structure during program execution.

Measuring path simulation by running the program is possible thanks to the applied modelling of the virtual CMM with all kinematic connections between the components, which allows the motion of a CMM virtual model as a system of rigid bodies [21, 25]. Detailed virtual CMM with all kinematic relation between moving components is defined as it is shown in Fig. 7. Slider connection is used for the all translatory movement (X, Y, Z). Such assembly enables the motion of the model in the range defined for each connection. This simulation of measuring path allows motion of movable segments of virtual CMM, with the probe at the end.

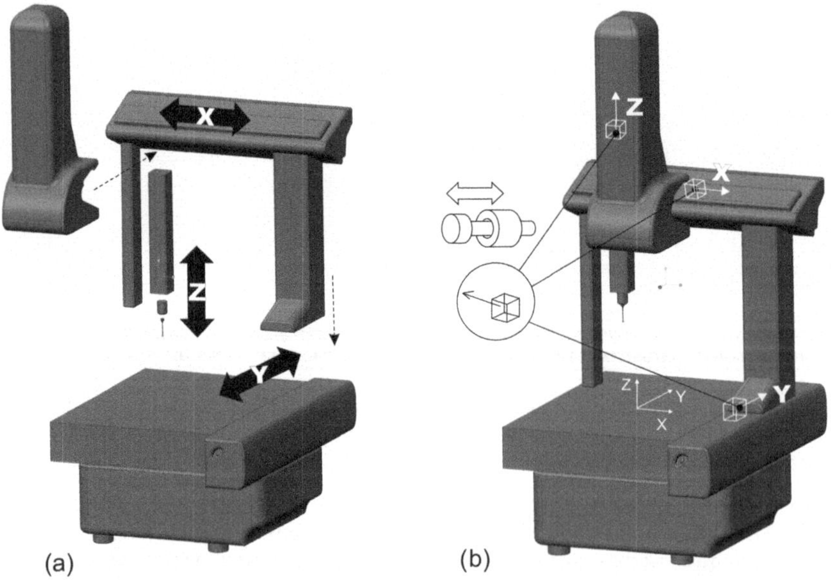

Fig. 7. CAD model of CMM for the simulation of measuring path

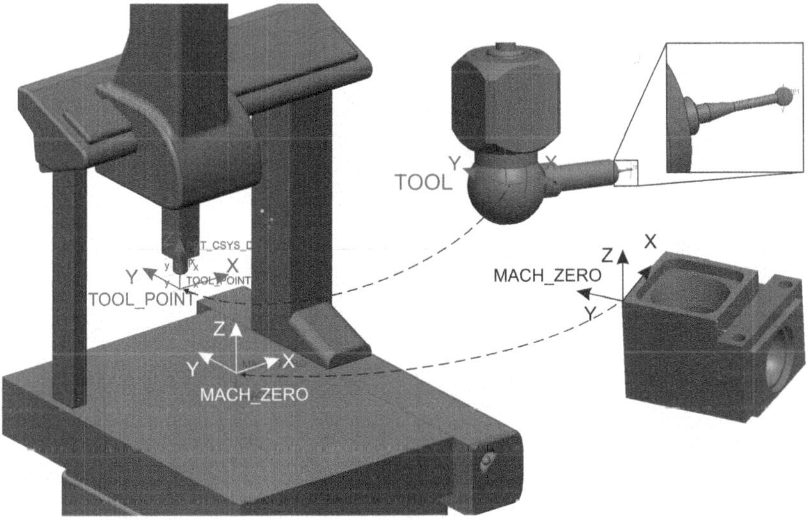

Fig. 8. Coordinate systems of the workpiece and probe for the simulation of measuring path

Besides already mentioned, as it can be seen on Fig. 8, it is necessary to define coordinate systems on:

- working table, market as MACH_ZERO,
- workpiece, market also as MACH_ZERO,
- head of sensor, market as TOOL_POINT, which axis is defined with same direction and orientation as the axis of coordinate system on workpiece, and
- probe, market as TOOL, if it is necessary to create own probe.

Matching the coordinate system MACH_ZERO of workpiece and coordinate system MACH_ZERO on working table enables the setting of the workpiece on the configured virtual CMM during the measuring simulation. Similar is also with matching of coordinate systems TOOL and TOOL_POINT during the setting of the probe on the head of sensor.

During the path simulation on Virtual CMM, Fig. 9, besides workpiece and probe that moves through measuring path, it is possible to create and load fixture that is of importance in verification of measuring path and detection of the possible collisions during the program execution. After the simulation on virtual CMM generated measuring path that is saved in CL file (DMIS program) is verified.

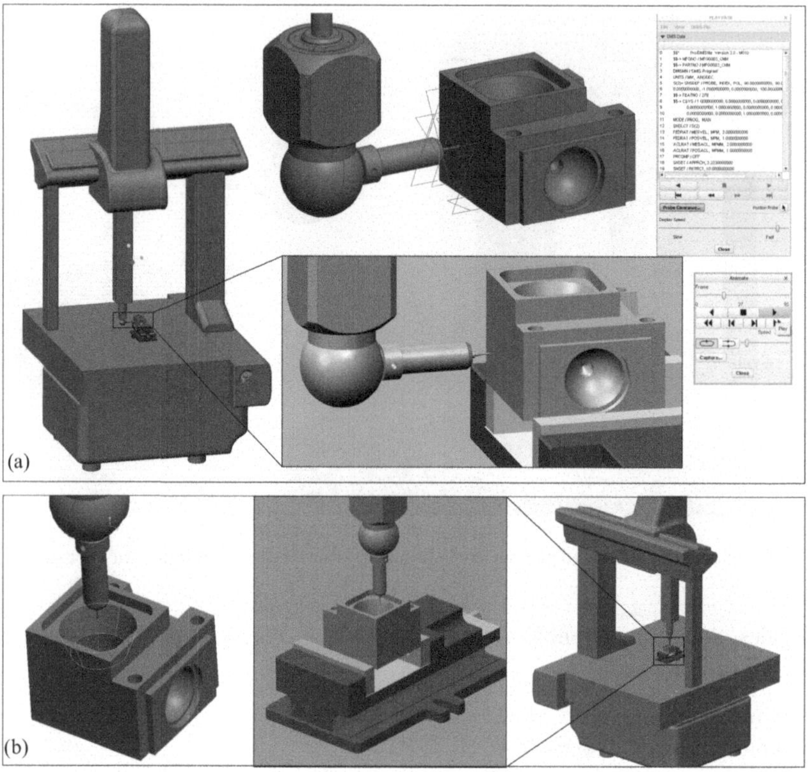

Fig. 9. Measuring path simulation on virtual CMM in CAD/CAM environment (a) plane, and (b) truncated cone

5 Conclusion

The CMMs are recognized as a flexible element of production metrology, and they are applicable for a wide range of metrological tasks. From other side, metrological tasks depend on geometrical and metrological complexity of workpiece and it is very important to verify measuring process before execution of measuring path on CMM. The verification of measuring path should check of collision between the measuring sensor and the workpiece and fixture in virtual environment.

In this paper, the verification of measuring path, based on modified Hammersly's algorithm for basic geometrical features of workpiece, is first visualized in MatLab and than through simulations in CMM module of PTC Creo. The simulation of the measuring path was realized on the configured virtual CMM in the CAD/CAM environment. Output from the simulation on virtual CMM generated measuring path is saved in CL file (DMIS program). When this file is generated, developing of a postprocessor leave the possibility for create of control data list for different CMMs. Proposed concept for CMM simulation and its output could be useful for other CMMs producer.

References

1. Weckenmann, A., Lorz, J.: Monitoring coordinate measuring machines by calibrated parts. J. Phys: Conf. Ser. **13**, 190–193 (2005)
2. Limaiem, A., ElMaraghy, A.H.: Automatic path planning for coordinate measuring machine. In: Proceedings of the 1998 IEEE, International Conference on Robotics and Automation, Leuven, Belgium, pp. 887–892 (1998)
3. Zhao, H., Kruth, J.P., Gestel, N.V., Boeckmans, B., Bleys, P.: Automated dimensional inspection planning using the combination of laser scanner and tactile probe. Measurement **45**, 1057–1066 (2012)
4. Ravishankar, S., Dutt, H.N.V., Gurumoorthy, B.: Automated inspection of aircraft parts using a modified ICP algorithm. Int. J. Adv. Manuf. Technol. **46**, 227–236 (2010)
5. Chang, H.C., Lin, A.C.: Automatic inspection of turbine blades using 5-axis coordinate measurement machine. Int. J. Comput. Integr. Manuf. **23**(12), 1071–1081 (2010)
6. Chang, H.C., Lin, A.C.: Five-axis automated measurement by coordinate measuring machine. Int. J. Adv. Manuf. Technol. **55**, 657–673 (2011)
7. Yau, H.T., Menq, C.H.: Automated CMM path planning for dimensional inspection of dies and molds having complex surface. Int. J. Mach. Tools Manufact **35**(6), 861–876 (2005)
8. Stojadinovic, S., Majstorovic, V.: An Intelligent Inspection Planning System for Prismatic Parts on CMMs. Springer, Switzerland (2019)
9. Stojadinovic, S., Majstorovic, V., Durakbasa, N., Sibalija, T.: Towards an intelligent approach for CMM inspection planning of prismatic parts. Measurement **92**, 326–339 (2016)
10. Stojadinovic, S., Majstorovic, V.: Developing engineering ontology for domain coordinate metrology. FME Trans. **42**(3), 249–255 (2014)
11. Stojadinovic, S., Majstorovic, V.: Towards the development of feature – based ontology for inspection planning system on CMM. J. Mach. Eng. **12**(1), 89–98 (2012)
12. Chen, Y.H., Wang, Y.Z., Yang, Z.Y.: Towards a haptic virtual coordinate measuring machine. Int. J. Mach. Tools Manuf **44**(10), 1009–1017 (2004)
13. Sładek, J., Gąska, A.: Evaluation of coordinate measurement uncertainty with use of virtual machine model based on Monte Carlo method. Measurement **45**(6), 1564–1575 (2012)
14. Ostrowska, K., Gąska, A., Sładek, J.: Determining the uncertainty of measurement with the use of a virtual coordinate measuring arm. Int. J. Adv. Manuf. Technol. **71**, 529–537 (2014)
15. Hu, Y., Yang, Q., Sun, X.: Design, implementation, and testing of advanced virtual coordinate-measuring machines. IEEE Trans. Instrum. Measur. **61**(5), 1368–1376 (2012)
16. Majstorovic, V., Stojadinovic, S., Sibalija, T.: Development of a knowledge base for the planning of prismatic parts inspection on CMM. Acta IMEKO **4**(2), 10–17 (2015)
17. Stojadinovic, S., Majstorovic, V., Durakbasa, N., Sibalija, T.: Ants colony optimization of the measuring path of prismatic parts on a CMM. Metrol. Measur. Syst. **23**(1), 119–132 (2016)
18. Lee, G., Mou, J., Shen, Y.: Sampling strategy design for dimensional measurement of geometric features using coordinate measuring machine. Int. J. Mach. Tools Manuf **37**(7), 917–934 (1997)
19. Slavkovic, N., Dimic, Z., Zivanovic, S., Milutinovic, M.: Kinematic modeling of 5-axis horizontal milling machine emulated from vertical articulated robot. FME Trans. **46**(1), 46–56 (2018)
20. Kanife, P.O.: Computer Aided Virtual Manufacturing Using Creo Parametric. Springer, Switzerland (2016)

21. Zivanovic, S., Glavonjic, M., Milutinovic, D.: Configuring a mini-laboratory and desktop 3-axis parallel kinematic milling machine. Strojniški Vjesnik J. Mech. Eng. **61**(1), 33–42 (2015)
22. Zivanovic, S., Kokotovic, B.: Configuring a virtual desktop 5-axis machine tool for machine simulation. In: Proceedings of the 12th International Conference on Accomplishments in Electrical and Mechanical Engineering and Information Technology DEMI 2015, Faculty of Mechanical Engineering Banja Luka, Banja Luka, pp. 255–262 (2015)
23. Slavkovic, N., Zivanovic, S., Milutinovic, D., Kokotovic, B.: Robot machining simulation in step-nc machine environment. In: Proceedings of the 13th International Conference on Accomplishments in Electrical and Mechanical Engineering and Information Technology DEMI 2017, Faculty of Mechanical Engineering Banja Luka, Banja Luka, pp. 43–50 (2017)
24. Stojadinovic, S., Slavkovic, N., Milutinovic, D.: Off-line programming and simulation of the cell based on the "MITSHUBISHI MOVEMASTER RV-M1" robot (on Serbian language). In: Proceedings of the 36th JUPITER Conference, University of Belgrade, Faculty of Mechanical Engineering, Belgrade, pp. 3.64–3.69 (2010)
25. Dimic, Z., Milutinovic, D., Zivanovic, S., Kvrgic, V.: Virtual environment in control and programming system for reconfigurable machining robot. Tehnicki Vjesnik **23**(6), 1821–1829 (2016)

Validation of Virtual CMM-Based Method for Uncertainty Estimation of Measurements Performed on Five-Axis Coordinate Measuring Machines

Adam Gąska(✉), Wiktor Harmatys, Piotr Gąska, Maciej Gruza, and Jerzy Sładek

Laboratory of Coordinate Metrology, Cracow University of Technology, al. Jana Pawla II 37, 31-864 Cracow, Poland
agaska@mech.pk.edu.pl

Abstract. According to guidelines of international standards it is necessary that all methods used in calibration and test laboratories for determination of measurement result and uncertainty should be validated. It is important especially in case of methods that are newly developed to fit some purpose. One of such methods is a virtual CMM-based method that was designed for uncertainty estimation of measurements performed using five-axis coordinate measuring machines. Different methodologies were chosen for validation of this approach. Theoretical aspects of them like basic rules of their implementation and selection of validation criterions were presented in this paper along with description of validation experiments and their results. The main finding of this research is that Virtual CMM prepared for five-axis coordinate measuring machines passed the validation carried out using all mentioned methodologies.

Keywords: Five-axis system · CMM · Uncertainty estimation · Validation

1 Introduction

International standards that regulate the functioning of accredited laboratories [1] indicates that laboratories should chose the current measurement methods described in the well-known normative documents. However, when there is a requirement to develop specific, fit for purpose methods, each time their proper functioning should be proven in the validation process. There are lots of factors that affect the proper organization of the validation process such as professional laboratory staff with extensive knowledge and measuring skills, adequate resources provided by the top management of the laboratory which include appropriate measuring machines with software and effective communication.

All non-standardized methods should be agreed with a potential client and validated. Validation of measuring methods, as the process of confirming that the chosen method used to perform a particular specified measurement is suitable for intended purpose, provides the reliability and consistency of the measurements. This is why it is

V. D. Majstorovic and N. Durakbasa (Eds.): IMEKOTC14 2019, LNME, pp. 39–46, 2019.
https://doi.org/10.1007/978-3-030-18177-2_4

so important to use correct validation methods both in scientific community and in production engineering [2].

Development of coordinate measuring technique requires validated methods for assessing the accuracy of measurements. On the other hand, the constant struggle to reduce time of all operations connected with production process, including quality inspection, necessitates formulation of new methods for uncertainty estimation, as the old ones require multiple repetitions of performed measurements. The most popular group of methods that are capable of quick determination of measurement uncertainty are simulation methods based on usage of virtual models of measurements. Virtual models have to be created separately for each model of CMM on which it is used. In order to allow their daily use it is necessary to previously run their validation, which require the choice of the most important parameters of the validation and appropriate validation techniques. Validation technique must provide the best laboratory and environmental conditions for carrying out the measurements so that the results of the validation parameters were reliable and unambiguous.

This paper propose two validation procedures for verification of virtual model of five-axis CMM developed in Laboratory of Coordinate Metrology at Cracow University of Technology. These two techniques of validation are based on comparison technique with another method and interlaboratory comparisons. Regarding the first one, the results obtained using virtual CMM-based method (which is under validation) are compared to results obtained utilizing method which use calibrated workpieces (called later in this paper the comparative method) [3, 4]. In case of interlaboratory comparisons the method based on checking the metrological compatibility was used [5–7]. Metrological compatibility was checked for results obtained using the two methods mentioned above. Details on implementation of these two validation procedures in relation to virtual model of five-axis CMM are given in next chapter.

2 Validation Models Chosen for Verification of Virtual CMM Model for Five-Axis Coordinate Measuring Machine

2.1 Model Based on Consistency Check

Both models presented in this chapter are based on statistical analysis of series of results obtained using the method that is under validation (in this case it is virtual CMM method for five-axis CMM) and the method that is widely accepted and has already been validated (in this case it is the method based on usage of calibrated workpieces).

The first validation model, which is the model based on consistency check, starts with the calculation of the weighted mean (1), so called Reference Value (RV):

$$RV = \frac{x/u^2(x) + y/u^2(y)}{1/u^2(x) + 1/u^2(y)} \tag{1}$$

where:

x, y-the mean values of the results obtained by calibrated workpieces method (x) and developed virtual CMM method (y), u(x), u(y) -uncertainties calculated according to the particular method

The next step of the model is to perform chi-squared test which is calculated as (2):

$$\chi^2_{obs} = \frac{(x - RV)^2 + (y - RV)^2}{u^2(x) + u^2(y)} \tag{2}$$

$$\Pr\{\chi^2(v) > \chi^2_{obs}\} > 0.05$$

where:
$v = N - 1$ - the degrees of freedom.

This test, if fails, assumes consistency of coordinate measuring methods used. In this case, next steps have to be performed. If the test rejects the hypothesis about consistency of results obtained using considered methods, the validation ends with negative result.

In next step, the standard uncertainty associated with reference value (RV) need to be calculated using (3):

$$\frac{1}{u^2(RV)} = \frac{1}{u^2(x)} + \frac{1}{u^2(y)} \tag{3}$$

where:
u(RV) is the standard uncertainty of determination of reference value

The most important is the validation acceptance interval for the results obtained with considered methods (4):

$$< RV - u(RV); RV + u(RV) > \tag{4}$$

The validation model assumes that if both intervals (5)

$$\begin{gathered} < x - u(x); x + u(x) > \\ < y - u(y); y + u(y) > \end{gathered} \tag{5}$$

have the common part with the validation acceptance interval then the validation ends with the positive result and the method that were under validation may now be considered as validated method.

2.2 Model Based on Metrological Compatibility Check

This model is based on checking if the following formula is satisfied (6):

$$\frac{|x_i - x_R|}{\sqrt{u^2(x_i) + u^2(x_R) - 2r(x_i, x_R)u(x_i)u(x_R)}} \leq \kappa \tag{6}$$

where:

$x_R, u(x_R)$ - the reference values with the standard uncertainties (obtained using calibrated workpieces method)

x_i, $u(x_i)$ - values with the standard uncertainties obtained using virtual CMM method

$r(x_i, x_R)$ - correlation coefficient, calculated using Eq. (7)

$$r(x_i, x_R) = \frac{\sum_{i=1}^{n} [(x_i - \overline{x})(x_R - \overline{x})]}{\{[\sum_{i=1}^{n} [(x_i - \overline{x})^2] \cdot [\sum_{i=1}^{n} [(x_R - \overline{x})^2]\}^{\frac{1}{2}}} \tag{7}$$

κ- the threshold determined on the basis of the measurements and calculations for coordinate metrology should be smaller than 1, 2 (basing on research presented in [8]).

The result of the test of metrological comparability gives the answer to the question whether the measurement results with their uncertainties obtained by a method being under validation $[x_i, u(x_i)]$ are sufficiently comparable with the set of results obtained from the reference method $[x_R, u(x_R)]$ and thus acceptable.

3 Performed Validation Measurements

Verification measurements included measurement of the Multi-Feature Check (MFC) standard. It is the complex shaped measuring standard for the purposes of measurement accuracy assessment and the determination of measurement uncertainty for nearly all features, dimensions and relations applicable to the coordinate measuring technique. The standard that was used has length of 200 mm and external diameter of 100 mm.

The following features were measured on MFC standard: distance between front planes, plane flatness, cylindricity and diameter of internal cylinder, parallelism deviation between front planes and angle between two planes (front plane and side plane).

The MFC standard was calibrated on PMM 12106 Leitz Messtechnik measuring machine with CMC (Calibration and Measurement Capability) for calibration of geometrical standards equal to 0,0006 + 0,0007·L mm (where L is measured length given in m).

The measurements were performed on five-axis CMM Zeiss WMM 850S equipped with PH20 probe head with TP20 STD module (Fig. 1). A measurement procedure for measurements of MFC consisted of: manual measurement of external cylinder using 16 points divided into 2 sections, each containing 8 points regularly distributed over the circumference of cylinder's section; measurement of side and front plane using 8 points that were regularly distributed over the planes; construction of manual part alignment, where the axis of cylinder is the main axis of constructed coordinate system, side plane is used to indicate the direction of second axis of coordinate system and its origin is at the intersection point of the front plane and the external cylinder's axis. Next step involved repeating of mentioned measurements in automatic mode, construction of part coordinate system using features measured in automatic mode and then automatic measurements of second front plane (also using 8 points) and chosen internal cylinder using similar points distribution as for external cylinder.

Fig. 1. Measurement of calibrated workpieces and MFC standard during validation of virtual model for five-axis CMM

The gauge block and the ring standard that satisfies the similarity conditions presented in [3] were used as reference objects during uncertainty estimation according to guidelines of comparative method. 20 measuring cycles were run, each consisted of measurements of both calibrated workpieces and MFC standard.

For measurements performed using comparative method the same number of points was used for each feature as in case of measurements of MFC and their distribution over the standard surface was similar for both measurements.

The results presented for developed virtual model are the results of single measurement (taken separately from measurements according to comparative method), while the corresponding uncertainty was determined using multiple simulation of measurement performed by virtual CMM software developed by authors of this paper.

After determination of results and standard uncertainties corresponding to them, the validation tests were performed using methodology described in Sects. 2.1 and 2.2. Obtained results for measurements of MFC and results of statistical analyses according to validation models are presented in Tables 1, 2 and Fig. 2.

4 Results

Results presented in Table 1 shows that for all presented measuring tasks the intervals that contains the true value of measured quantity have the common part with the validation acceptance interval so the validation process according to procedure presented in Sect. 2.1 ends with positive result.

Table 1. Results of verification of developed virtual model using validation model based on consistency check. Results for angle given in °, for other features/relations in mm

Characteristic/ relation	Calibrated worpieces method		Virtual CMM		Statistical consistency check		
	x	u(x)	y	u(y)	VAI	<x – u(x); x + u(x)>	<y – u(y); y + u(y)>
Plane – plane distance	199,6208	0,0006	199,6219	0,0009	<199,6206; 199,6216>	<199,6202; 199,6214>	<199,6210; 199,6228>
Internal diameter	59,9652	0,0005	59,9656	0,0003	<59,9652; 59,9658>	<59,9647; 59,9657>	<59,9653; 59,9659>
Cylindricity	0,0023	0,0007	0,0024	0,0008	<0,0018; 0,0029>	<0,0016; 0,0030>	<0,0016; 0,0032>
Flatness	0,0008	0,0006	0,0009	0,0006	<0,0004; 0,0012>	<0,0002; 0,0014>	<0,0002; 0,0014>
Parallelism	0,0034	0,0011	0,0036	0,0007	<0,0030; 0,0041>	<0,0023; 0,0045>	<0,0029; 0,0043>
Plane – plane angle	89,9824	0,0023	89,9806	0,0019	<89,9799; 89,9828>	<89,9801; 89,9847>	<89,9787; 89,9825>

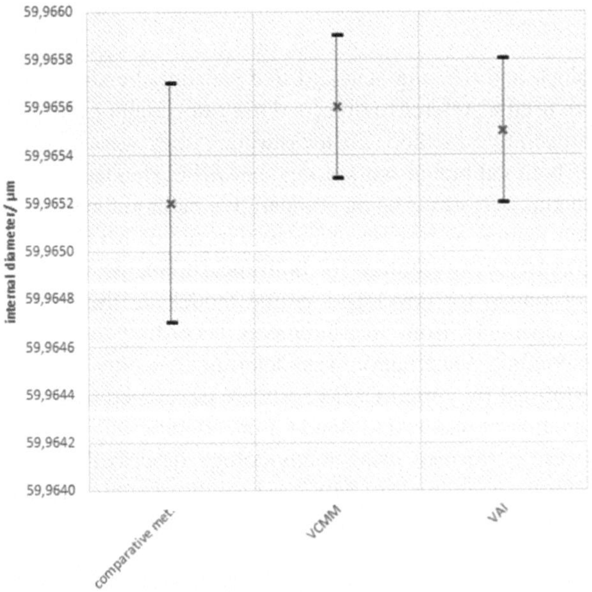

Fig. 2. Results of internal diameter measurement with uncertainty estimated using considered methods compared to validation acceptance interval

In case of results presented in Table 2, the correlation coefficient between results produced by two compared methods was calculated using 20 results of measurements performed using comparative method and 20 simulation results randomly chosen out of total 1000 simulation trials.

As can be observed for all performed measuring tasks the value of validation threshold calculated using left side of the inequality (6) was smaller than 1, 2 so this inequality is a true statement for all considered measuring tasks.

Table 2. Results of verification of developed virtual model using validation model based on metrological compatibility check. x_i, x_R, $u(x_i)$ and $u(x_R)$ given in mm.

Characteristic/relation	Calibrated worpieces method		Virtual CMM		Validation threshold
	x_R	$u(x_R)$	x_i	$u(x_i)$	κ
Plane – plane distance	199,6208	0,0006	199,6219	0,0009	1,13
Internal diameter	59,9652	0,0005	59,9656	0,0003	0,78
Cylindricity	0,0023	0,0007	0,0024	0,0008	0,11
Flatness	0,0008	0,0006	0,0009	0,0006	0,13
Parallelism	0,0034	0,0011	0,0036	0,0007	0,17
Plane – plane angle	89,9824	0,0023	89,9806	0,0019	0,72

On basis of presented results, developed virtual model may be regarded as working properly and providing correct uncertainty values for measurements performed on five-axis measuring systems.

5 Conclusions

This paper presents two different methodologies for validation of virtual CMM-based uncertainty estimation method for five-axis CMM. According to both of performed validation procedures the virtual model of five-axis coordinate machine should be regarded as giving results of evaluated measurements and associated standard uncertainties comparable to those produced by calibrated workpieces method (that is widely used all around the world). On that basis, the developed model may be regarded as validated. Thus, as was mentioned in introduction, confirm that this virtual model used to evaluate a common measuring tasks known from geometrical dimensioning and tolerancing framework is suitable for intended purpose, and provides the reliability and consistency of the measurements.

Validation methods according to which the measurements presented in this paper were done may be used in analogous way for verification of correct functioning of other virtual models that would be developed in the future (or of virtual models that already exist but have not been experimentally validated yet). It is of huge importance that all virtual models, created in different scientific disciplines, in addition to purely theoretical verification (usually based on a comparison of the results produced by the model being under verification with the mathematical/simulation model of the process/phenomenon) undergone also the practical one, which may be based on the procedure presented in this paper.

References

1. ISO/IEC 17025: General requirements for the competence of testing and calibration laboratories (2017)
2. Sładek, J.: Accuracy of coordinate measurements, Publishing House of Cracow University of Technology, Cracow 2011. ISBN 978-83-7242-643-7
3. ISO/TS 15530-3: Geometrical product specifications (GPS) - coordinate measuring machines (CMM): technique for determining the uncertainty of measurement – part 3: use of calibrated workpieces or measurement standards (2011)
4. Weckenmann, A., Lorz, J.: Monitoring coordinate measuring machines by calibrated parts. J. Phys: Conf. Ser. **13**, 190–193 (2005)
5. Campanelli, M., Kacker, R., Kessel, R.: Variance gradients and uncertainty budgets for nonlinear measurement functions with independent inputs. Meas. Sci. Technol. **24**, 025002 (2013)
6. Possolo, A., Elster, C.: Evaluating the uncertainty of input quantities in measurement models. Metrol. **51**(3), 339 (2014)
7. Kacker, R.N., Kessel, R., Sommer, K.D.: Metrological compatibility and statistical consistency. In: Proceedings 10th International Symposium on Measurement and Quality Control, Osaka, Japan, 05–09 September 2010
8. Gromczak, K., Gąska, A., Kowalski, M., Ostrowska, K., Sładek, J., Gruza, M., Gąska, P.: Determination of validation threshold for coordinate measuring methods using a metrological compatibility model. Meas. Sci. Technol. **28**, 1 (2017)

Metrological Approach for Testing Performance of Optical 3D Measurements Systems

Bojan Acko(✉) and Rok Klobucar

Faculty of Mechanical Engineering,
University of Maribor, 2000 Maribor, Slovenia
bojan.acko@um.si

Abstract. For quality control purposes, many manufacturing industries perform dimensional metrology checking processes that often necessitate the use of precision optical 3D measurement instruments, such as fringe projection systems, laser scanners and other similar non–contact systems. The typical measurement accuracies of commercially available instruments are down to few micrometers. In order to assure traceability of measurements to the SI, the meter, these instruments are normally calibrated by using different precision measurement standards, such ball bars and special 3D set-ups. With these optical standards, a range of different kinds of optical measurement instruments, together with their associated internal reference scales and image processing algorithms can be evaluated and then verified.

Recent research work in the field 3D optical artefact calibration is presented in this book. In addition, a tetrahedron ball standard and a method for calibrating 3D optical measurement systems are presented. The outcome of the presented research is a calibration procedure with approved measurement uncertainty that has already been accredited by the national accreditation body.

Keywords: Calibration · Traceability · Optical CMMs

1 Introduction

With the initial development of coordinate measuring machines (CMMs) in the 1970s, 3D metrology has become established in industry. Since that time, industry together with scientific metrology institutes have developed test procedures which enable machine manufacturers as well as end users to evaluate the performance of different types of CMMs [1, 2].

The specified acceptance and re-verification procedures have meanwhile been established in many national and international guidelines [3]. However, traceable measurements on CMMs are only possible by using a well–known substitution method based on calibrated artefacts and by employing advanced virtual CMM techniques [4, 5].

Apart from tactile measurements, traceability of optical 3D measurements [6, 7] is still an open issue, as a qualified statement of the task specific measurement uncertainty, requested by international standards like ISO 9001 [8] can hardly be given. Traceable standard reference artefacts and procedures for both calibration and

V. D. Majstorovic and N. Durakbasa (Eds.): IMEKOTC14 2019, LNME, pp. 47–61, 2019.
https://doi.org/10.1007/978-3-030-18177-2_5

verification of optical 3D systems practically do not exist. Some forms of equipment verification tests are performed by producers of measuring systems, but these employ non–validated procedures in-conjunction with in–house standards. Unfortunately no accredited or national laboratories are involved in the traceability chain. Thus demonstration of the conformance of a piece of measuring equipment to meet specifications according to ISO 14253-1 [9] is consequently not possible. This leads to increasing costs as it is not possible to distinguish reliably between acceptable and non–acceptable parts, especially where tolerances are small compared to the measurement uncertainty.

For the above reasons, the consortium of the European project iMERA Plus JRP T3.J2.2 NIMTech project [10] decided to develop different types of 3D artefacts and corresponding procedures for verifying the freeform measurement capability of optical and tactile co–ordinate measuring systems. A range of reference artefacts has been developed, allowing the performance of optical–based 3D measurement systems, such as fringe projection, laser scanners and other similar non–contact systems to be verified against a set of known surface conditions. The purpose of these artefacts was to demonstrate the dimensional measurement capability of selected optical–based 3D measurement technologies to measure specific forms and various surface conditions, rather than to be universal standards [11].

All the standards were designed in accordance with industrial lead requirements. They are now available from the respective national metrology institutes (NMIs) and can be supplied with appropriate calibration data. High precision specifications and associated metrological characteristics were confirmed by measuring all the artefacts on different tactile and optical CMMs using the facilities from all the three participating NMIs.

These developments were performed by three project partners, namely National Physical Laboratory from United Kingdom (NPL), Physikalisch–Technische Bundesanstalt from Germany (PTB), and Laboratory for Production Measurement from the University in Maribor, which is representing the Metrology Institute of the Republic of Slovenia (MIRS/UM-FS/LTM).

This article is presenting the tetrahedron standard developed by MIRS/UM-FS/LTM and its application in assuring traceability of optical 3D fringe projection measurement systems.

2 Standards and Procedures

2.1 Documented Standards

From around 1994 onwards, ISO 10360-2 series part 3, "Geometrical Product Specifications (GPS) acceptance and re-verification tests for coordinate measuring machines (CMM)" [3] has been available to assist in verifying the performance of such machines. Until 2011 there were 6 parts of this document, each part specialising in different technical areas. For example, Part 1 describes fundamental CMM "Vocabulary" and Part 6 describes "Estimation of errors in computing Gaussian associated features". This standard does not cater for optical based coordinate measuring systems, such as those employing laser triangulation or fringe projection scanners [7]. In 2011, ISO 10360-2 Part 7 "CMMs equipped with imaging probing systems" was introduced.

In 2002, a German guideline VDI/VDE 2634 [12] was introduced. This guide relates to optical–based 3D scanning systems and currently consists of three parts:

- Part 1: "Imaging systems with point-by-point probing";
- Part 2: "Optical systems based on area scanning";
- Part 3: "Multiple view systems based on area scanning";

This VDI/VDE guideline defines a particular way of measuring a reference artefact, which is typically used to define a spatial length or simple forms (sphere, plane) to a high accuracy. Although much more suited to optical systems than the ISO 10360-2, parts 2 and 3 of the VDI/VDE 2634 guideline are more relevant to surface scanning. However the guidelines do not cover performance verification of freeform surface measuring systems.

2.2 Standards of Measurement

A variety of standards or artefacts can be used for calibrating and testing performance of CMMs [13, 14]. Commercially available standards, like gauge blocks, balls, internal and external cylinders, step gauges, etc. can be used on their own or in different combinations. In addition, different kinds of special 2D and 3D artefacts have been developed. Most of them consist of basic standards in different spatial or planar configurations. In most cases, such artefacts are equipped with external and internal balls.

A ball is the most common element currently used for determining metrological characteristics of optical 3D measurement systems such as fringe projection, laser scanners, photogrammetric and other similar non-contact systems. Artefacts [13] are produced in the form of a plane and a single ball or a ball-bar with two or more balls attached as illustrated in Fig. 1.

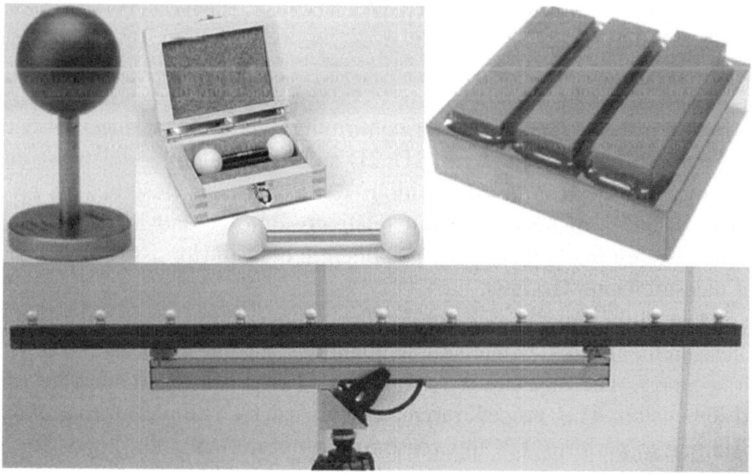

Fig. 1. Existing artefacts for testing optical CMMs [11, 13]

Verification of these non-contact measurement systems is complex and useful guides such as the VDI/VDE 2634 2 series [12] describe methods to demonstrate capability using test artefacts (see Fig. 2) with prismatic features, such as spheres, ball-bars and planes. However, the guides do not extend to fully address performance verification when freeform surfaces are to be measured using optical-based techniques. In order to verify most parts of the measurement volume, the verification artefact needs to be measured in several positions within the measuring volume. VDI/VDE 2634 states at least three arbitrary positions need to be chosen for the measurements, but it is recommended to use five to seven positions. Especially for multiple view scanning systems, the total number of scans resulting from at least five different sensor positions for at least three artefact positions is a minimum. Thus the number of measurements required can dramatically increase in number, from typically 15, to as many as 40 data sets. Thus this type of verification approach can be extremely time consuming and result in high costs.

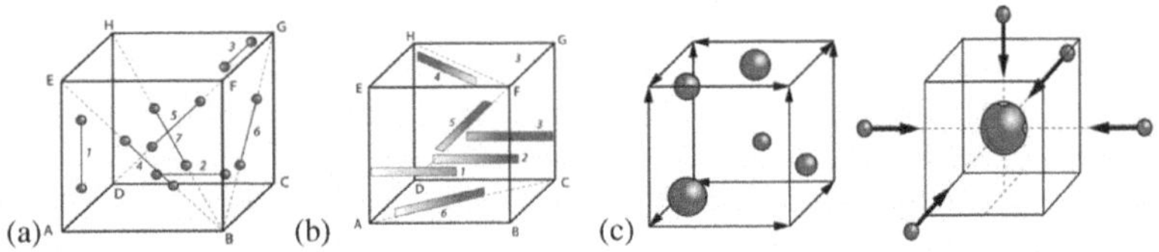

Fig. 2. Verification tests according to VDI/VDE 2634: (a) determination of sphere spacing error, (b) determination of flatness error, (c) determination of probing error

2.3 Existing Procedures

Most commonly three types of procedures are used for assuring traceability of 3D measurements:

- performance verification tests [2, 15];
- task related calibration; and
- Virtual CMM [4, 5, 12];

Performance tests serve as a tool for confirming measurement uncertainty declared by a CMM manufacturer. A variety of 1D, 2D and 3D standard artefacts are employed for performing such tests. The application of such tests is designed to be straight forward and is aimed to be carried out relatively quickly and in a financially economical way. Therefore, they are widely applied in industry for assuring traceability of industrial measurements [16, 17].

Task related calibrations, which employ the comparator principle, are often used for specific measurement tasks, where simple quantities need to be measured or calibrated to a high accuracy. The measurand is compared with a calibrated standard of similar form and dimension. This procedure cannot be used for complex forms and is quite time consuming to perform. For this reasons it is not normally employed for assuring traceability of precision industrial measurements.

The Virtual CMM [5, 18] represents the highest level of assuring traceability of coordinate measurements. It is the only method for performing calibrations using CMMs, in strict accordance with the definition of the term 'calibration'. Significant random and systematic errors of a CMM are modelled to characterize the kinematic behavior of the CMM and its probing process. Therefore, all input parameters have to be determined traceable to SI units. For instance these are geometrical errors of the CMM's guides or probing errors. The methods for determining the input parameters are relatively time consuming and as a consequence also quite expensive. However, often the Virtual CMM is used by calibration laboratories, which calibrate complex 2D or 3D standards.

3 Tetrahedron Artefact for Testing Performance of Optical CMMs

Experiences of MIRS/UM-FS/LTM in traceability of optical measurements were limited to 1D [19] and 2D artefacts before joining the project [10]. Tetrahedronal verification artefacts that were developed during the project were the first 3D artefacts designed by this laboratory. The purpose of these artefacts is for testing the performance of fringe projection and similar 3D optical measurement devices with both single scan and multiple scan measuring capabilities [11].

In order to minimize verification time and still comply with standard requirements (e.g. VDI/VDE 2634), a new artefact was designed by combining the advantages of both ball-cube and ball-bar artefacts [11]. Normally, balls to the rear side of a ball cube are hidden from the measurement sensor.

By moving the ball from the back side to the interior of the artefact, this problem is resolved. This new spatial artefact is applicable for verifying single, as well as multiple-view optical-based area scanning systems. When comparing only three, in contrast to seven ball-bar measuring positions [12] with a single positioning of this spatial artefact, extensive time reductions in verification periods can be achieved. Because of the exposed position of the balls, the artefact can successfully replace a single-bar artefact and all the necessary multiple repositioning.

3.1 Design of the Artefact

The developed artefact has the geometric shape of an irregular tetrahedron. It consists of six tubes, three long and three short ones in a ratio of 2.3 : 1, and four ceramic spheres serving as probing elements (Fig. 3). Such design was chosen in order to have a spatial standard with a minimum number of probing elements, which can be scanned by an optical scanner in a single scan. The original idea was to have all four balls pointing out of the tetrahedron. However, after performing some virtual probing tests, it was decided to move one of the balls to the interior of the tetrahedron in order to enable the tested scanner to see all balls at the same time.

The artefact was constructed in two sizes. The height of the small artefact is 260 mm, while the height of the larger artefact is 1050 mm. The diameters of probing balls employed are 20 mm for the small artefact and 30 mm on the larger artefact. The main purpose of making the artefact in two different sizes was to enable the testing of different measurement ranges commonly available with commercial fringe projection measurement systems. The cameras on these systems can be set in various configurations and thus their metrological characteristics can change when operated over different measurement ranges. To support multi-scanning measurement systems, the artefact is additionally equipped with removable targets as shown on the right hand picture in Fig. 3.

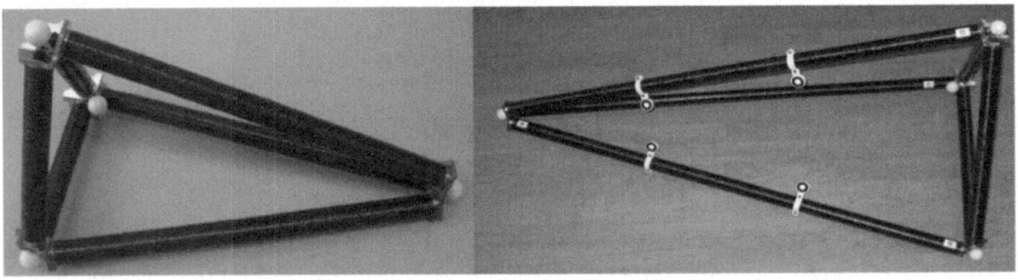

Fig. 3. Small (left-hand picture) and large tetrahedron artefacts [11]

The tubes are constructed from composite materials consisting of carbon fibers in an epoxy matrix.

Ceramic spheres with lambertain surfaces, which offer desirable light scattering properties for optical systems, are glued on to stainless steel ball holders which also serve as joining elements between the tubes. These joints are attached to the composite tubes using epoxy glue and examples are shown in Fig. 4.

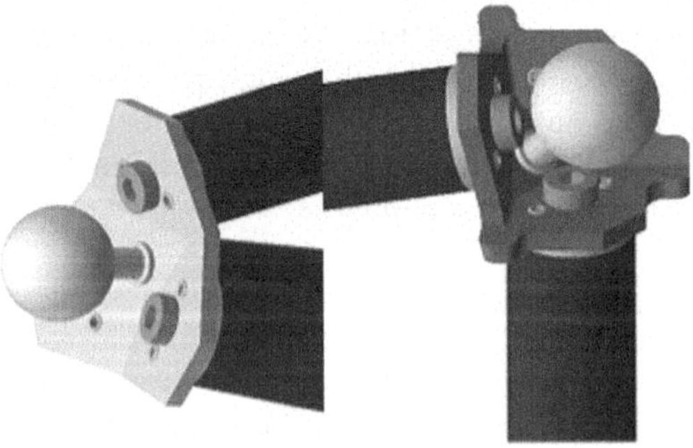

Fig. 4. Two details of the bar joint with a ball holder [11]

4 Calibration of the Artefact

4.1 Measurement Standards and Traceability

The artefact is being calibrated with the coordinate measuring machine (CMM) Zeiss UMC 850 [20]. The traceability of this machine is assured through periodic performance tests by using a set of long gauge blocks 125 mm to 500 mm. The standard is a link of the traceability chain presented in Fig. 7.

The standards are not calibrated periodically following a predefined calibration interval, but before each performance test in which they are used.

4.2 Calibration Procedure

Distances between the centres of the six corner ceramic spheres are measured by means of the ZEISS UMC 850 CMM. Distances between the spheres A-B, A-C and B-C have a nominal value of 506 mm, while distances between spheres A-D, B-D and C-D have a nominal value of 1060 mm.

Special stylus configuration is assembled for this calibration. The system consists of 4 styli, so that spheres can be reached from each direction. Figure 5 presents the example of the stylus system position for probing sphere C.

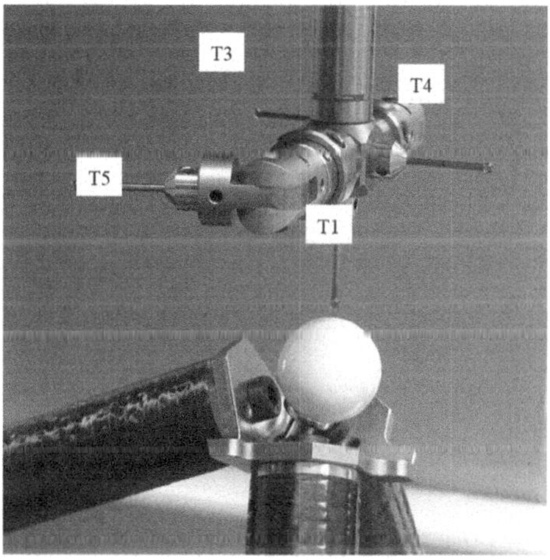

Fig. 5. Stylus system in sphere C position for probing

As indicated in Fig. 6, standard is laid on three prisms $(40 \times 35 \times 120)\ mm^3$, with steel corners glued to prisms, assuring tension-free fixation. Spheres are reachable for styli from all directions.

Each ball is probed in 25 points (5 groups of 5 probing points). Standard's coordinate system shall be put in the centre of the sphere A (bottom left corner, Fig. 11). Spatial rotation is carried out on the plane through centres of spheres A, B and C. Planar rotation is performed around the line through the centres of spheres A and B.

Measurements of the distances between sphere centres are carried out five times and arithmetical mean of altogether six distances (A-B, A-C, B-C, A-D, B-D, and C-D) between sphere centres is calculated.

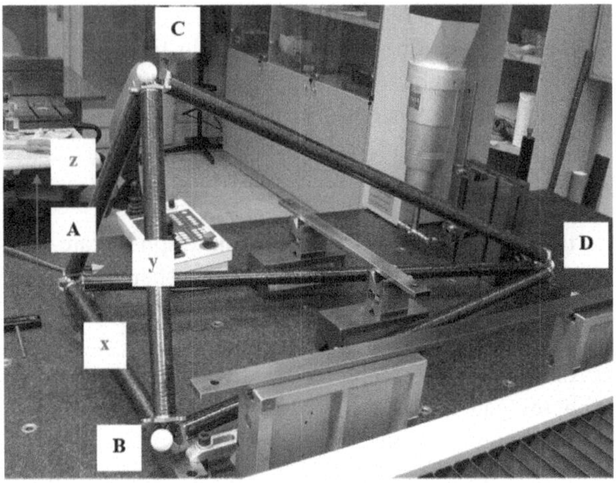

Fig. 6. Position of the bigger standard in the measuring volume

Calibration uncertainty representing best calibration and measurement capability (CMC) is equal to the uncertainty of the CMM performance test and is:

$$U = 3,2\,\mu\text{m} + 5,0 \cdot 10^{-6} \cdot L, k = 2 \tag{1}$$

where:
U - expanded measurement uncertainty at a level of confidence $\approx$ 95%
L - measured length
k - coverage factor.

5 Performance Test of a Fringe Projection System by Using the Tetrahedron Standard

The procedure [11, 21] developed in the Laboratory for Production Measurement in the frame of the EU metrology project iMERA Plus JRP T3.J2.2 NIMTech, is applicable for testing optical measurement machines for quality parameter "length measurement error" in one or more images. It is suitable for testing fringe projection and photogrammetric measurement systems.

5.1 Measurement Standards and Traceability

Tetrahedron measurement standards described in Chapter 3 are used for the performance test. Ceramic spheres have a diameter of 20 mm (small standard) resp. 30 mm (large standard) and diffusely scattering surface. Roughness of the spheres is negligibly small in comparison with the requested precision of the test. Large standard is additionally equipped with four mark plates (see Fig. 3) in order to allow performance testing in multiple scan mode. The standard balls are well visible by all optical instruments, spraying is not necessary.

The standards are calibrated by using a tactile CMM, while the CMM is calibrated in our laboratory by using gauge clocks of dimensions 125 mm to 500 mm. The traceability chain with corresponding measurement uncertainties in all calibration stages is presented in Fig. 7.

5.2 Test Procedure – Single Scan

Small standard (see Sect. 3.1) is normally selected for measurement ranges from (100 × 100 × 100) mm^3 to (500 × 500 × 500) mm^3. However, it can also be used for bigger ranges, if the customer wants to check accuracy within smaller portions of the measurement range. For ranges above (500 × 500 × 500) mm^3 we normally select the bigger standard.

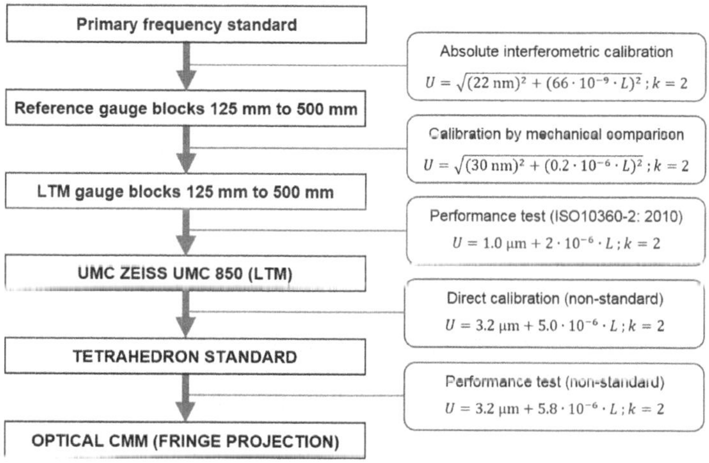

Fig. 7. Traceability chain for the performance test of a fringe projection systems

The standard shall be placed in accordance with the sketches in Fig. 10, as follows:

- The standard should face the camera with the bar AB comprising angle in vertical projection of approx. 15° to the optical axis of the camera.
- Focus of the camera should be in the centre of the standard.
- Camera should comprise a vertical angle of approx. 40°.

The standard should be scanned accordingly to machine's standard procedure. Therefore, a machine operator is required to be present at the test (Fig. 8).

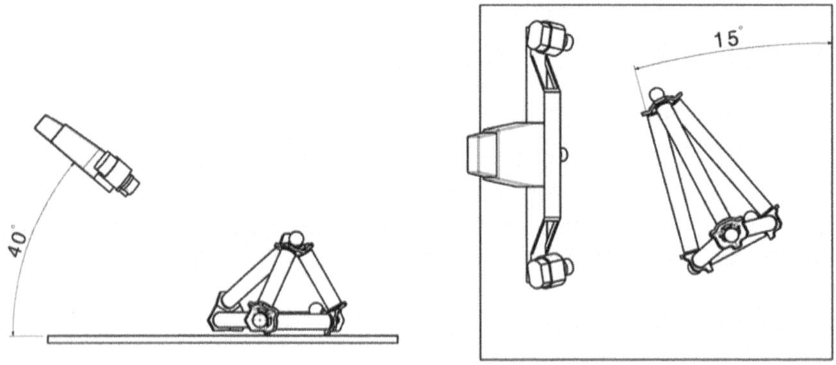

Fig. 8. Position of the standard [11, 21]

Primitives (spheres) should be evaluated by using Best-Fit Sphere procedure with $1-\sigma$ deviation from ideal geometric primitive. Six Point to Point distances between all spheres centres (AB, AC, AD, BC, BD, CD) are to be measured.

5.3 Test Procedure – Multiple Scan

For measurements with more images, large standard (with mark plates) is to be used. The number of images is determined by including all distances between spheres in the measurement. Standard is positioned according to exact instructions in the procedure. Normally, two or three different positions are applied. Primitives (spheres) should be evaluated in the same way as in the single scan procedure.

Measurement results are to be compared with calibrated values of distances between sphere centres.

In accordance with EN ISO 10360-2 [3], the limiting value of the length measurement error MPEE is stated either as a length-dependent quantity $\pm(A + L/K)$, as a maximum value $\pm B$, or as a combination of the two. It must be complied with over the whole measuring volume of the optical 3D measuring system and under all admissible conditions of measurement. For the limiting value for the length measurement error to be completely stated, the operating and ambient conditions referred to must also be given.

When comparing the length measurement error E with its limits MPEE, the expanded uncertainty of measurement U of the test method is also to be taken into account:

$$|E| \leq |\mathrm{MPEE}| - U \text{ for the manufacturer} \tag{2}$$

$$|E| \leq |\text{MPEE}| + U \text{ for the customer} \tag{3}$$

The quality parameter length measurement error is complied with if – taking account of the uncertainty of measurement – no length measurement error exceeds the limit for the maximum permissible length measurement error MPEE as regards its amount. If this limiting value is exceeded maximally once, the measurement in which the limit has been exceeded must be repeated three times. In these repeat measurements the limit must not be exceeded again. Otherwise, the acceptance test is not successful.

The limit for the quality parameter is to be complied with under all conditions permitted by the manufacturer. This applies in particular with regard to the surface properties of the artefacts and the filter parameters.

5.4 Measurement Uncertainty

Mathematical Model of Measurement. Mathematical model of measurement for the presented calibration (test) task can be expressed as follows:

$$e = L_{\text{om}} - L_{\text{m}} \cdot (1 + \alpha_{\text{m}} \cdot \theta_{\text{m}}) \tag{4}$$

where:

e - deviation (measurement result) at 20 °C
L_{om} - length reading on the optical measurement machine
L_{m} - length of the standard
α_{m} - linear temperature expansion coefficient of the standard
θ_{m} - temperature deviation of the standard from 20 °C

Standard Uncertainties of the Input Value Estimates and Combined Standard Uncertainty of Measurement. Combined standard uncertainty [21–25] is expressed with the uncertainties of the input values by the following equation:

$$u_c^2(e) = c_{\text{Lom}}^2 u^2(L_{\text{om}}) + c_{\text{Lm}}^2 u^2(L_{\text{m}}) + c_{\alpha\text{m}}^2 u^2(\alpha_{\text{m}}) + c_{\theta\text{m}}^2 u^2(\theta_{\text{m}}) \tag{5}$$

where c_{i} are partial derivatives of the function (4):

$$c_{\text{Lom}} = \partial \text{f} / \partial L_{\text{om}} = 1 \tag{6}$$

$$c_{\text{Lm}} = \partial \text{f} / \partial L_{\text{m}} = -(1 + \alpha_{\text{m}} \cdot \theta_{\text{m}}) \approx -1 \text{ at } \theta_{\text{mmax}} = \pm 1\,^{\circ}\text{C} \tag{7}$$

$$c_{\alpha\text{m}} = \partial \text{f} / \partial \alpha_{\text{m}} = -\theta_{\text{m}} \cdot L_{\text{m}} \tag{8}$$

$$c_{\theta\text{m}} = \partial \text{f} / \partial \theta_{\text{e}} = -\alpha_{\text{m}} \cdot L_{\text{m}} \tag{9}$$

Standard uncertainties of influence (input) values are calculated (estimated) for applied equipment and method as well as for supposed measurement conditions (Table 1).

(a) *Uncertainty of optical measurement machine reading* $u(L_{om})$

Uncertainty of optical measurement machine reading consists of uncertainty due to rounding the measured value, and uncertainty due to of repeatability of the measurement.

- Optical measurement machine's resolution of 0,01 mm causes the error interval of ±5 μm (due to rounding). With the rectangular distribution, standard uncertainty of machine's reading is:

$$u(L_L) = (5\ \mu\text{m})\Big/\sqrt{3} = 2,2\ \mu\text{m}$$

For resolution 0,0001 mm, $u(L_L)$ = 0,3 μm.

- Repeatability of measurement has to be established during the calibration by measuring the distance AC five times. Average distance is calculated, as well as deviations e and average deviation $\bar{e}$.
- Standard uncertainty of measurement machine reading $u(L_{om})$ is:

$$u(L_{om}) = \sqrt{u(L_L)^2 + (\bar{e}/\sqrt{3})^2} \quad (10)$$

(b) *Uncertainty of the standard's length* $u(L_m)$

According to calibration certificate, uncertainty of the standard's length is:

$$u(L_m) = 1,6\ \mu\text{m} + 2,5 \cdot 10^{-6} \cdot L \quad (11)$$

(c) *Uncertainty of the standard's linear temperature expansion coefficient* $u(\alpha_m)$

Standard's linear temperature expansion coefficient, as established by extensive tests, is $2,2 \cdot 10^{-6}\ °C^{-1}$, interval of $\pm 1 \cdot 10^{-6}\ °C^{-1}$ is estimated. Standard uncertainty at supposed rectangular distribution is:

$$u(\alpha_m) = (1 \cdot 10^{-6}\,°C^{-1})/\sqrt{3} = 0,58 \cdot 10^{-6}\,°C^{-1}$$

(d) *Uncertainty of the standard's* temperature deviation $u(\theta_m)$

Temperature deviations are estimated to be ±1 °C. Standard uncertainty at assumed rectangular distribution is:

$$u(\theta_m) = 0,58\,^\circ\mathrm{C}$$

Table 1. Uncertainty budget in optical CMM performance test

Value X_i	Standard uncertainty	Distribution	Sensitivity coefficient	Uncertainty contribution
L_{om}	(to be calc.)	Rectangular	1	(to be calc.)
L_m	1,6 μm + 2, 5·10^{-6}·L	Rectangular	−1	1,6 μm + 2, 5·10^{-6}·L
α_m	0,58·10^{-6} °C^{-1}	Rectangular	1 °C·L	0,58·10^{-6}·L
θ_m	0,58 °C	Rectangular	2,2·10^{-6} °C^{-1}·L	1,3·10^{-6}·L
		Total: $\sqrt{(\sqrt{(1,6\,\mu\mathrm{m})^2 + u(L_{om})^2})^2 + (2,9 \cdot 10^{-6} \cdot L)^2}$		

For the CMC calculation, supposed resolution of the optical measurement machine is 0, 1 μm and the repeatability of 5 consecutive measurements is 0. CMC (expanded measurement uncertainty with the level of confidence 95%) is then:

$$U = 3,2\,\mu\mathrm{m} + 5,8 \cdot 10^6 \cdot L; k = 2 \tag{12}$$

6 Conclusions

In order to assist optical 3D metrology requirement, our laboratory joined the consortium of the iMERA Plus JRP T3.J2.2 NIMTech project, within which we have developed and presented a number of verification artefacts for different traceability purposes. The presented tetrahedron standards offer quicker solution for performance testing of optical 3D instruments compared with currently available standards. Its design is based on an extensive study of metrology tasks of 3D optical devices, such as sheet forming tools, sheet parts, freeform pipes (exhaust systems, ...), housings of home appliances, car bodies etc. and is therefore suited for efficient evaluation of performing single scan and multiple scan tasks. Further investigations will be focused into the long–term stability and the sensitivity of the artefacts when used in harsh environmental conditions. Also some new materials with lower mass, better surface properties and even better stability should be examined as well. Improved designs based on experiences through application are also expected. The calibration and verification test procedures have already been developed, but should be subject to further investigation and development.

Acknowledgments. The authors acknowledge the financial support from the Slovenian Research Agency (research core funding No. P2-0190), as well as from Metrology Institute of the Republic of Slovenia (funding of national standard of length; contract No. C3212-10-000072). The research was performed by using equipment financed from the European Structural and Investment funds (Measuring instrument for length measurement in two coordinates with sub-micrometre resolution; contract with MIRS No. C2132-13-000033).

References

1. Acko, B.: System for assuring traceability of industrial measurements in Slovenia - present state and development strategy. In: Proceedings of Congress International de Métrologie, pp. 673–676, Bordeaux, France (1999)
2. Curran, E., Phelan, P.: Quick check error verification of coordinate measuring machines. J. Mater. Process. Technol. **1207**(13), 155–156 (2004)
3. ISO 10360-2: General product specifications (GPS) - acceptance and reverification tests for coordinate measuring machines (CMM), part 2: CMMs used for measuring linear dimensions, Geneva, Switzerland (2009)
4. Härtig, F., Trapet, E., Wäldele, F., Wiegand, U.: Traceability of coordinate measurements according to the virtual CMM concept. In: Proceedings of the 5th IMEKO TC-14 Symposium on Dimensional Metrology in Production and Quality Control, pp. 245–254, Zaragoza (1995)
5. Ramu, P., Yagüe, J.A., Hocken, R.J., Miller, J.: Development of a parametric model and virtual machine to estimate task specific measurement uncertainty for a five-axis multi-sensor coordinate measuring machine. Prec. Eng. **35**, 431–439 (2011)
6. Rodger, G., McCarthy, M.B., Flack, D.R.: A review of industrial capabilities to measure free-form surfaces. NPL report DEPC-EM 014 (2007)
7. Rajoub, B.A., Burton, D.R., Lalor, M.J.: A new phase-to-height model for measuring object shape using collimated projections of structured light. J. Opt.: Pure Appl. Opt. **7**, S368–S375 (2005)
8. ISO 9001: Quality management systems. requirements. International Organization for Standardization, Geneva, Switzerland (2015)
9. ISO 14253-1: geometrical product specifications (GPS) - inspection by measurement of workpieces and measuring equipment - part 1: decision rules for proving conformance or non-conformance with specifications, International Organization for Standardization, Geneva, Switzerland (1998)
10. Euramet. http://www.euramet.org/index.php?id=nimtech. Accessed 15 Nov 2018
11. Acko, B., McCarthy, M., Haertig, F., Buchmeister, B.: Standards for testing freeform measurement capability of optical and tactile coordinate measuring machines. Meas. Sci. Technol. **23**, 1–23 (2012)
12. VDI/VDE 2634 - 1,2,3: Optical 3D measuring systems - imaging systems with point-by-point probing. VDI, Beuth Verlag, Berlin (2002)
13. KOBA. https://www.koba.de/en/26-products-en/artifacts-for-3-coordinate-measuring-machines-en.html. Accessed 10 Nov 2018
14. Zeiss. https://mx.probes.zeiss.com/en/Machine-Accessories/Calibration-Artifacts/category-383.html?force_sid=0ugeq7tbgks7sk8k3bvs52lt97%20. Accessed 10 Nov 2018
15. Salsbury, J.G., Morse, E.P.: Measurement uncertainty in the performance verification of indicating measuring instruments. Precis. Eng. **36**(2), 218–228 (2011)

16. Cox, M.G., Harris, P.M.: Measurement uncertainty and traceability. Meas. Sci. Technol. **17**, 533–540 (2006)
17. Beges, G., Drnovsek, J., Pusnik, I., Bojkovski, J.: Measurement uncertainty, traceability and evaluation of test results in testing laboratories. Meas. Sci. Technol. **13**, 565–572 (2002)
18. Härtig, F., Rost, K., Goch, G.: Large gear material standard for the traceability of gears for transmission manufacturing. In: International Conference on Gears (Garching) VDI-Berichte, 2108, vol. 2 (2010)
19. Druzovec, M., Acko, B., Godina, A., Welzer-Druzovec, T.: Simulation of line scale contamination in calibration uncertainty model. Int. J. Simul. Model. **7**(3), 113–123 (2008)
20. Godina, A., Acko, B.: Calibration of the standard for optical measuring instruments, SOP 37: E-1, Faculty of Mechanical Engineering, Laboratory for Production Measurement, Maribor (2014)
21. Godina, A., Acko, B.: Performance verification of noncontact measuring systems. SOP 36, E-1. Faculty of Mechanical Engineering, Laboratory for production measurement, Maribor (2014)
22. EA-4/02: Evaluation of the Uncertainty of Measurement in Calibration. European Accreditation guidance document (2013)
23. ISO Guide: ISO Guide to the Expression of Uncertainty in Measurement ISO Guidance Document. Geneve, Swizerland (2008)
24. Kacker, R., Jones, A.: On use of Bayesian statistics to make the guide to the expression of uncertainty in measurement consistent. Metrologia **40**(5), 235–248 (2003)
25. Grabe, M.: Estimation of measurement uncertainties – an alternative to the ISO guide. Metrologia **38**(2), 97–106 (2001)

Determination of Resistance Spot Welding Parameters to Guarantee Certain Strength Values Including Regression Analysis

Cem Yurci[1](✉), Anil Akdogan[1], and M. Numan Durakbasa[2]

[1] Mechanical Engineering Department, Yildiz Technical University, Istanbul, Turkey
yurci_cem@yahoo.com, nomak@yildiz.edu.tr

[2] Industrial Metrology and Adaptronic Systems, Institute of Production, Engineering and Photonic Technologies, TU Wien, Vienna, Austria
numan.durakbasa@tuwien.ac.at

Abstract. In today's manufacturing world spot welding and its quality control take especially in the sheet metal industry a big place. According to that, a sheet metal part group has been taken from an automotive supplier firm. Several spot welding parameters (force, current and time) have been applied to these parts. The parameter limits have been apartly applied to the certain test parts. After applying the chisel tests, the spot welding region dimensions have been measured. The strengths of these regions can be defined with tensile tests, too. A relationship between these determined nugget diameters, strengths, calculated empirical strength values and welding parameters can be created. Finally, the relationships can be explained with mathematical formulas, too. These examinations have been done to give limit values for a simulation research to find a relationship between distortions after spot welding and spot welding parameters. It has been checked and confirmed on the production line of the supplier firm that these limit parameter values have guaranteed the welding strength. Verbal observations for the relationships and exceptions by those will be achieved, too.

Keywords: Spot welding parameters · Quality inspection · Strength

1 Introduction

Sheet metal industry is very important branch of the industry. Resistance spot welding is a crucial application in sheet metal industry especially in the combination of sheets. Sheets are joined mostly with spot welding. By spot welding a nugget occurs by the joint point of sheets. The nugget and form of this nugget is important by defining the strength and endurance of the weld joint. By testing of weld joint destructive and non-destructive testing methods are used. Ultrasonic testing can be counted in non-destructive testing in quality inspection methods. By ultrasonic testing ultrasonic waves are sent through spot welding region. By this, weld nugget, penetration, internal defects and outflow can be controlled. According to Zhang [1] et al., destructive testing methods are chisel test, peel test, tension test and tensile shear tests. Bend test, fatigue test, combined tension and shear test are other destructive testing methods. Inspecting

V. D. Majstorovic and N. Durakbasa (Eds.): IMEKOTC14 2019, LNME, pp. 62–73, 2019.
https://doi.org/10.1007/978-3-030-18177-2_6

the spot point size and their mechanical properties visually is very hard. Because of that the easiest method for that is to tear the weld joint and to measure the nugget diameter. By chisel test the experience of the operator is very important. He can feel or hear whether the weld region is brittle or not. The aim here is to detect cold regions which do not include welds. Weld nugget can be determined when the weld joint is opened.

By peel test one of the sheets is hold with a clamp and the other is rolled. So the weld joint is torn. If the weld region is brittle the sheets are torn easily. If there is a good joint a button occurs on one of the sheets. This test can be applied on coupons or on specimens cut from related parts.

Besides, tension and tension shear testing have been applied. These tests are applied according to the standards DIN EN ISO 14272 [2] and DIN EN ISO 14273 [3]. The specimens in these tests have been produced according to these standards. Their dimensions have been extracted from these standards and from Zhou et al. [4]. There are many FEM work examples including and investigating tensile and shear testing. For example, Kulkarni [5] makes shear strength prediction by Ansys. Akkaş [6] investigated the effect of the parameter resistance spot welding time on tensile shear loading on a steel sheet example. Chao [7] investigated combined tensile/shear loading and the stress distribution by that. Bandgar et al. [8] researched the effect of welding parameters in this study and the sheet thickness on shear strength with experiments and FEM, too. Raut et al. [9] worked about the optimization of spot welding process parameters for maximum tensile strength with the help of Taguchi method. In addition, one sample of Taguchi studies can be seen in Durakbasa et al.'s work [10]. Cakmakci et al. [11] investigate the relationship between quality engineering and Taguchi methodology. This study is based on the doctoral thesis of Yurci [12]. Besides, Yurci et al. [13] have another study about resistance spot welding investigating its effects on distortions.

2 Experimental Procedure and Set Up Conditions

An assembly group including some sheet metal parts has been taken from TOFAS TURK Company (the FIAT automotive producer) to investigate their distortions after spot welding. This group consists of four parts (Fig. 1).

Fig. 1. Sheetmetal parts from TOFAS TURK Company

During the spot welding of singular parts, the welding parameters have been attained approximately. Then ultrasonic tests have been applied to determine their welding quality. To determine the spot welding parameters it was necessary to generate Tensile Test and Tensile Shear Test between probable parameter limit values. The observed assembly group will be both scanned and analysed and modelled closest according to the actual process in the Simufact software. So distortions will be achieved. For that, an experiment set will be organized and limit parameter values have to be determined. Because of these two reasons, it has been decided to apply tensile and tensile shear tests to the specimens having the same material with the assembly group. Meanwhile, the previous experiments in TOFAS TURK Company will be considered and used, too. These tests have been implemented in an automotive firm ERMETAL Company in Bursa-Turkey which produces sheet metal parts.

Before realizing the mentioned tests, the sheets having the same material and quality with the singular parts of the assembly group have been searched.

Table 1. Tensile Shear Tests experiment setup table

Experiment information			Material information (1 Group = 81 specimens)		
Welding region (parts' names)	Experiment levels (I, t, F)	Number of experiment repetitions	'1.8 mm FEE 340 F ZNT/F/10/2S'	'1.6 mm FEE 220 BH ZNT/F/10/2S'	'3 mm FEE 340 F ZNT/F/14/2S'
Rinforzo + Scatolamento	3 × 2 × 2	3	X X		
Rinforzo + MontanteAttacco	3 × 2 × 2	3	X	X	
Rinforzo + Scatolamento + MontanteAttacco (three sheets region)	3 × 2 × 2	3	X X	X	
Scatolamento + MontanteAttacco	3 × 2 × 2	3	X	X	
Montante DX + MontanteAttacco	3 × 2 × 2	3		X	X
Total specimen number (1 specimen: 110 × 40 mm)			216	144	36
Total sheet requirement (mm^2)			950400 mm^2	633600 mm^2	158400 mm^2
Total sheet requirement ≈ (m^2)			0.9504 m^2	0.6336 m^2	0.1584 m^2

The singular parts' material is the HSLA (High-strength low-alloy) steel and their thickness and quality are so:

- 1.8 mm FEE 340 F ZNT/F/10/2S
- 1.6 mm FEE 220 BH ZNT/F/10/2S
- 3 mm FEE 340 F ZNT/F/14/2S

FEE 340 F-ZNT/F/2S steel is in the class 'High Strength Low Alloyed Zinc Coated (galvanized) Steels by Continuous Hot-Dip Process for Cold Forming'. Besides, the FEE 220 BHZNT/F/2S steel is in the 'Continuously Hot- Dip Zinc Coated (Galvanized) Bake-Hardening High Yield Strength Steels For Cold Forming' class.

Here for example, in the FEE 340 F ZNT/F/10/2S steel, ZNT/F shows the zinc plating type, 2S shows that plating is both-sided, F shows that the steel is a cold rolled product and 10 shows that the minimum plating thickness is 10 μm.

Tensile shear tests have been applied according to the experiment organization in Table 1 which has been implemented with the experiment setup Full Factorial. Specimen drawing has been given in Fig. 2 with its dimensions. Dimensions have been attained according to the DIN EN ISO Standard 14273 [3].

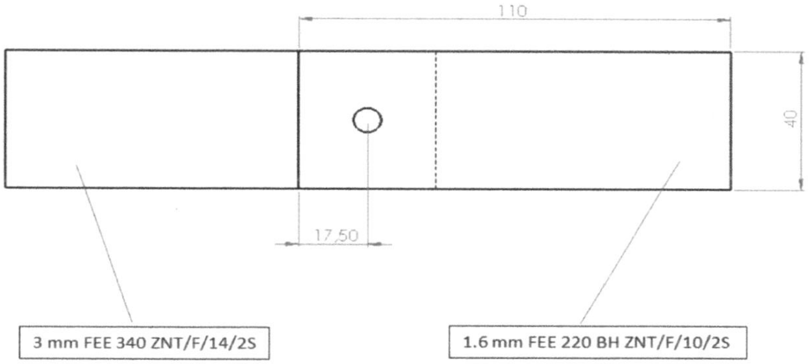

Fig. 2. Tensile Shear Test specimen dimensions example

In all of the tests, three parameters were the welding current (illustrated with "I"), the welding time (illustrated with "t") and the electrode force (illustrated with "F"). The level of the current in the experiments is 3; the level of welding time and the electrode force is 2. The welding current values are in A, the welding time values are in periods (frequency: 50 Hz) and the electrode force values are in N (in Table 3).

The experiments which were organized according to Full Factorial method were carried out in the mixed order randomly. However, their order for each experiment group was determined with the software Minitab. The parameter values for these tests were applied apartly for each welding part combination.

Tensile tests have been applied according to the experiment organization in Table 2 which has been implemented again with the experiment setup Full Factorial. The difference from tensile shear tests is that here the number of experiment repetitions is 2. Specimen drawing has been given in Fig. 5 with its dimensions. Dimensions have been attained according to the DIN EN ISO Standard 14272 [2].

The unfolded sheets shown in Fig. 3 have been bent and specimens (Fig. 4) ready for tension have been prepared.

At the beginning, while finding proper welding parameter limits we have only the parameter values applied in the work in TOFAS Turk. Because of that some parameter combinations have been tried. Destructive Chisel test (Fig. 5) and peel test have been applied to the specimens after several combination trials. Here the purpose is to see that a weld nugget has occurred and to test that the diameter of this nugget is between certain limits. It has been accepted that the nugget diameter can change between 3,5 $\sqrt{t}$ and 6 $\sqrt{t}$ according to DIN EN ISO 14373 [14]. If these nugget limits are provided the parameter values providing these nuggets will be our parameter limit values for tensile shear and tensile testing. After these tests our parameter limits for the works in the

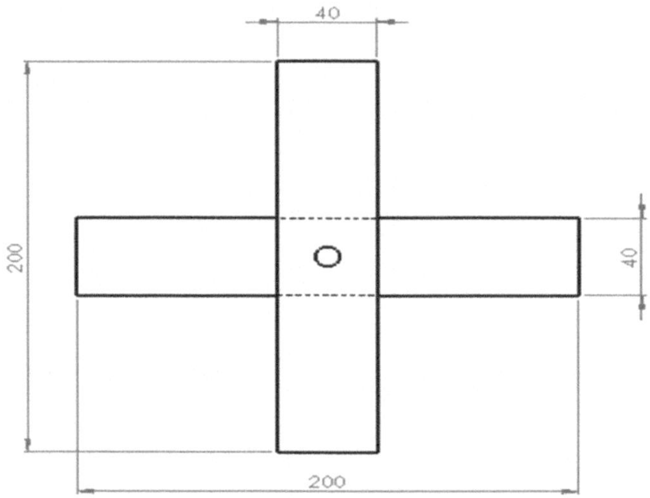

Fig. 3. Tensile Test specimen dimensions

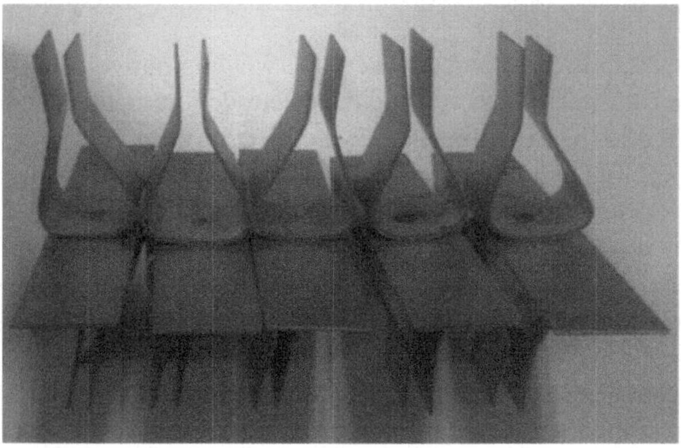

Fig. 4. Tensile test part specimens for three parts region

software Simufact will be determined. "t" is here the sheet thickness. The minimum nugget diameter is calculated approximately as 4,5 mm for our specimens. But this value has been accepted as 6 mm after the recommendations of the ERMETAL firm and its engineers.

Before applying the welding, to fixture the parts during welding a specific apparatus has been designed. The apparatus under the welding electrode can be seen in Fig. 6. In the experiments 'G0-20-22-50-8-8-22c5' type electrode has been used according to DIN EN ISO 5128 [15].

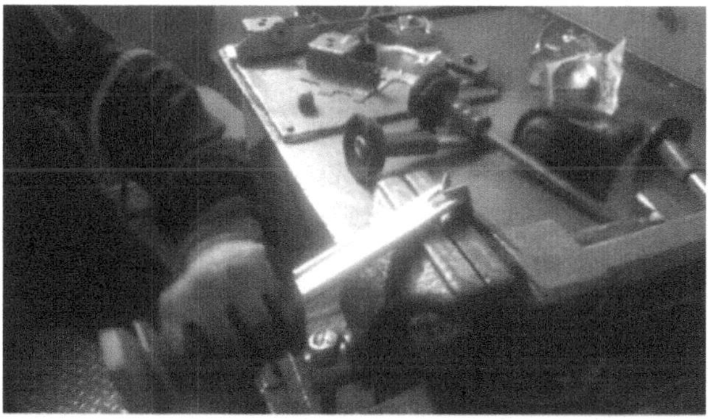

Fig. 5. Application of the destructive testing for specimens to determine welding parameter limit values

Table 2. Tensile Tests experiment setup table

Experiment information		Specimens number according to the material			
Welding region (parts' names)	Experiment levels (I, t, F)	Number of experiment repetitions	'1.8 mm FEE 340 F ZNT/F/10/2S'	'1.6 mm FEE 220 BH ZNT/F/10/2S'	'3 mm FEE 340 F ZNT/F/14/2S'
Rinforzo + Scatolamento	3 × 2 × 2	2	48		
Rinforzo + MontanteAttacco	3 × 2 × 2	2	24	24	
Rinforzo + Scatolamento + MontanteAttacco (three parts region)	3 × 2 × 2	2	48	24	
Scatolamento + MontanteAttacco	3 × 2 × 2	2	24	24	
Montante DX + MontanteAttacco	3 × 2 × 2	2		24	24
Total specimen number (1 specimen: 200 × 40 mm)		144		96	24

Fig. 6. Fixture apparatus for spot welding of testing specimens

3 Results and Discussions

The result for the combination of Scatolamento and Rinforzo for the Tensile Shear Test is given in the Table 3.

Table 3. The result for the combination of Scatolamento and Rinforzo for the Tensile Shear Test

I (A)	t (Per.50 Hz)	F (N)	Pressure (Mpa)	Tensile strength (N)	Nugget diameter (mm)	Calculated empirical strength (N)	Result
10000	14	3200	0.22	17658.56	8.09	16724.15667	Ok
12000	9	3500	0.25	19717.57	9.01	18614.53386	Ok
10000	9	3200	0.22	19945.42	9.91	20473.92126	Ok
8000	14	3500	0.25	15012.83	No nugget occuring	–	
12000	9	3500	0.25	20380.75	11.65	24068.73690	
12000	14	3500	0.25	21673.76	11.64	24048.07704	
8000	14	3200	0.22	16616.88	No nugget occuring	–	
12000	14	3500	0.25	21717.06	12.1	24998.43060	
8000	14	3500	0.25	11135.43	No nugget occuring	–	
10000	9	3500	0.25	18267.38	8.67	17912.09862	Ok
10000	9	3200	0.22	18111.90	8.62	17808.79932	Ok
10000	9	3500	0.25	18489.09	8.87	18325.29582	Ok
10000	14	3200	0.22	20451.51	11.71	24192.69606	
10000	14	3500	0.25	19924.96	10.24	21155.69664	
10000	14	3500	0.25	20688.65	11.98	24750.51228	
12000	14	3200	0.22	21520.80	11.23	23201.02278	
8000	14	3500	0.25	8054.97	No nugget occuring	–	
8000	14	3200	0.22	7977.47	No nugget occuring	–	
12000	9	3200	0.22	19865.42	9.74	20122.70364	Ok
10000	9	3500	0.25	16521.01	No nugget occuring	–	
8000	9	3500	0.25	0.46	No nugget occuring	–	
10000	9	3200	0.22	16516.9	No nugget occuring	–	
8000	9	3200	0.22	1.01	No nugget occuring	–	
12000	9	3200	0.22	20527.05	10.87	22457.26782	

(continued)

Table 3. (*continued*)

I (A)	t (Per.50 Hz)	F (N)	Pressure (Mpa)	Tensile strength (N)	Nugget diameter (mm)	Calculated empirical strength (N)	Result
8000	14	3200	0.22	7889.55	No nugget occuring	–	
8000	9	3500	0.25	0.83	No nugget occuring	–	
12000	9	3200	0.22	19829.86	9.51	19647.52686	Ok
12000	14	3200	0.22	21534.24	11.42	23593.56012	
12000	14	3200	0.22	21105.87	10.95	22622.54670	
10000	14	3500	0.25	20176.82	11.93	24647.21298	OK
8000	9	3500	0.25	11902.62	No nugget occuring	–	
8000	9	3200	0.22	1.04	No nugget occuring	–	
12000	9	3500	0.25	20378.66	9.97	20597.88042	
10000	14	3200	0.22	20338.09	10.65	22002.75090	
12000	14	3500	0.25	21384.11	11.50	23758.83900	
8000	9	3200	0.22	12623.73	No nugget occuring	–	

The calculated empirical strength value shown in the table above is calculated according to the formula in DIN EN ISO 14373 [14].

$$Ps = 2,6 * t * d_w * R_m \tag{1}$$

d_w: Welding nugget diameter (mm)
P_s: Tensile shear strength of the welding joint region (N)
R_m: Tensile strength of the steel (Mpa)
t: Sheet thickness (mm)

Tensile strength of the materials have been found firstly with obtaining their Rockwell hardness values and then with necessary transformations from SAE 1958 Rockwell Hardness Transformation Table of ERMETAL. According to that for FEE 340 the hardness value is HRB 71 and for BH 220 it is HRB 50. So for FEE 340 the tensile strength is 45 kg/mm^2 (441,45 N/mm^2) and for BH 220 it is 32,5 kg/mm^2.

As it is seen in the table above at the experiment stages by which the tensile strength obtained with tensile testing is bigger than the calculated empirical strength value OK is written which shows that the strength of the weld joint can be accepted. The formula according to the standard above has been given with a tolerance %20.

Anyway it is impossible to give a formula including the specimens with the same thickness and plating. According to that, other experiment stages different from stages by which no nugget occurred can be accepted. As it is seen no nugget or poor nugget has occurred at the current value 8000 A. Besides by some current values 10000 A no nugget has occurred, too. The reason for that is that the electrode tip is worn with the time and its resistance decreases, so its heat energy input is low, too. All the results will be compiled and commented. After that, essential acceptations and calculations will be made. So, the input limit values for Simufact will be obtained. In this work the welding time values are being applied with 3 pulses.

3.1 Regression, Taguchi and ANOVA (Analysis of Variance) Analyses in Minitab

The software Minitab 16 was used for these analyses. The result parameter for these analyses is the measured tensile strength.

The linear regression analysis for Table 3 gave the conclusion regression equation:

$$\text{Tensile Strength} = -26928,2 + 3,30038\ \mathrm{I} + 686,454\ \mathrm{t} + 0,539196\ \mathrm{F}$$

The p value of F is 0, 906 and also is bigger than 0,05 (with a confidence level 95%. Because of that when F is neglected the new equation is so:

$$\text{Tensile Strength} = -25121,9 + 3,30038\ \mathrm{I} + 686,454\ \mathrm{t}$$

The other results are so: R-Sq = 68,55% R-Sq(adj) = 66,65% R-Sq(pred) = 62,30%

To increase the R-sq values the square values of the parameters I, t and F are included and so regression calculations are calculated. After that the new equation is so:

$$\text{Tensile Strength} = -141831 + 26,9105\ \mathrm{I} + 686,454\ \mathrm{t} + 0,539196\ \mathrm{F} - 0,00118051\ \mathrm{I}\hat{\ }2$$

Because p-value of F is high F is neglected and the new equation is so:

$$\text{Tensile Strength} = -140024 + 26,9105\ \mathrm{I} + 686,454\ \mathrm{t} - 0,00118051\ \mathrm{F}$$

By adding square values of the parameters, R values are so increased and other results are so: R-Sq = 79,17% R-Sq(adj) = 77,22% R-Sq(pred) = 73,64%

Taguchi analysis gave these results shown in Figs. 7 and 8.

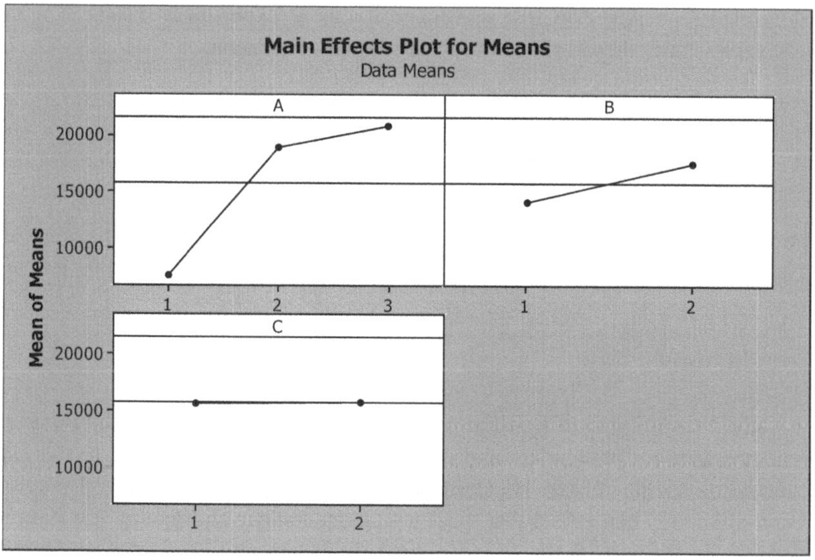

Fig. 7. The Taguchi results for Table 3; Main Effects Plot for Means

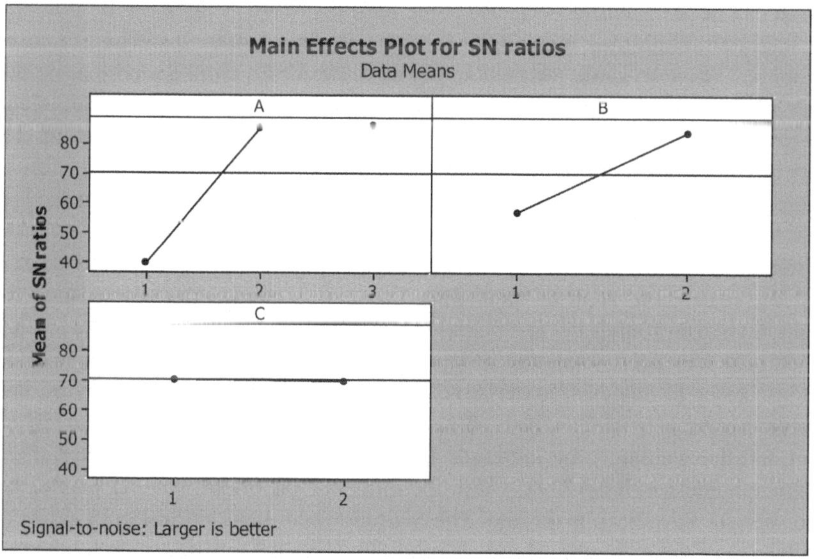

Fig. 8. The Taguchi results for Table 3; Main Effects Plot for SN ratios

The SN ratios for 'Larger is better' analysis show that the tensile strength is increased when the current and the welding time are in a bigger magnitude. The tensile strength is decreased when the electrode force is high.

Additionally, an ANOVA test was made for these experiment results. The calculated SS values gave these influence factors for the parameters:

For the welding current: 72,86%
For the welding time: 6,31%
For the electrode force: 0,014%
For the error: 20,81%

The SN ratios and ANOVA test both show, too that the influence of the changing of the electrode force on the tensile strength is low.

4 Conclusions

In this study, the aim was to find the parameter values for inspecting the relationship between spot welding parameters and distortions in a work. According to that, specimens according to the related standards have been prepared and several parameters have been applied. Then chisel test, peel test, tensile shear and tensile test have been implemented. In these tests the found nugget diameters and properties and strengths have been considered. So necessary parameter limits have been found.

After making related result tables for every part combination some observations can be made. By some stages of the experiments by which no nugget or poor nugget occurred can be accepted too because some tensile strength values can be obtained by some stages although they are smaller than the empirical strengths. This is caused by the empirical strength value formula's tolerance. Besides it can be seen that no nugget or poor nugget has occurred at the minimum current values. Besides by some current values one bigger than minimum current values no nugget has occurred again. This is caused by the worn electrode tip with the passing time. During that, its resistance decreases because of the increasing contact of the tip, so its heat energy input decreases, too.

After the conducted tests, regression, Taguchi and ANOVA analyses have been made. By regression analysis equations between welding parameters and tensile strength have been found. By neglecting some parameters and adding square values of parameters it has been seen that R-square values of functions have increased and functions closer to analysis points have been obtained. By Taguchi analysis, the SN ratios for 'Larger is better' analysis showed that the tensile strength increases when the current and the welding time increase. The tensile strength decreases when the electrode force increases. By ANOVA test, the influence factors of welding parameters on tensile strength have been inspected. The SN ratios and ANOVA test both showed that the influence of the changing of the electrode force on the tensile strength is very low. The biggest influence factor by that category belongs to the welding current.

This study will be followed with the application of the found spot welding parameter limits to the Simufact works to find a relationship between parameters and distortions after spot welding. The works in this study can be broadened in the future for other materials, specimen type and dimensions and test types.

References

1. Zhang, H., Senkara, J.: Resistance Welding Fundamentals and Applications. CRC Press Taylor & Francis Group, Boca Raton (2012)
2. DIN EN ISO 14272. Resistance welding – destructive testing of welds – specimen dimensions and procedure for cross tension testing of resistance spot and embossed projection welds; German version (2016)
3. DIN EN ISO 14273. Resistance welding – destructive testing of welds – specimen dimensions and procedure for tensile shear testing resistance spot and embossed projection welds; German version (2016)
4. Zhou, M., Hu, S.J., Zhang, H.: Critical specimen sizes for tensile-shear testing of steel sheets. Weld. Res. Suppl., 305–313 (1999)
5. Kulkarni, P.P.: Shear strength prediction of multi-spot welded lap shear specimen through experimentation and validation by FEM. Int. J. Eng. Res. Appl. **4**(4), 80–87 (2014)
6. Akkaş, N.: Welding time effect on tensile-shear loading in resistance spot welding of SPA-H weathering steel sheets used in railway vehicles. Acta Phys. Pol. **A131**(1), 52–54 (2017)
7. Chao, Y.J.: Ultimate strength and failure mechanism of resistance spot weld subjected to tensile, shear, or combined tensile/shear loads. Trans. ASME **125**, 125–132 (2003)
8. Bandgar, R.T., Dhawale, P.A., Vyavahare, R.T.: Parametric study of multi-spot welded lap shear specimen for shear strength and validation through finite element analysis. Int. J. Sci. Eng. Technol. Res. **4**(9), 329–3236 (2015)
9. Raut, M., Achwal, V.: Optimization of spot welding process parameters for maximum tensile strength. Int. J. Mech. Eng. Robot. Res. **3**(4), 506–517 (2014)
10. Durakbasa, M.N., Akdogan, A., Vanli, A.S., Bulutsuz, A.G.: Optimization of end milling parameters and determination of the effects of edge profile for high surface quality of AISI H13 steel by using precise and fast measurements. Measurement **68**, 92–99 (2015)
11. Cakmakci, M., Durakbasa, N.M., Karasu, M.K., Bas, G., Gurel, U.: The importance of quality control within the relationship between the quality engineering and Taguchi methodology. Key Eng. Mater. **637**, 27–35 (2015)
12. Yurci, C.: Determining of cumulative dimension deviations in multi-part sheet metal constructions by means of measurements and analyses and development of an algorithm. Doctoral thesis, Mechanical Engineering Department, Yildiz Technical University, Istanbul (2017)
13. Yurci, C., Akdogan, A., Durakbasa, N.M.: Investigation of effects (welding sequence, fixturing, welding points) on distortions after spot welding for determining individual and cumulative tolerances. IFAC PapersOnLine **49**(29), 30–35 (2016)
14. DIN EN ISO 14373. Resistance welding – procedure for spot welding of uncoated and coated low carbon steels; German version (2015)
15. DIN EN ISO 5821. Resistance welding – spot welding electrode caps; German version (2009)

Development of a 3-Axis Displacement Measuring Heterodyne Interferometer System Usable with a He-Ne Laser of Either 3.76 MHz or 20 MHz Split Frequency

Kyu Sik Yoon[1], Eun Ji Jeong[1], Don Young Jeong[2], and Chu-Shik Kang[2,3(✉)]

[1] EPIR Co. Ltd., 2802, Building A, Heungdeok IT Valley, 13, Heungdeok 1-ro, Giheung-gu, Yongin-si, Gyeonggi-do 16954, Republic of Korea
[2] Korea Research Institute of Standards and Science, 267 Gajeong-ro, Yuseong-gu, Daejeon 34113, Republic of Korea
cskang@kriss.re.kr
[3] University of Science and Technology, 217 Gajeong-ro, Yuseong-gu, Daejeon 34113, Republic of Korea

Abstract. We present a 3-axis displacement measuring heterodyne interferometer system developed for use with a He-Ne laser at the wavelength of 633 nm having a split frequency of either 3.76 MHz or 20 MHz. The system consists of an optical interferometer, a signal processing board, and a software.

The interferometer can simultaneously measure the displacements of three axes, from which the distance, tilt angles of a moving object can be determined in real time. A digital logic circuit which calculates the displacement values from electric signals converted from 3 optical interference signals, was developed. To demonstrate the measured results graphically, we developed a graphical-user-interface software. In addition, a feedback function through a high speed communication bus was added to the system so that it can be applied for motion control of a moving stage requiring scanning and alignment of a thin film coated substrate such as a silicon wafer. The performance of the interferometer was checked by comparing the measurement values of the displacement of a moving stage obtained simultaneously by a commercial laser interferometer and the developed interferometer.

Keywords: Laser interferometer · 3-axis · Split frequency · Optical signal processing

1 Introduction

Laser interferometers are widely used in various equipment for measuring displacement in nanometer precision. In particular, laser interferometer system is an essential part of lithographic equipment or metrology and inspection equipment used in semiconductor manufacturing processes. The equipment for fabricating or inspecting thin-film patterns on silicon wafers requires measurements of multiple degrees-of-freedom (DoF) such as height, tilts, and rotation, as well as displacements.

V. D. Majstorovic and N. Durakbasa (Eds.): IMEKOTC14 2019, LNME, pp. 74–80, 2019.
https://doi.org/10.1007/978-3-030-18177-2_7

For a motion stage that performs 6 DoF transfer and alignment, at least 6-axis measurement is required. Also, transformation from 6-axis displacement measurement data into 6 DoF coordinate system is necessary because the optical axes are not parallel to each other and are not perpendicular to the moving object carrying a silicon wafer.

To perform displacement measurement and feedback for high-speed stage movement, the optical signal processing and the displacement computing should be performed at a high sampling frequency.

In this paper, we report on the current status of the developed 3-axis laser interferometer that can be used for 6 DoF measurements, which uses a Xilinx Zynq 7000 series board, which is suitable for a signal processing and computer board requiring large calculation and high speed communication. The performance of the 3-axis laser interferometer was evaluated by measuring motion of a moving stage, where a Renishaw's single axis laser interferometer system was also installed and used for the measured simultaneously.

2 Interferometer Design and Components

2.1 Interferometer Displacement Measurement System

Figure 1 shows the overall layout of the 3-axis laser interferometer. The developed 3-axis laser interferometer and a single axis laser interferometer from Renishaw [1] were placed side by side and the measurement object was placed on a single-axis linear motor stage. We used a Keysight 5517D He-Ne laser [2] with 3.76 MHz split frequency and a Zygo ZMI 7702 He-Ne laser [3] with 20 MHz split frequency, widely used in semiconductor equipment as the measurement light source.

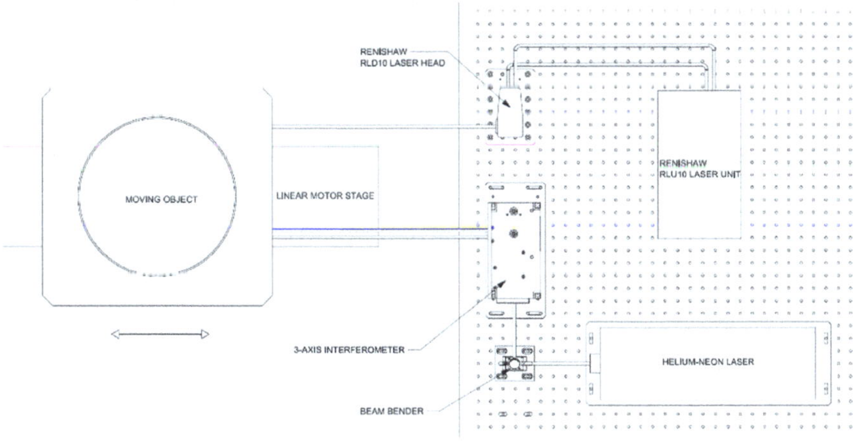

Fig. 1. Displacement measurement system for moving object using 3-axis interferometer

2.2 The 3-Axis Interferometer

Figure 2 shows the optical layout of the developed a 3-axis interferometer.

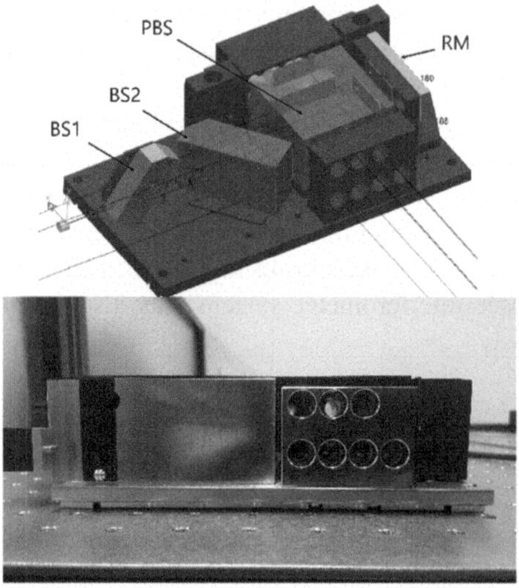

Fig. 2. Optical layout and photo of the 3-axis interferometer. BS1/BS2: beam splitter, PBS: polarizing beam splitter, RM: reference mirror

The entrance of the light source is arranged with two prisms for separating three beams of a single He-Ne laser beam. The three beams are separated into polarized vertical light and horizontal light, which are sent to the corner cube and the moving object, respectively. The measurement beams reflected from the moving object are superimposed on the reflected beams from the corner cube, respectively, and sent to the three optical connectors.

2.3 Optical Signal Processing

We developed an optical signal processing board with four photodiode receivers for 3-axis interferometer measurements. The optical signal transmitted by the optical cable is converted to a low level current through the photodiode. This current is amplified by the voltage and converted into a square wave of 3.3 V. Four square waves, one from the reference and three from measurement beams, are transmitted to the field programmable gate array (FPGA) input.

2.4 FPGA Logic Design

A digital logic circuit which calculates the displacement values from electric signals converted from 3 optical interference signals, was developed using an FPGA board of Xilinx Zynq 7000 series. To demonstrate the measured results graphically, we developed a graphical user interface software that can run on the Linux operating system of the ARM CPU included in the Zynq chip and the remote PC. A screenshot of the software is shown in Fig. 3.

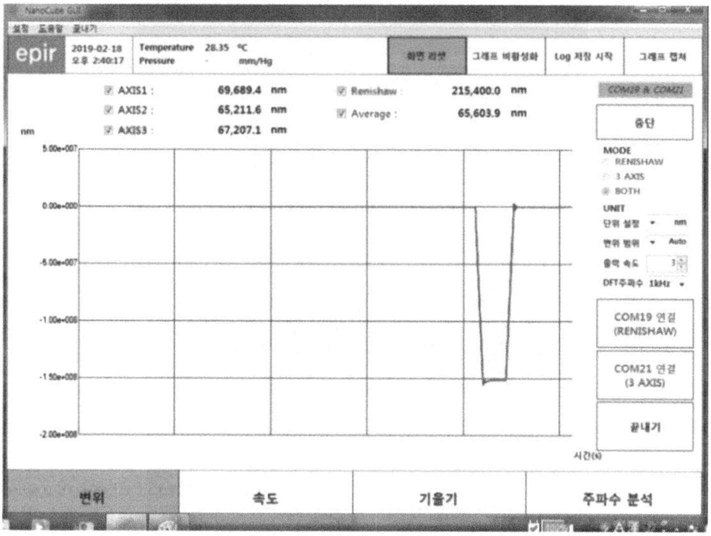

Fig. 3. Captured image of the graphical-user-interface software

In addition, a feedback function through a high speed communication bus was added to the system so that it can be applied for motion control of a moving stage requiring scanning and alignment of a thin film coated substrate such as a silicon wafer. The FPGA logic counts the number of square wave pulses and calculates rotation and tilts as well as 3-axis displacements.

3 Test Results

3.1 Optical Signal Processing for Two Split Frequencies

Split frequencies of two He-Ne lasers (Keysight 5517D and Zygo 7702) were measured using the developed optical signal processing board. The measurement results of the optical signal processing board showed correct split frequencies of 3.76 MHz and 20.0 MHz, for the Keysight and Zygo laser, respectively. Figure 4 shows the measured square waves of the reference frequencies of the two laser heads.

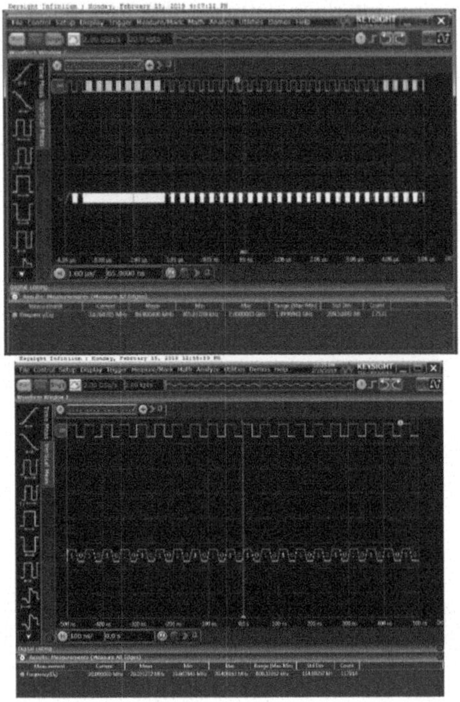

Fig. 4. Optical signal square wave of 3.76 MHz (left) and 20 MHz (right).

3.2 Measuring Single Axis Movement

Using the setup shown in Fig. 1, the displacements of a stage moving at 0.2 m/s were measured by the two laser interferometers. The results are shown in Fig. 5. Our 3-axis

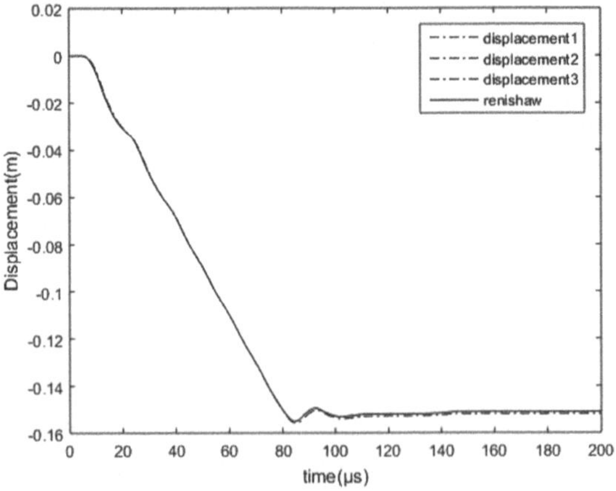

Fig. 5. Measurement results of single-axis displacement (dot lines: 3-axis interferometer, solid line: Renishaw single-axis interferometer)

laser interferometer and Renishaw's 1-axis laser interferometer both performed measurements at a 1 kHz sampling frequency. Three results measured with a 3-axis interferometer agree well with each other.

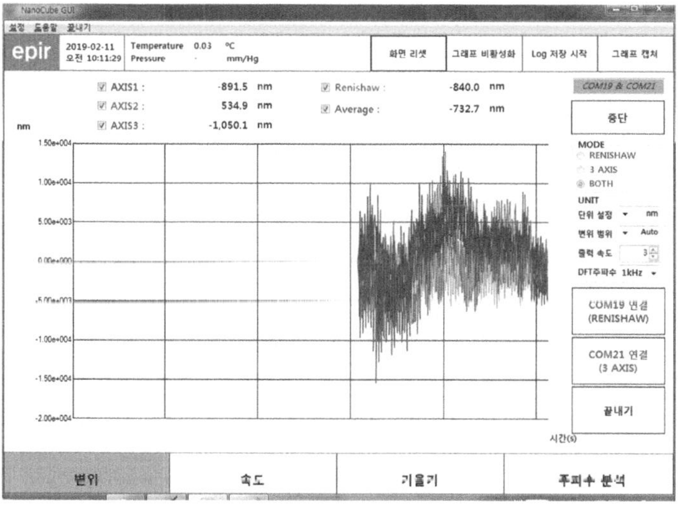

Fig. 6. Comparison of data measured by the 3-axis interferometer (blue, red, green) and the Renishaw interferometer (Orange)

Figure 6 shows the displacement data measured by the developed 3-axis interferometer and the Renishaw interferometer, when the stage is not moving. The two interferometers show slightly different values because the scaling factor is not yet calibrated as well as the wavelength of the laser not compensated. It can be seen that the four graphs are comparable and move in the same manner. The reason why the measured values fluctuate quite much is that the interferometers and the stage are installed on a different platform. So the fluctuation shows the relative vibration between the stage and the optical table where the interferometers are located.

4 Conclusion

We developed a 3-axis heterodyne laser interferometer system consisting interferometer optics module, a signal processing board, and a measurement software. It is designed for use with a He-Ne laser at the wavelength of 633 nm having a split frequency of either 3.76 MHz or 20 MHz, and is capable of 6 DoF measurements with high sampling frequency of 1 kHz. Through future works, it is expected to be applied to equipment for fabricating or inspecting thin-film patterns on silicon wafers.

Acknowledgments. This work was supported by the Industrial Strategic Technology Development Program (10079332, Development of 6 degrees of freedom laser interferometer position measurement system with 40 nm repeatability for wafer stage) funded by the Ministry of Trade, Industry & Energy (MOTIE, Korea).

References

1. https://www.renishaw.com/en/xl-80-laser-system–8268
2. https://www.keysight.com/en/pd-1000001477%3Aepsg%3Apro-pn-5517D/laser-head?nid=-536900395.536880855&cc=US&lc=eng
3. https://www.zygo.com/?/met/markets/stageposition/zmi/laserheads/

Hard Gauge Visualization – Effective Tool for MMR Verification Discussion

Marcin Berta[1] and Zbigniew Humienny[2(✉)]

[1] DMG MORI Polska Sp. z o.o., Fabryczna 7, 63-300 Pleszew, Poland
[2] Institute of Machine Design Fundamentals, Warsaw University of Technology, Narbutta 84, 02-524 Warsaw, Poland
zbigniew.humienny@pw.edu.pl

Abstract. The concept of author's application *Geometrical tolerancing* that supports teaching and vocational training on the geometrical product specifications rules and verification methods is briefly presented. The maximum material requirement (MMR) modifier defined in the ISO 2692 changes the classical meaning of the geometrical tolerance symbols. The classical concept of a tolerance zone is replaced by the concept of a gauge. The idea of the maximum material modifier is based on functional requirement – assembleability. The paper contains a few screenshots from the interactive animations that are employed to demonstrate the MMR concept as well as relevant verification method. The sequence of developed scenes effectively help to understand and distinguish the role of the MMR modifier for the tolerance features and the datum features. For the datum features application of the MMR modifier produces an additional mobility for a group of all features considered as a pattern. This extra mobility depends on the datum feature actual mating sizes as well as on the 3D configuration of the datums and their precedence. It is demonstrated that employed animations clearly show that the bonus tolerance when the datum feature deviates from the MMC is virtually allowable rarely – only for specific configurations of a datum features geometry.

Keywords: Geometrical tolerancing · MMR · E-learning

1 Introduction

The role for tolerancing, mathematical modelling of geometrical specifications and manufacturing, uncertainty in areas of design, manufacturing and metrology as well as challenges introduced by new processes are discussed in [1]. Manufacturing always produces workpieces that are not perfect and which differ from the nominal geometry and from workpiece-to-workpiece.

After the designer has the functional concept of nominal workpiece geometry his main task is to specify how far an actual workpiece may by from the optimum. The ISO GPS system [2] supplies a number of Geometrical Product Specification (GPS) tools that may be used by a designer to set explicitly, or implicitly as default, the limits for production imperfections. The design and engineering students shall be equipped with good knowledge and understanding of GPS relevant to the conventional Technical Product Specifications (TPS) as well as the Model Based Definition (MBD) [3]. This

V. D. Majstorovic and N. Durakbasa (Eds.): IMEKOTC14 2019, LNME, pp. 81–88, 2019.
https://doi.org/10.1007/978-3-030-18177-2_8

paper presents a part of application *Geometrical tolerancing* that has been developed to visualize the great advantages of maximum material requirement (MMR) applied to toleranced feature and restrictions for benefits of the MMR specification for datum features in a datum system.

The purpose of maximum material requirement expressed by Ⓜ modifier is to make it possible to accept parts based on their functionality i.e. assembleability and avoid rejecting them with respect to criteria that fail to take into account the actual function. The MMR is extremely useful when a producer intends to accept all parts that fulfil the assumed function (assembleability) regardless of separately evaluated size deviations and geometrical deviations. This is due to the fact that parts assembly depends on the combined effect of the size and geometrical deviations. The idea of the maximum material modifier employs a material hard gauge that can be produced and implemented physically to verify MMR specification.

2 Application *Geometrical Tolerancing*

The first animations for the application *Geometrical tolerancing* that supports teaching and vocational training on the geometrical product specification rules and verification methods were developed a few years ago [4] and from that time the application is permanently enriched and developed [5]. Animations are extensively used in the application *Geometrical tolerancing* because we share opinion on their overall positive effect to memorizing and understanding over static graphics [6].

Below the concept of the application *Geometrical tolerancing* is briefly presented and then the paper is focused on the animation that reflects design of a gauge for the MMR for position tolerance of three hole pattern in the disc with respect to datum system establish by the primary datum A – plane, the secondary datum B – hole with MMR and the tertiary datum C – groove with MMR (Fig. 3).

The application *Geometrical tolerancing* starts from the *Main Window* (Fig. 1) where a user can left-click on one of the 14 *Tolerance symbol buttons*, the *Datums button*, any *Modifier button* or the *Size button* to open the *Case selection* window with the list of cases of its applications.

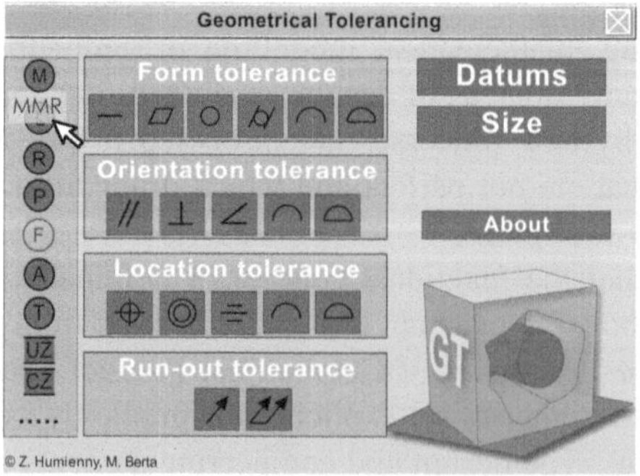

Fig. 1. *Main window* of the application *Geometrical tolerancing* – the *MMR button* description is displayed adjacent to the button.

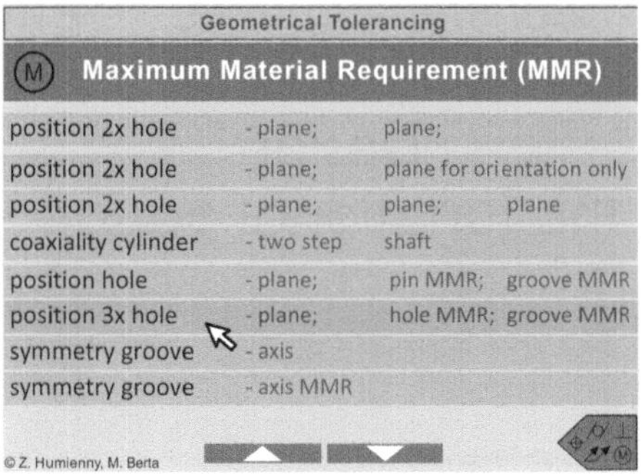

Fig. 2. *Case selection window* with scrolled list of *Line buttons* for the MMR tolerance applications

The click event opens the *Case selection window* where different applications of the maximum material requirement modifier (Fig. 2) are listed. The *Case selection window* may be scrolled by buttons that are placed on this window bottom. Currently 14 cases of the MMR specifications are discussed and visualized in the application. Selection of the particular case by click event on the *Line button* opens the *Definition window* (Fig. 3) with the relevant tolerance indicator and datum indicators attached to a workpiece. The short description of the MMR concept is placed on the right side of the *Definition window*. This information shall refresh a user the concept of the MMR and help her/him to solve task of a gauge design that is assigned to this animation. It shall be underlined that the presented case is more advanced, so it is not listed at the first page of animations devoted to the MMR issue and it assumed that a user have already studied the simpler cases. The idea of the presented animation is to guide a user through the design process of a gauge dedicated to given workpiece and thanks to it help a user to find whether she/he understands what are implications of the MMR application for a toleranced feature and for datum features. The click event on the *Explanation button* initiates series of scenes developed to assist a user in the gauge design and the gauge usage (Fig. 4).

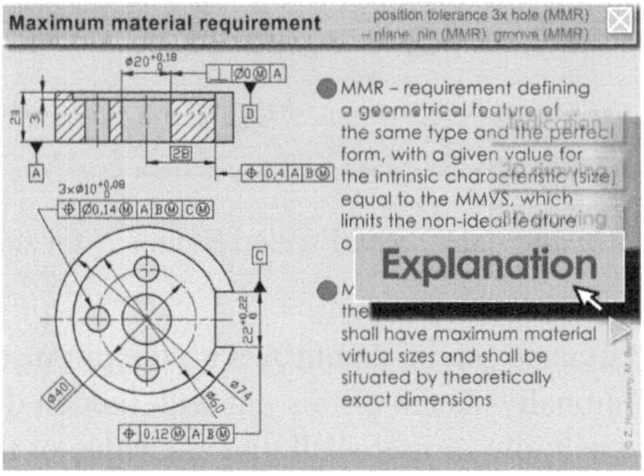

Fig. 3. *Definition window* of MMR specification. The click on the *Explanation button* at pull down menu available on its right edge transfers a user to the next window.

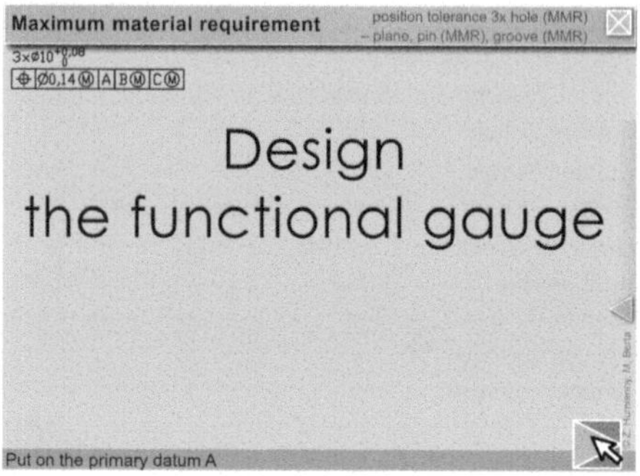

Fig. 4. *Explanation window* – the first scene. The arrow cursor dragged and held over the *Forward button* activates the *Prompt line*

A short help makes the task easier for a user. The *Forward button* that is available on the bottom right corner of each *Explanation Window* is used to release next scenes in the *Explanation Window.* The command displayed in the *Prompt line* in the window bottom when the arrow cursor is dragged and held over the *Forward button* (e.g. *Put on the primary datum A,* Fig. 4) informs what action will be demonstrated in the next scene/animation released by click on the *Forward button.* In this scene the click event on the *Forward button* triggers placement of the toleranced disc on a plane – the datum A is established (Fig. 5). Now a user has to indicate the successive step in the gauge designing – the pin of the constructed gauge shall be inserted into disc central hole to establish the datum B. The user shall give the pin diameter that is equal to the maximum material virtual size for the central hole with perpendicularity requirement.

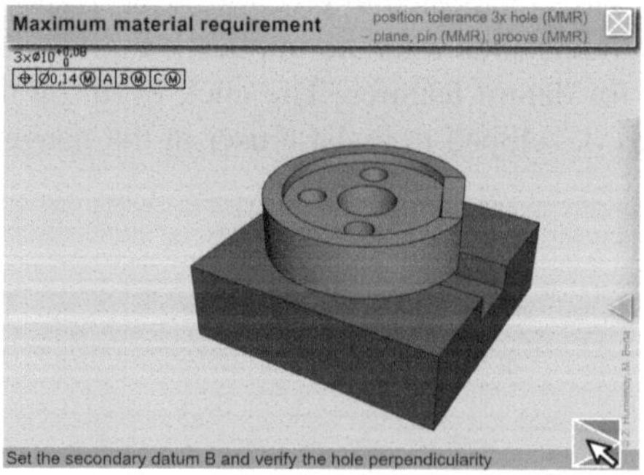

Fig. 5. *Explanation window* – the second scene. Datum A for the disc is established

The toleranced disc situated on the datum A with the pin inserted in its central hole is shown in Fig. 6. Additionally sketch of this pin with hidden diameter is presented in bottom left corner of the window. A user shall give the value of the pin diameter and by click event she/he can disclose the hidden diameter to verify correctness of the given answer. In next scenes the block is inserted into the groove to establish the datum B. A user also has opportunity to verify whether the proposed width of the block is correct.

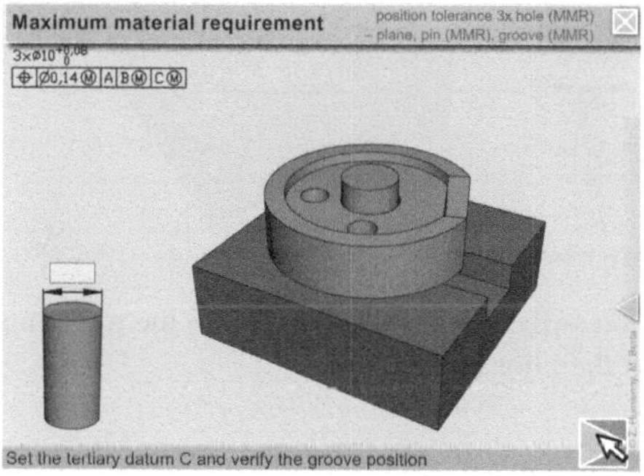

Fig. 6. *Explanation window* – the third scene. A user has opportunity to verify whether the proposed pin diameter is correct.

The final step of the gauge designing is shown in Fig. 7. The datum system A B C is established and three pins are inserted to the pattern of three disc holes to verify their position tolerances. Unfortunately the pins do not fit into their pilot holes that are in the gauge base. So it seems that the examined disc is a scrap. On the other hand due to the MMR modifier specified after the datum B and the MMR modifier specified after the datum C the examined disc may be moved with respect to both datums. The command *Rotate/translate the disc* displayed in the *Prompt line* encourage a user to click on the *Forward button* to see results of such displacement (Fig. 8). Next the verification of the second disc is visualized in the similar way – the only difference is, that all inserted verification features are shown at one animated sequence in the seventh scene (Fig. 9).

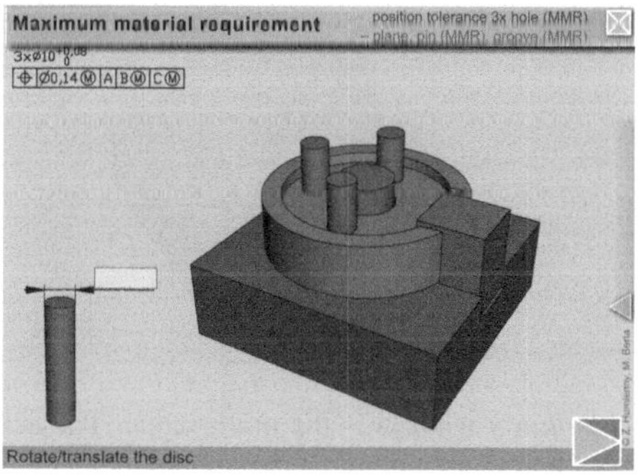

Fig. 7. *Explanation window* – the fifth scene. The three pins do not fit into their pilot holes

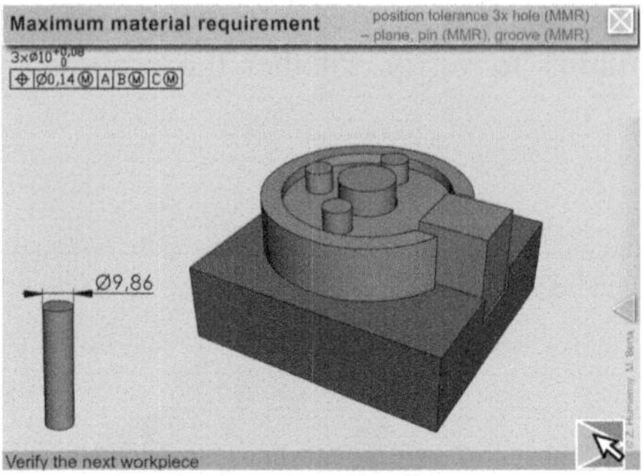

Fig. 8. *Explanation window* – the sixth scene. Thanks to the rotation of first inspected disc the three pins have fallen into the pilot holes.

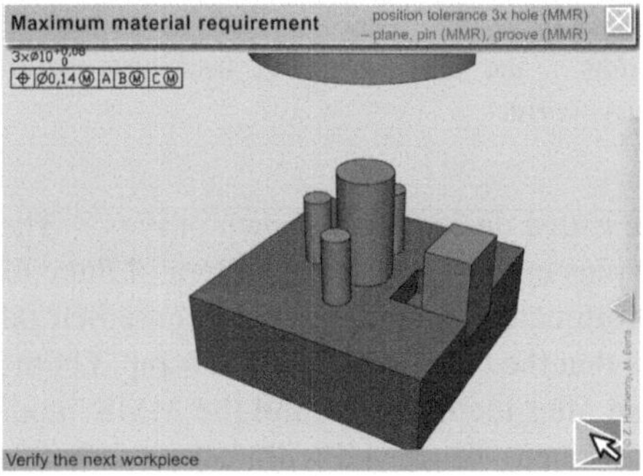

Fig. 9. *Explanation window* – the seventh scene – the second disc verification. Again the three pins do not fit into their pilot holes

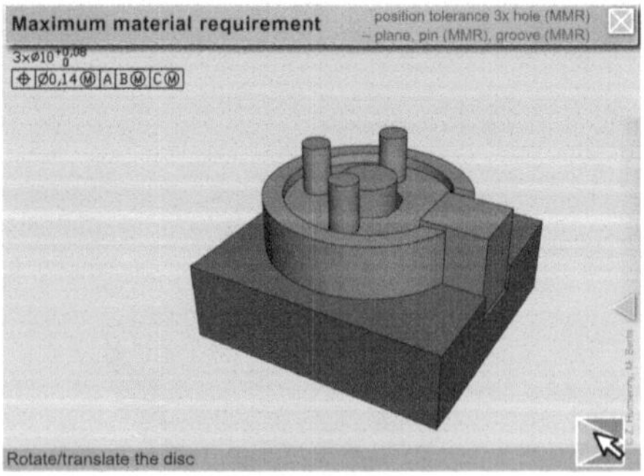

Fig. 10. *Explanation window* – the ninth scene. The actual gauge.

Unfortunately again the three pins do not fit into their pilot holes. This time the command *Rotate/translate the disc* displayed in the *Prompt line* activates translation of the examined disc and finally the pins fall into the holes.

The two types of available disc displacements are presented in the animation discussed above. So, the natural question arises as to how large additional deviations can occur when the groove – the datum C – deviates from the maximum material condition. Unfortunately large number of less skilled engineers and technicians without sufficient imagination claim during vocational training that additional bonus tolerance for pattern of three holes is equal to the difference between actual groove width and its maximum material size.

It is not true, the disc rotation consumes the clearance between the gauge block and the groove that is available at the radius 30 mm with respect to the disc axis (Fig. 3). The three holes are located on the radius 20 mm with respect to the disc axis, so when the grove is at its least material condition i.e. has the width 22,22 mm the three hole pattern can only rotate clockwise or counter clockwise by 0,22 × 20/30 mm. It is also often stated by participants of vocational training that the groove with 22,22 mm width (0,22 mm wider) allows an additional deviation of the position of the three hole pattern by 0,22 mm in a direction perpendicular to its symmetry plane. This is also not true, the displacement in this direction is limited by an actual diameter of the central hole. If the hole is in the maximum material condition, i.e. has the diameter of 20 mm (like in the first examined disc – Fig. 7), it is not possible to translate the disc in any direction. In extreme case, when the central hole is in the state of the least material condition, i.e. has the diameter of 20,18 mm, the additional disc translation is limited to 0,18 mm (see the second examined disc – Fig. 9).

The presented animation helps to clarify many confusions related to the misunderstanding of the MMR specification and to highlight the differences between the use of the MMR for a toleranced feature and datum features.

The actual gauge is shown in the last but one scene of the *Explanation window* (Fig. 10). The four pins and the block are fixed in the gauge base. The central pin is slightly longer to make the gauge easier to use. The verification takes a short while, but remember to accept a disc its face shall be in full contact with the gauge plane.

3 Conclusions

Despite its importance and potential benefits the academic education on the geometrical dimensioning and tolerancing methods is insufficient and knowledge of the modern geometrical product specification tools in industry is unsatisfactory. Specification of the MMR modifier allows design engineers to document functional intent more accurately and more completely.

Unfortunately the consequences of the MMR application for toleranced feature and the MMR application for datum features in a datum system are not well distinguish by many designers and metrologist. This observation have encouraged us to develop the presented animation within the application *Geometrical tolerancing*.

The application *Geometrical tolerancing* is created in the Flash, the tool that can integrate images, text, animations, sound and video. To optimize the application we decided to employ images, text and animations. Good reception of the application during university lectures as well as at vocational training in industry stimulates us to expand it.

The application is constantly being developed to show the latest provisions of new editions of the international standards that have been published recently as well as to reflect experience gained by authors during standardisation works and cooperation with industry. The inquires over the reception and understanding of GPS symbols, rules and principles during vocational trainings shows the positive impact of implementation of the application *Geometrical Tolerancing* in the courses.

The undisputed advantage of Geometrical Tolerancing application is that it combines usage of multimedia instructional tools to present information on geometrical tolerancing rules with demonstration of selected verification techniques for the particular requirements. In this paper verification by hard gauge that is very suitable for the MMR application in mass production is shown. For other specifications verification by classical, manually operated measuring equipment or by coordinate measuring systems are presented in the application *Geometrical Tolerancing*.

References

1. Morse, E., Dantan, J.-Y., Anwer, N., Soderberg, R., Moroni, G., Qureshi, A., Jiang, X., Mathieu, L.: Tolerancing: managing uncertainty from conceptual design to final product. CIRP Ann. Manuf. Technol. **67**, 695–717 (2018)
2. Charpentier, F.: Handbook for the Geometrical Product Specifications of Products: The ISO-GPS standards. AFNOR (2016)
3. Garland, N., Glithro, R., Wade, R.: The challenges facing education in engineering drawing practice. In: International Conference on Engineering & Product Design Education. Oslo & Akershus University College of Applied Sciences, Norway (2017)
4. Humienny, Z., Berta, M.: Using animations to support the understanding of geometrical tolerancing concepts. Technisches Messen **82**(9), 422–431 (2015)
5. Berta, M., Humienny, Z.: XXII world congress of the international measurement confederation (IMEKO 2018). IOP Conf. Ser.: J. Phys.: Conf. Ser. **1065** (2018)
6. Berney, S., Betrancourt, M.: Does animation enhance learning? A meta-analysis. Comput. Educ. **101**, 150–167 (2016)

Research of Traceability of Unit of Length in Metrology System of Bosnia and Herzegovina

Almira Softić[1], Hazim Bašić[1], Samir Lemeš[2(✉)], and Nermina Zaimović-Uzunović[2]

[1] University of Sarajevo, Vilsonovo setaliste 9, 71000 Sarajevo, Bosnia and Herzegovina
{softic,basic}@mef.unsa.ba
[2] University of Zenica, Fakultetska 1, 72000 Zenica, Bosnia and Herzegovina
slemes@unze.ba, nzaimovic@mf.unze.ba

Abstract. The actual status of metrology infrastructure of B&H on the state level and possibilities for improvement in future was discussed in this paper. The Institute of Metrology of Bosnia and Herzegovina (IMBiH) is a Government authority on the state level performing the functions of conformity assessment. Institute also has a role in harmonization of B&H legislation with the EC Directives. Since 2009. IMBiH has been full member of the European Association of National Metrology Institutes (EURAMET). Basic adopted laws in force on metrology in B&H are Law of metrology as well as Law on Measurement Units. The metrology system in B&H is highly decentralized, with a lack of correspondence between different levels of authorities. Accredited laboratories and other laboratories occupied with length measurement make their own traceability chain using laboratories in region and without inclusion of IMBiH. Laboratory for length in IMBiH is responsible for the realization and maintenance of the National Standard for Length and transferring unit of the length to lower rank standards and calibration of measuring instruments. Unit of length, in the Laboratory for length is realized through Iodine stabilized He-Ne laser of wavelength 633 nm. The laser is realized according to the definition of meter, i.e. according to the requirements of the International Recommendation for Practical Realization of the meter (CIMP 1997/2001 'Mise en pratique'). This laser can be used for calibration of other lasers with wavelength of 633 nm including laser interferometers for length measurements.

Keywords: Metrology infrastructure · Bosnia and Herzegovina · Length measurement

1 Introduction

Metrology is a key tool for fundamental research and innovation in all areas and enables the availability of accurate and reliable measurements relevant to industry, science, ecology, politics and everyday life. Development of technology, and therefore science without a well-established measurement system is not possible. National

V. D. Majstorovic and N. Durakbasa (Eds.): IMEKOTC14 2019, LNME, pp. 89–97, 2019.
https://doi.org/10.1007/978-3-030-18177-2_9

metrology system is of primary importance for any state, since it is the prerequisite for the development of other branches of metrology as well as new technologies, and for the development of the society as a whole. The basic tasks of a national metrology institution are providing a measurement traceability of the national measurement standards and delivering measurement traceability to the users and their measurement needs in various segments (science, environment, health, food, agriculture, transport, telecommunications, industry, trade, taxation, judicial authorities, and police) [1, 2].

In metrology system, it is necessary to provide traceability, the chain of comparisons ensures that national standards are traceable to the unit of measure in BIMP, and with the national standard conforming to the secondary (reference) standards, and with these industrial standards. Finally, industrial standards harmonize the end-user measuring instruments. Measuring of length and availability of length standards are crucial for the existence and development of a modern and technically developed society [2]. The development of technology also directly affects the improvement of the general quality of life, without meter and metrology, this would not be possible, and so metrology can be as an interface between science and technology that enables them to be stable.

The beginnings of metrology in B&H date from Ottoman Empire, when for almost five centuries a system of measures and measurement existed. European units of measurement and metrological system in our country brings Austro-Hungarian - one of the countries that signed and ratified the Metre Convention is in Budapest on 1875 [3].

The first laws on metrology – Law on measures for B&H were adopted 1911, and political and territorial changes and forming new Kingdom of Yugoslavia gives the new laws and rules in field of metrology. After the Second World War, the production and reconstruction of the former Yugoslavia also contributes to the changes and the emergence of new laws – Law of measuring units and measures in 1961 and 1974.

After the war years (1992–1995) in Bosnia and Herzegovina, with the help of the international community, major changes in the system of measurement units and metrology becomes and brought a whole set of laws and acts to the future organization of standardization, metrology, accreditation and intellectual property in Bosnia and Herzegovina.

The future organization of standardization, metrology, accreditation and intellectual property in Bosnia and Herzegovina was adopted in November 2000, by the then High Representative for B&H. Bosnia and Herzegovina participate in CIPM MRA since 2011, with signatory of IMBIH and participating national laboratories. For the past five years, the Institute for Metrology of Bosnia and Herzegovina has published a total of 59 lines of CMCs in 24 different fields and subdivisions, in which way it has achieved a stabile status in international metrology organizations. Internationally recognized Calibration and Measurement Capabilities (CMC) in the BIPM key data comparison KCDB have Laboratories of IMBIH: Mass and related quantities (pressure, density), Chemistry, Temperature and humidity, Electrical quantities, Time and frequency [3].

2 The Current State of Metrological Infrastructure in Bosnia and Herzegovina

According to the National Law of Metrology, main task of IMB&H is forming National laboratories for realization of basic SI units in B&H, as well as assuring international traceability up to the highest metrology level, Fig. 1. Based on allocation of needs in Bosnia and Herzegovina, IMB&H strongly started with process of establishing large number of national laboratories. Active laboratories of IMBIH are: Laboratory for Mass, Laboratory for Chemistry, Laboratory for Pressure and Vacuum, Laboratory for Volume, Laboratory for Density and Viscosity, Laboratory for Temperature and Humidity, Laboratory for Electrical Quantities, Laboratory for Time and Frequency, Laboratory for Ionizing Radiation and Laboratory for Verification. Some of these laboratories demonstrated their competencies at the international level by fulfilling the requirements for proving test and calibration skills, which resulted with published Calibration and Measurement Capabilities CMCs in their field in the KCDB – BIPM [3, 10].

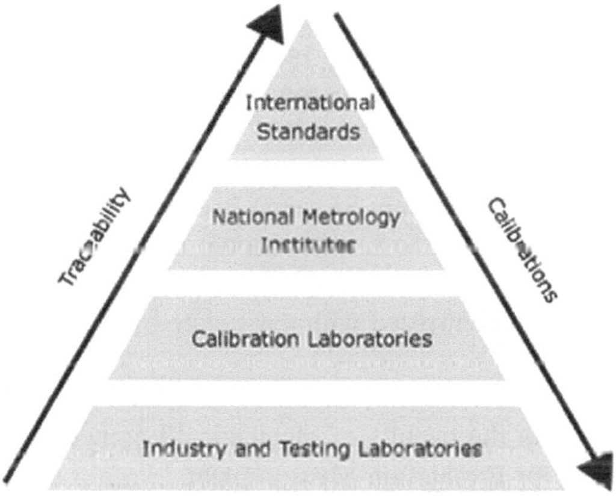

Fig. 1. Traceability chain for length

Furthermore, it is obvious is that the national Laboratory for the length is not on the list of active laboratories, which implies that length and measurement of length is one of the missing areas of activity of IMB&H [13].

Of course, this doesn't mean that the length and the length measurement is not being carried out in B&H, but international traceability and standards in this area are not as well covered as in other areas of IMB&H activity. The situation in the area of definition and dissemination of the unit of length is still in the regulation stage. Namely, IMB&H is in the possession of the primary standard for a unit of length and meter (m) is realized in the Laboratory for length through Iodine stabilized He-Ne laser of wavelength 633 nm [4, 11]. The laser is realized according to the definition of meter, i.e. according to the requirements of the International Recommendation for Practical Realization of the meter (CIMP 1997/2001 'Mise en pratique').

The traceability chain continuity should be from the primary IMB&H standard to lower levels, however conditions have not ensured yet due to the lack capacity and equipment that would provide this.

3 Comparison with Metrological Infrastructure of Slovenia

Bosnia and Herzegovina is a potential candidate for EU membership. This means that on the one side, the country is interested in joining the Union and on the other side there is interest of the Union in the country's accession. Bosnia and Herzegovina is at an early stage in achieving the capacity to cope with competitive pressure and market forces within the Union. The state is at an early stage regarding its approximation with European standards and has established institutions for standardization, metrology, accreditation and market surveillance [2]. Because of the complex constitutional and legal conformation, the metrology infrastructure of B&H still does not function in an optimal way. Slovenia, as one of the countries in the neighborhood, is a member of the EU and represents a good and harmonized approach to the regulation of metrology infrastructure [5]. In accordance with this fact and the lack of adequate regulation of metrological unity and dissemination of the unit of length in B&H, the accredited laboratory Lotric Control d.o.o has established its chain of traceability of the unit of length through the National Metrology Institute of Slovenia - MIRS.

The National Metrology System in Slovenia is entrusted to the Metrology Institute of the Republic of Slovenia (MIRS) as the national metrology institute (NMI), which provides traceability for selected units of SI System [5]. For certain SI Units, national standards are recognized and entrusted with a mandate by MIRS to other legal entities, Table 1.

The national standard of length of Slovenia is within the Institute of Production Mechanical Engineering of the Faculty of Mechanical Engineering at the University of Maribor, in Laboratory for Production Measurement.

The Laboratory for Production Measurement is ensuring traceability of the national standard of length in the primary laboratory of the Nederland's Meetinstituut, the Netherlands, (for laser interferometer), in the primary laboratory of the Physikalisch-Technische Bundesanstalt (PTB), Germany (for large length standards up to 1000 mm), and in the national laboratory of BNM-LNE, France (for small length standards up to 100 mm). In view of reducing measurement uncertainty, the Laboratory also participates in inter-laboratory comparisons. The dissemination of the value by means of national standards conducts following the accredited procedures for comparison calibration, to the laboratory's working standards, which are then used for calibration of customers' standards. The comparison measurement is performed by a comparison measuring instrument (comparator). The length of the calibrated standard (gauge block) compares to the length of the reference standard, which is traceable to the primary standard of length (international level). The comparison measuring instruments (a comparator up to 100 mm, and a comparator up to 1000 mm) operate on the principle of differential measurement of deviations by two inductive pick-offs [5].

Table 1. National measurement standard system of Slovenia [5].

Field	Recognized institution
Time and frequency	Slovenian Institute of Quality and Metrology, Metrology
Length	University of Maribor, Faculty of Mechanical Engineering, Laboratory for Production Measurement
Electricity	Slovenian Institute of Quality and Metrology, Metrology
Ionizing radiation	Jožef Stefan Institute, Low and Medium Energy Physics F2
Amount of substance/Organic Compounds, in particular Fatty Acids, Sterols, Biophenols, Tocopherols, Waxes, Triacylglycerols, Stigmastadienes/in Biological Materials and Food	Science and Research Centre, Laboratory of the Institute for Oliveculture
Amount of substance/Bioanalysis of Nucleic Acids/GMOs and Microorganisms	National Institute of Biology, Department of Biotechnology and systems biology
Amount of substance/Inorganic Non-metals and their Compounds, Manganese, Loss on Ignition, insoluble and main Components/in Mineral Binders and Mortars	Slovenian National Building and Civil Engineering Institute, Laboratory for Cements, Mortars and Ceramics
Amount of substance/*Chemical trace Elements/in the Organic and Inorganic Materials*	Jožef Stefan Institute, Environmental Sciences O2
Amount of substance/Inorganic Metalloids and their Compounds, sum Parameters, pH, Toxicity/in Water	National Institute of Chemistry, Laboratory for Environmental Sciences and Engineering
Thermodynamic temperature	University of Ljubljana, Faculty of Electrical Engineering, Laboratory of Metrology and Quality
Pressure	Institute of Metals and Technology, Laboratory of pressure metrology
Humidity	University of Ljubljana, Faculty of Electrical Engineering, Laboratory of Metrology and Quality

4 Dissemination of Unit of Length in Bosnia and Herzegovina

Standard ISO 17025:2018 demands that the laboratories have to establish and maintain metrological traceability of all measuring results by means of documented unbroken chain of calibrations linking them to an appropriate reference [2, 14]. In B&H exist several laboratories dealing with length measurements, and some of them are accredited according to standard ISO 17025 by national accreditation body in B&H named

BATA [6, 12]. Most of laboratories are part of larger production plants, and in order to achieve correctness and the accuracy of their production have formed laboratories and adapted to their needs, Table 2. Naturally, globalization of the world market put strict demands on products and services, so these B&H laboratories have to meet these conditions.

Table 2. Laboratories for length measurement in B&H

No.	Laboratory	Accreditation ISO 17025	Field of accreditation	Traceability chain
1.	BNT-Factory of hydraulics and machines, Novi Travnik - Calibration Laboratory	Yes	Length, mass, pressure	"ORAO", Bijeljina, Laboratory – Metrological laboratory
2.	JSC for production and overhaul "ORAO", Bijeljina, Laboratory – Metrological laboratory	Yes	Length, mass, electricity and magnetism and other quantities	DMDM Belgrade Serbia
3.	LOTRIC CONTROL, Mostar-Laboratories	Yes	Length, mass, humidity, thermometry	MIRS Slovenia
4.	Mechanical Engineering Faculty, Sarajevo – Laboratory for production measuring technique	No	Length, angle, roughness	DKD Laboratory Germany*

*expired

Due to the fact that there is no traceability chain for a unit of length to lower levels, i.e. accredited laboratories, at the state level, these four laboratories were forced to find their own ways to meet the requirements of the standard and establish continuity of traceability chain for length measurement.

"ORAO" Bijeljina in B&H, is factory for production and overhaul of airplane engines, with the most contemporary equipment and highly trained stuff [7]. Laboratories are very important potential of ORAO and they are primary developed for overhaul process of turbojet engines. Laboratory for length measurement is accredited according to standard ISO 17025 by BATA, and they maintained their traceability chain for unit of length according to the National Metrology Institute Serbia (DMDM), Fig. 2.

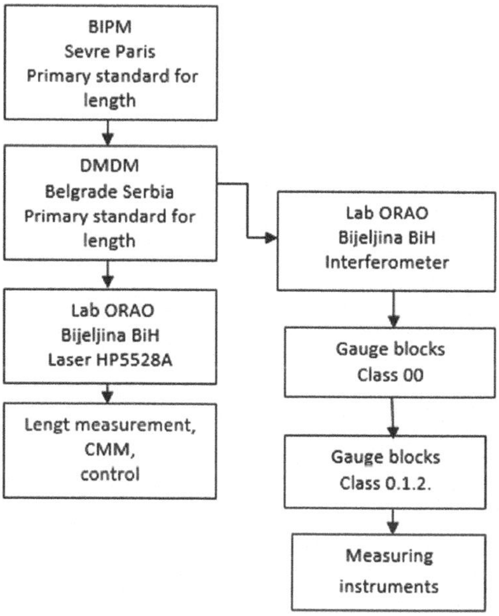

Fig. 2. Traceability chain for length of laboratory "ORAO" [7].

As one can see from Fig. 2, the Laboratory of ORAO, unlike the laboratory Lotric control, has established its chain of traceability of the unit of length through the neighboring state of Serbia. Serbia has a stable metrology system with Directorate of Measures and Precious Metals (DMDM) as National Metrology Institute (NMI) and national authority on legal metrology with responsibilities for control of precious metals articles [8]. As NMI, the DMDM has task to maintain national measuring standards of SI units, perform their international comparisons and disseminate legal units of measurement by calibration of measuring standards and instruments.

For the realization and maintenance of the standard of the unit of length and other related sizes at the NMI of Serbia, the Laboratory for dimensional measures (LDM) is responsible. The national etalon of the unit of length (meter) was achieved through two frequency-stabilized helium-neon (He-Ne) lasers stabilized on pairs of dual-atomic isotope molecules 127 (127I2) at a wavelength of 633 nm. Lasers are made by definition, i.e. in accordance with the requirements of the international recommendation for the practical realization of the meter [8, 9]. With these lasers, other lasers with a wavelength of 633 nm are installed, including laser interferometers for measuring the length.

Laboratory measurement and calibration capabilities (CMCs) can be found in the database of the International Bureau for Weights and Measures Appendix C of the KCDB [10], Fig. 3.

The international traceability for all measurements made in the LDM Serbia is ensured through participation in international inter-comparisons as well as in various international projects. LDM participates in these activities mainly at the regional level,

Calibration and Measurement Capabilities

Length, Serbia, DMDM (Directorate of Measures and Precious Metals) KCDB

Calibration or Measurement Service			Measurand Level or Range			Measurement Conditions/Independent Variable		Expanded Uncertainty						
Class	Instrument or Artifact: Measurand	Instrument Type or Method	Minimum value	Maximum value	Units	Parameter	Specifications	Value	Units	Coverage Factor	Level of Confidence	Is the expanded uncertainty a relative one?	NMI Internal Service Identifier	Comments
Laser radiation	Frequency stabilized laser: vacuum wavelength	Optical beat frequency	633	633	nm			0.04	fm	2	95%	No	1	Approved on 22 March 2005
Laser radiation	Frequency stabilized laser: absolute frequency	Optical beat frequency	474	474	THz			24	kHz	2	95%	No	2	Approved on 22 March 2005
Laser radiation	Frequency stabilized laser: vacuum wavelength	Optical beat frequency	633	633	nm			1E-09		2	95%	Yes	3	Approved on 22 March 2005
End standards	Gauge block: central length *L*	Interferometry, exact fraction	0.5	100	mm			Q[20, 0.2*L*], *L* in mm	nm	2	95%	No	4	Approved on 22 March 2005
End standards	Gauge block: central length *L*	Mechanical comparison to gauge block	0.5	100	mm			Q[50, 0.5*L*], *L* in mm	nm	2	95%	No	5	Approved on 22 March 2005
Angle by circle dividers	Optical polygon: face angle	Goniometer	0	360	°			0.2	"	2	95%	No	7	Approved on 22 March 2005
Angle by circular dividers	Rotary table, heads and goniometers: angle	Standard polygon and autocollimator	15	360	°			0.2	"	2	95%	No	8	Approved on 22 March 2005
Angle instruments	Autocollimator: error of indicated angle	Goniometer	0	according to the autocollimator	"			0.2	"	2	95%	No	9	Approved on 22 March 2005
Angle artefacts	Angle block: included angle	Goniometer	0	360	°			0.2	"	2	95%	No	10	Approved on 22 March 2005
Surface texture	Depth standard: (ISO 5436-1 type A), depth *d*	Stylus instrument	0.01	50	µm	Depth	*d*	Q[15, 15*d*], *d* in µm	nm	2	95%	No	11	Approved on 12 September 2016
Surface texture	Spacing standard (ISO 5436-1 type C): amplitude parameters	Stylus instrument	0.01	15	µm	Average parameters	*Ra*, *Rq*	Q[10, 30*Ra*], *Ra* in µm	nm	2	95%	No	12	Approved on 12 September 2016

The BIPM key comparison database, June 2018 1/2

Fig. 3. Length, Serbia, DMDM, KCDB [10].

through active participation in EURAMET projects, the European Association of National Metrology Institutes [10].

5 Conclusion

Bosnia and Herzegovina is at an early stage in achieving the capacity to cope with competitive pressure and market forces within the EU. Through this research, it has been established that there are solid laboratory capacities of four laboratories for length measurement that successfully meet the needs of the B&H market. Main problem is dissemination of the unit of length on the state level, which is not established. All mentioned laboratories made their own traceability chain for unit of length through different laboratories in Europe (three European countries) and that is main problem for optimal functioning of metrology system in B&H in field of length measurement. The Slovenian metrology system was observed as an example of good and strategic approach.

IMB&H cannot adequately achieve the goals, because the complexity and scope of metrology requirements require great investment. Because of the complex constitutional and legal conformation of B&H, the metrology infrastructure of B&H still does not function in optimal way.

One of the transitional arrangements for setting up a chain of traceability per unit length in B&H could be the announcement of one of the accredited laboratories in B&H for the national laboratory for the length and the establishment of a single traceability chain for the whole country [14–16]. This would prevent the establishment of different chains of traceability of the unit length on the state level and the establishment of a unique one. Of course, further development and expansion of the unit length base by IMB&H would establish a unique distributed metrology system, ensuring the proper establishment of traceability of state-level and dissemination to low levels in B&H. Cooperation and coordination with the metrology institutes of EU needs further improvement and a country-wide development strategy remains to be adopted.

References

1. Aganovic, A.: Organization and quality system management of the national metrology institute in respect with international requirements and national metrology infrastructure. Master thesis, University of Zenica (2013). (in Bosnian)
2. Seferović, E., Bašić, H.: Metrology and measuring results data analysis. Mechanical Engineering Faculty Sarajevo, Sarajevo (2005). (in Bosnian)
3. Softic, A.: Traceability of unit of the length research using circular intercomparation. Ph.D. thesis, University of Sarajevo (2014). (in Bosnian)
4. http://www.met.gov.ba. Accessed 20 Feb 2019
5. Bosnjakovic, A.: Measuring uncertainty calculation of iodine stabilized He-Ne laser with wavelength at 633 nm. Master thesis, University of Sarajevo, Sarajevo (2013). (in Bosnian)
6. http://www.mirs.si. Accessed 16 Feb 2019
7. http://www.bata.gov.ba. Accessed 15 Jan 2019
8. Group of Authors: Measurements in theory and practice. In: PoMaCoM, Proceedings. University of Zenica, Zenica (2007)
9. http://www.dmdm.rs. Accessed 10 Feb 2019
10. http://www.bipm.fr. Accessed 20 Feb 2019
11. http://kcdb.bipm.org/. Accessed 17 Feb 2019
12. Bošnjaković, A., Bašić, H.: The method of calibration of He-Ne laser using the NPL iodine stabilized He-Ne laser with wavelength at 633 nm. In: 18th International Research/Expert Conference Trends in the Development of Machinery and Associated Technology – TMT 2014, Budapest, Hungary, pp. 293–296 (2014). ISSN 1840-4944
13. Softić, A., Zaimovic-Uzunović, N., Bašić, H.: Proficiency testing and interlaboratory comparisons in laboratory for dimensional measurement. J. Trends Dev. Mach. Assoc. Technol. **16**(1), 115–118 (2012)
14. Bašić, H., Softić, A.: Metrology infrastructure of Bosnia and Herzegovina in the context of European integration. In: IN-TECH 2017, International Conference on Innovative Technologies, Ljubljana, Slovenia, pp. 145–148 (2017)
15. Bašić, H., Softić, A.: The proficiency testing analyses for dimensional laboratories using the gauge blocks. In: 19th International Research/Expert Conference 'Trends in the Development of Machinery and Associated Technology – TMT 2015, Barcelona, Spain, pp. 369–372 (2015). ISSN 1840-4944
16. Softić, A., Bašić, H., Zaimović-Uzunović, N.: Implementation methodology scheme for proficiency testing in length measurement. In: 9 Research/Expert Conference with International participation, QUALITY 2015, Neum, B&H, pp. 195–200 (2015). ISSN 1512-9268

Surface Topological Investigation of Seal Bushing by Using Chromium Plating Process

P. Demircioglu[1], I. Bogrekci[1(✉)], M. N. Durakbasa[2], N. Demir[1], and U. Kose[3]

[1] Faculty of Engineering, Department of Mechanical Engineering, Aydın Adnan Menderes University, 09010 Aydın, Turkey
{pinar.demircioglu, ibogrekci, neslihan.demir}@adu.edu.tr

[2] Institute for Production Engineering and Laser Technology, Department of Interchangeable Manufacturing and Industrial Metrology/Production Metrology and Quality, Vienna University of Technology (TU Wien), BA-09, 1060 Vienna, Austria
durakbasa@ift.tuwien.ac.at

[3] EYS METAL Research and Development Center, 09010 Aydın, Turkey
ukose@e-y-s.com.tr

Abstract. According to utilization functions and environments; various plating methods have been applied on metallic manufactured products to improve some properties such as friction, wear resistance and corrosion; besides mechanical properties such as stiffness, ductility, toughness, fatigue strength, etc. The chromium plating has been commonly used plating technology at manufacturing sector providing a coating of wear resistant chromium with a micro-scaled thickness on metal products. The chromium plating is to gain high toughness, resistance of corrosion, low coefficient of friction and improve some properties of material. Chromium acid solutions consisting of one or more catalytic anions are used for chromium plating. The aim of this study is to investigate the surface roughness parameters for coated and uncoated seal bushing surfaces. The material of seal bushing is AISI 420 steel, which is included in martensitic stainless steels and has quite high tensile strength. After the heating process was applied to the seal bushing, chromium plating was implemented with different currents and periods on the surface of the material. The chromium coating was successfully deposited on a seal bushing surfaces. Experimental data was collected from three different types of the bushing surfaces. After coating process, the finishing methods, such as grinding etc. were implemented. The results indicated that coated surfaces had higher surface roughness values than the uncoated surfaces. According to the results of Wilcoxon test, there were a significant difference between Workpieces B (cutting+turning+chromium plating) and C (cutting+turning+chromium plating+grinding); Workpieces A (cutting+turning) and B.

Keywords: Bushing · Chromium plating · Surface roughness · Topology

V. D. Majstorovic and N. Durakbasa (Eds.): IMEKOTC14 2019, LNME, pp. 98–106, 2019.
https://doi.org/10.1007/978-3-030-18177-2_10

1 Introduction

In manufacturing industry, various manufacturing processes have been used and diversified as turning, grinding, cutting, welding, milling, grinding, plating etc. The chromium plating, which is the one of these methods, has been used in various manufacturing fields such as automotive, aerospace, petro-chemistry, which are used for maintaining wear resistance of parts and enhance the material properties [1, 2]. Chromium coating is remarkable for high resistance to obtain high hardness and wear resistance as well as low coefficient of friction resistance to cold welding [3]. Chromium plating is per-formed with a chromium salt solution and a passive anode made of lead. Trivalent chromium solutions are limited to low deposition rates and are mainly used for the deposition of thin decorative coatings. Hexavalent chromium solutions provide higher deposition rates than trivalent chromium solutions and are used for the production of thick wear-resistant hard chromium coatings [4].

Surface roughness is one of the surface texture measurement and is quantified by the vertical deviations of the real surface from its ideal. If the surface comes along rough, the deviations are high; if smooth, the deviations are low. Roughness is typically considered the high frequency, short wavelength component of *a* measured texture [5]. In measuring surface roughness, average surface roughness is commonly used, which is denoted as R_a. R_a is the arithmetic average value of departure of the profile from the mean line throughout the sampling length. R_a is also an important factor in controlling machining performance [6].

Surface roughness can provide various knowledge about plating, such as the bath stability: stable baths generate less rough coatings, free of nickel particles adsorbed on the surface of the coating [7]. De Mello et al. analyzed influence of hard chromium plating on surface texturing and found out a significant influence of plating on topological behavior [8].

2 Material and Method

Bushing has been processed by different manufacturing methods for purpose of higher surface quality. In the first manufacturing operation, raw material of bushing, AISI 420 stainless steel, has been operated for cutting and turning. In this process, the aim of turning is to smooth out bushing and it is significant to reach the required cutting parameters to obtain high cutting performance [9]. After these operations, chromium plating has been operated to intend to improve some mechanical properties, good aesthetic appearance and superior resistance to corrosion [10]. The advantages of chromium plating are the high level of hardness, resistance to corrosion and wear, or low coefficient of friction of the coatings. Due to the deposition process, these coatings are extensively micro-cracked and present a crystallographic texture [11].

Surface quality of bushing with the chromium plating has been improved when compared with the uncoated surfaces. To improve the surface finish, grinding operation was conducted. Grinding is a machining process utilizing hard, abrasive particles as the material removal medium, and usually regarded as a key step to obtain the high dimensional accuracy and fine surface roughness of the workpieces [12]. After grinding, semi-finished product has been ready for assembling.

Many factors cause the traverse and longitudinal waviness produced by the random nature of the turning, the plating and the grinding. Consequently, surface topography becomes a complicated issue. Surface roughness is a typical parameter to quantify the surface quality [13]. Surface roughness is defined as the finer irregularities of the surface texture that usually result from the inherent action of the machining process [14]. The surface roughness is presented by the arithmetic mean value R_a, the root mean-square-average R_q, and the maximum roughness height R_t.

In this study, it was sought the answer which workpieces, manufactured by related process, have higher surface quality. So that there are three types of workpieces manufactured by related processes and measured the surface roughness by using scanning type confocal laser microscope. The roughness of bushing surfaces was evaluated. The bushing, which is investigated surface roughness value in this study, is used in high-pressure pump for avoiding corrosion etc.

2.1 Materials

The raw material of workpieces is AISI 420, martensitic stainless steel, which offered moderate corrosion resistance in comparison to the austenitic and duplex grades and used for prolonging lifetime of product achieved with a surface treatment and coating, as long as corrosion resistance is sustained, and adhesion is acceptable [15].

2.2 Bushing Manufacturing Processes

These workpieces used for characterization of surface roughness. The workpieces had the manufacturing processes, respectively, turning, chromium plating and grinding. The workpieces to be measured are shown in Fig. 1.

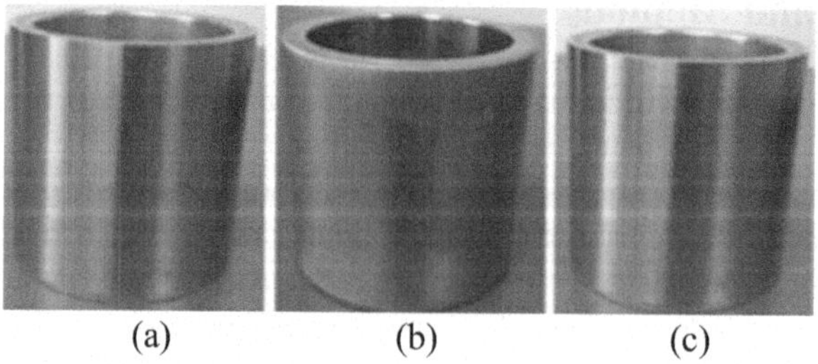

Fig. 1. (a) Workpiece A applied turning (b) Workpiece B applied chromium plating (c) Workpiece C applied grinding

The evaluation process of flow steps is presented in Fig. 2 consisting of two main phases, respectively, manufacturing process of workpieces (Step 1).

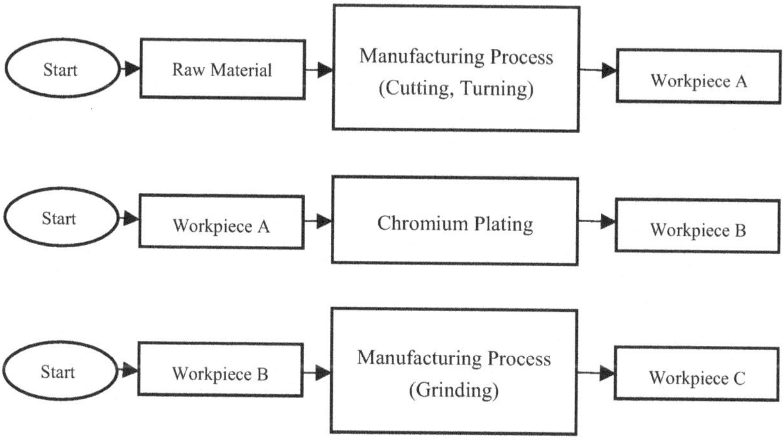

Fig. 2. Manufacturing process of workpieces (Step 1)

2.3 Surface Roughness Characterization

The evaluation of surface roughness parameters is configured in Step 2. The aim is to compare the surface roughness values for three different types of workpieces following step in Fig. 3.

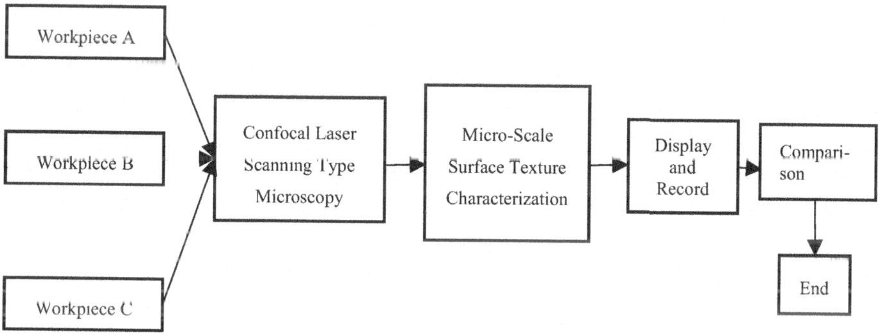

Fig. 3. Measurements for characterization of surface roughness (Step 2)

2D surface profile parameter R_a as amplitude parameter (average of ordinates) and 3D surface profile parameter S_a as surface average deviation are listed in Table 1. The surface roughness of workpieces of three different bushing were investigated [17, 22].

For surface roughness measurements, a rotating (Nipkow disc) confocal microscope (NanoFocus - µsurf) [14, 20] was used. Each workpieces from all types was measured for surface roughness. The most commonly used roughness parameters in this experimental study are used as R_a and S_a.

Table 1. 2D/3D surface profile parameters [18–22]

Equation with description	Graphical/Schematical Representation
Arithmetical mean ddeviation of the assessed profile Ra $$Ra = \frac{1}{l}\int_{0}^{l} \lvert Z(x)\rvert \, dx$$ Arithmetic mean of the absolute ordinate values *Z(x)* within a sampling length.	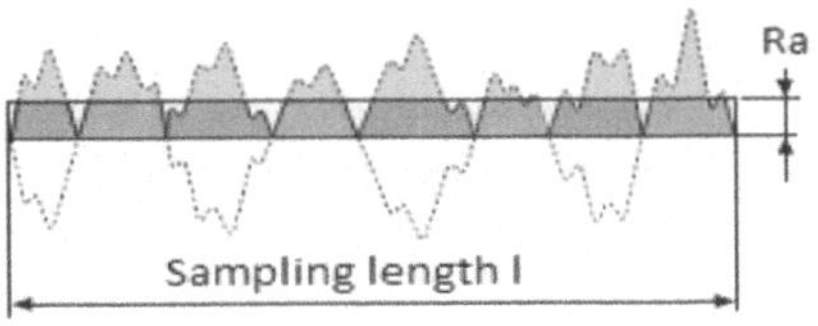
$$Sa = \iint_{a} \lvert Z(x,y)\rvert \, dx\, dy$$ Sa, The average deviation of the surface	

3 Results

3.1 2D Surface Roughness Measurement Results

There are different techniques used for surface metrology analysis. This paper presents the investigation of coated and uncoated surface quality using the scanning type confocal laser microscope.

The roughness measurement results from the scanning type confocal laser microscope are given at Table 2. Roughness values of Workpiece B (cutting+turning

Table 2. Surface roughness measurement results in terms of R_a

Roughness measurement results R_a (μm)	Workpieces		
	A	B	C
1	0.063	1.610	0.084
2	0.097	1.780	0.098
3	0.069	1.730	0.093
4	0.122	1.730	0.161
5	0.122	0.829	0.106
6	0.065	0.619	0.113
7	0.102	1.610	0.110
8	0.113	1.190	0.096
9	0.094	0.781	0.142
Mean	0.094	1.320	0.111
Std. Dev.	0.022	0.442	0.024

+chromium plating) turned out to be higher than those of Workpieces A (cutting +turning) and C (cutting+turning+chromium plating+grinding).

The main reason of high roughness values indicated that the manufacturing process consisted of the chromium plating. The roughness deviation verified that chromium plating increased the surface roughness values.

3.2 3D Surface Roughness Measurement Results

3D profiles of the measured workpieces are given in Fig. 4.

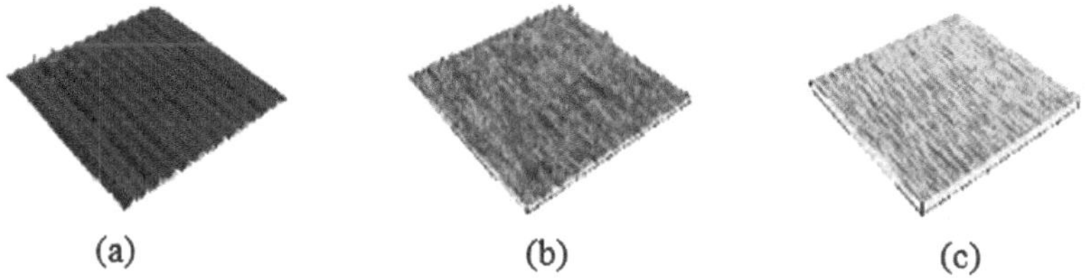

Fig. 4. 3D profiles of (a) Workpiece A (cutting+turning) (b) Workpiece B (cutting+turning +chromium plating) (c) Workpiece C (cutting+turning+chromium plating+grinding)

There are four groups of parameters for 3D (areal) surface roughness characterization based on statistic, respectively, amplitude, spatial, hybrid and functional [22, 23].

In the present study, it was used 3D amplitude height parameters such as S_q, S_p, S_v, S_z and S_a for the 3D evaluation. The results of surface roughness measurement in terms of 3D Height Parameters are given in Table 3.

S_q, Root mean square height of the surface
S_p, Maximum height of peaks
S_v, Maximum height of valleys
S_z, Maximum height of the surface
S_a, Arithmetical mean height of the surface

Table 3. Surface roughness measurement results in terms of height parameters.

Height parameters	Workpieces		
	A	B	C
S_a (μm)	1.294	1.577	0.205
S_q (μm)	1.469	2.098	0.279
S_p (μm)	11.186	21.111	6.688
S_v (μm)	3.897	6.303	2.447
S_z (μm)	15.084	27.422	7.632

S_a values indicates that coating increased the surface roughness, while grinding the coated surface decreased the surface roughness for Workpieces A, B and C. This phenomenon happens for all values of S_q, S_p, S_v, S_z and S_a.

3.3 Statistical Results

In order to find the significant difference between different manufacturing operations and coating/noncoating, the statistical analyses were conducted using SPSS statistical software. Friedman and Wilcoxon Tests were carried out for surface roughness values of Workpieces A, B and C.

Every measurement of surface roughness was compared using Friedman Test, regarding significance (p) value is 0.05. According to Friedman test results in Table 4, significance value is 0.001 and lower than 0.05, there was a significant difference between surface roughness measurement of Workpieces A, B and C.

Table 4. Friedman test results for comparing Workpieces A, B and C.

Number of measurement	9
Chi-Square	14.889
Significance value	.001

In order to compare these significant differences between them, Wilcoxon Test was applied. It was found that Workpieces A–B and B–C had significant changes, which mean that there are effects between these workpieces on surface topography. According to results of Wilcoxon Test in Table 5, the followings can be inferred:

Table 5. Wilcoxon test results for comparing Workpieces A, B and C.

		N	Mean rank	Sum of ranks
Workpiece_A - Workpiece_B	Negative ranks	9[a]	5.00	45.00
	Positive ranks	0[b]	.00	.00
Workpiece_C - Workpiece_B	Negative ranks	9[d]	5.00	45.00
	Positive ranks	0[e]	.00	.00

- There was a significant difference between Workpieces B and C. Workpiece C (cutting+turning+chromium plating+grinding) had lower roughness than Workpiece B (cutting+turning+chromium plating).
- There was a significant difference between Workpieces A and B. Workpiece A (cutting+turning) had lower roughness than Workpiece B had.

4 Conclusion

This study was performed by the researchers of different academic organizations and industry (R&D center approved by Ministry of Industry and Technology). The methodology to define the surface roughness characterization of the bushing surface parameters using the scanning type concofal microscope revealed an advantage for more precise and accurate results.

According to these results, chromium plating deteriorated the surface roughness, while improving some properties as friction, wear resistance and corrosion, besides mechanical properties. However, the coated surfaces had higher surface roughness. After plating process, finishing process ought to be implemented on the workpiece such as grinding etc.

The continuation of this study is to be the surface investigation of the bushings after a series of process and their coating performances and issues such as the optimization of the bushing surfaces. The future studies can be the investigation of surface coating and mechanical properties of the material.

References

1. Kirihaara, S., Umeda, Y., Tashiro, K., Honma, H., Takai, O.: Development of NI-W alloy plating as a substitution of hard chromium plating. Trans. Mater. Res. Soc. Japan **41**(1), 35–39 (2016)
2. Aydın, Z., Aldıç, G., Çimenoğlu, H.: An investigation on the mechanical properties of the hard chromium layer deposited by brush plating process on AISI H13 steel. Arch. Mater. Sci. Eng. **65**(2), 87–92 (2014)
3. Chauhan, K.S.: Effect of substrate texture on electroplating. In a Report Submitted in Partial Fulfilment of the Requirement for the Degree of Bachelor of Technology in Metallurgical and Materials Engineering. Department of Metallurgical and Materials Engineering National Institute of Technology Rourkela (2010)
4. Batchelor, A.W., Lam, L.N., Chandrasekaran, M.: Materials Degradation and its Control by Surface Engineering. World Scientific Publishing Co. Pte. Ltd., Singapore (1999)
5. Li, H.: A study of surface roughness in the metal forming process. Ph.D. thesis, School of Mechanical, Materials and Mechatronics Engineering, University of Wollongong (2012)
6. Öktem, H., Erzurumlu, T., Kurtaran, H.: Application of response surface methodology in the optimization of cutting conditions for surface roughness. J. Mater. Process. Technol. **170**, 11–16 (2005)
7. Bonin, L., Castro, C.C., Vitry, V., Hantson, A.L., Delaunois, F.: Optimization of electroless NiB deposition without stabilizer, based on surface roughness and plating rate. J. Alloys Compounds **767**, 276–284 (2018)
8. De Mello, J.D.B., Goncalves, J.L., Costa, H.L.: Influence of surface texturing and hard chromium coating on the wear of steels used in cold rolling mill rolls. Wear **302**, 1295–1309 (2013)
9. Sánchez Lasheras, F., Garcia Nieto, P.J., de Cos Juez, F.J., Vilan Vilan, J.A.: Evolutionary support vector regression algorithm applied to the prediction of the thickness of the chromium layer in a hard chromium plating process. Appl. Math. Comput. **227**, 164–170 (2014)

10. Pina, J., Dias, A., Francois, M., Lebrun, J.L.: Residual stresses and crystallographic texture in hard-chromium electroplated coatings. Surface Coatings Technol. **96**, 148–162 (1997)
11. He, B., Wei, C., Ding, S., Shi, S.: A survey of methods for detecting metallic grinding burn. Measurement **134**, 426–439 (2019)
12. Hecker, R., Liang, S.Y.: Predictive modeling of surface roughness in grinding. Int. J. Mach. Tools Manuf. **43**, 755–761 (2003)
13. Gupta, M.K., Sood, P.K.: Surface roughness measurements in NFMQL assisted turning of titanium alloys: an optimization approach. Friction **5**(2), 155–170 (2017)
14. Dalibon, E.L., Charadia, R., Cabo, A., Trava-Airoldi, V., Brühl, S.P.: Evaluation of the mechanical behaviour of a DLC film on plasma nitride AISI 420 with different surface finishing. Surface Coatings Technol. **235**, 735–740 (2013)
15. Osanna, P.H., Durakbasa, M.N., Kräuter, L.: Industrial metrology and interchangeable manufacturing under the viewpoint of nanotechnology and nanometrology. Bulgarian Acad. Sci. **59**, 60–73 (2008)
16. Nano Focus – μsurf. http://www.nanofocus.com/products/usurf/usurf-explorer/
17. ISO 4287:1997 – Geometrical Product Specifications (GPS) – Surface texture: Profile method – Terms, definitions and surface texture parameters. (Last Reviewed in 2015)
18. ISO 4288:1996 – Geometrical Product Specifications (GPS) – Surface texture: Profile method – Rules and procedures for the assessment of surface texture. (Last Reviewed in 2013)
19. ISO 25178-2:2012 – Geometrical product specifications (GPS) – Surface texture: Areal - Part 2: Terms, definitions and surface texture parameters
20. Demircioglu, P.: Topological evaluation of surfaces in relation to surface finish. In: Hashmi, M.S.J. (ed.) Comprehensive Materials Finishing, vol. 3, pp. 243–260. Elsevier, Oxford (2017)
21. Bogrekci, I., Demircioglu, P., Sucuoglu, H.S., Hacıyusufoglu, A.F.: Determining surface topography for cylinder liner surfaces using 2D fast Fourier transform. Sigma J. Eng. Nat. Sci. **35**(2), 311–322 (2017)
22. Manesh, K.K., Ramamoorthy, B., Singaperumal, M.: Numerical generation of anisotropic 3D non-Gaussian engineering surfaces with specified 3D surface roughness parameters. Wear **268**, 1371–1379 (2010)
23. Demircioglu, P., Bogrekci, I., Durakbasa, N.M.: Micro scale surface texture characterization of technical structures by computer vision. Measurement **46**, 2022–2028 (2013)

A General Concept of Measurements of Form Deviations of 3D Rotary Elements with the Use of the Adaptive Strategy

Krzysztof Stępień(✉), Dariusz Janecki, and Stanisław Adamczak

Kielce University of Technology, Al. 1000-lecia P.P. 7, 25-314 Kielce, Poland
kstepien@tu.kielce.pl

Abstract. Measurements of form deviations of 3D elements can be conducted with various strategies. Such strategies differ in a number and distribution of sampling points located on an investigated surface. Low number of sampling points can bring about the situation that some surface irregularities cannot detected by a measuring system. An application of higher density of sampling points, in turn, results in significant lengthening of measurement time, which is undesirable. This is why authors propose to develop a novel, adaptive measurement strategy. Proposed adaptive strategy consists of two stages: a preliminary measurement and additional measurements. During the preliminary measurement an investigated area is scanned along a preselected trajectory. If measurement results show that there is a significant change of sensor readings in a certain fragment of an investigated surface, then we conduct additional measurements in the area where the large change of sensor readings occurred. The paper presents the state-of-the-art on contemporary measurement strategies of 3D elements and fundamentals of the adaptive strategy proposed by authors.

Keywords: Measurement · Form deviation · Adaptive strategy

1 Introduction

Rotary elements constitute a very significant and numerous group of machine parts Such machine parts are used in various branches of engineering industry (for example in bearing, automotive or power industry). Such parts are usually cylindrical, spherical, barrel- and saddle-shaped or conical ones. Usually, rotary elements should be of very high accuracy. Therefore it is very important to apply a reliable method of measurements of their form deviations.

Usually, investigating form deviations of rotary parts is based on an analysis of results of 2D measurements. In other words, the investigation is performed in the following way: measurements of roundness deviations are conducted in a few presumed cross-sections of the part.

It is less common to measure 3D parameters of rotary parts. In industrial practice, measurements of 3D parameters are practically limited to an evaluation of out-of-cylindricity. Measurements of cylindricity measurements are usually performed with special-purpose radial systems. Typical example of such systems are Talyrond

V. D. Majstorovic and N. Durakbasa (Eds.): IMEKOTC14 2019, LNME, pp. 107–113, 2019.
https://doi.org/10.1007/978-3-030-18177-2_11

instruments by Taylor Hobson. Nowadays, radial systems, called also radius-change ones, can be used to measure roundness and cylindricity deviations. In addition, some radius-change instruments can also be used to measure flatness of the cylinder face. Radius-change instruments are characterized by high accuracy, however, their area of applications is limited, as they usually can be used to measure roundness, cylindricity or flatness of the face, as it was mentioned above.

Nowadays, in the area of metrology of geometrical quantities dynamic development of coordinate metrology can be observed. It contributes to increasing the measurement accuracy of coordinate measuring machines. This is why such type of machines is more and more common to measure form deviations, if tolerances of the part are significantly higher than the maximum permissible error of the machine.

Considering the problem of measurements of 3D parameters of rotary parts it is noteworthy that it is very important to apply proper measurement strategy. Term "measurement strategy" denotes here distribution of sampling points on the surface under investigation. Thus, this term is closely related to the path along which the sensor moves on the surface to be measured. It is obvious that applied measurement strategy should allow dense coverage of investigated surface with a grid of sampling points. On the other hand, it should be noted that the higher number of sampling points the longer measurement time. Therefore, sometimes selection of proper measurement strategy is not an easy task.

2 State-of-the-Art on Measurement Strategies of 3D Features

Measurement strategies of 3D parameters described in the scientific literature usually relate to measurements of cylindrical workpieces. In general, they can be divided into three main groups: uniform sampling strategies, strategies fitted to measured surfaces on the basis of predicted pattern of irregularities and so-called adaptive strategies.

2.1 Uniform Sampling Strategies

Uniform sampling strategies are most common in industrial practice. Usually measurements of form deviations of cylindrical parts are performed with the use of the circumferential and the generatrix strategy. The strategy that allows obtaining more accurate information about investigated surfaces is so-called "bird-cage" strategy. This strategy is simply a combination of measurements in cross- and longitudinal sections. Due to computational problems "bird-cage" strategy has not been available in the software of measuring systems for a long time. At the moment, there are some measuring systems that allow application of this strategy. The most important advantage of the "bird-cage" strategy is that it permits dense covering of measured surface with a grid of sampling points. Therefore such results are most reliable. The major drawback of the "bird-cage" strategy is that it is very time consuming [1].

Single points strategy is less common to measure form deviations of cylindrical parts. The reason is that low number of sampling points does not allow obtaining accurate information about investigated surface.

The strategy that was not described in the standard ISO 12180 is the helical trace strategy. This strategy is a compromise solution between the strategies of cross-sections and longitudinal sections and it is quite often applied under industrial conditions.

In general, it should be noted that the strategies mentioned above allow uniform sampling. It means that the whole surface of the specimen is scanned in the same way. Thus, for the case of circumferential, generatrix and "bird-cage" strategy all the linear distances between all cross-sections and angular distances between longitudinal sections are the same. The helical trace strategy can be regarded as a uniform sampling strategy, too, as the parameter that describes sampling density, which is pitch angle, remains constant during the measurement. Application of the single points strategies also requires approximate uniform distribution of sampling points on the surface under investigation. It can be performed with the use of specific numerical methods, for example by applying so-called Hammersley/Halton sampling [2].

Uniform sampling strategies are very useful if surfaces under study are characterized by regular pattern of form errors. However, sometimes significant irregularities are located only in a certain area of the surface. Such area should be then sampled with the use of many more points than other areas of the surface. This is the reason why research activities are conducted that are focused on development of a strategy that would denser sampling in areas where significant surface defects are predicted.

2.2 Strategies Based on Predicted Location of Irregularities

These strategies are fitted to the surface under study on the basis of predicted distribution of surface irregularities. If information about the machining process of the surface is available, then one can predict most probable location of the irregularities of the surface. For example, cylindrical surface of workpieces machined by turning that were placed in a three-jaw chucks are very often characterized by a triangular error (i.e. the third harmonic component is dominant in circumferential profile). Additionally, workpieces that were placed in lathe centers are usually characterized by a saddle-shaped error. Strategies fitted to predicted model of irregularities can be designed on the basis of data of preliminary measurements or on the basis of assumed pattern of form errors. An example of the former approach is a method of harmonics fitting described in work [3]. This method applies statistical evaluation of characteristics of individual Fourier components of the profile in order to establish as small set of sampling points data as possible that allows reliable evaluation of form errors of the surface under study. The latter approach is given in work [4]. In this method, a preliminary model of the shape of the surface is assumed. The model is described mathematically by the linear combination of the set of base functions (for example, polynomials, Fourier components or eigenfunctions). The model is then used to design the grid of sampling points. After the measurement, on the basis of the values in presumed sampling points, coefficients of assumed model of the linear combination are calculated (taking into account uncertainty of fitting). The coefficients are then used to reconstruct the image of the whole surface under study.

2.3 Adaptive Strategies

Adaptive strategies, described in the scientific literature, are iterative ones. In the case of application of such types of strategies, the sampling is conducted in a few stages. The first stage is a preliminary measurement. The preliminary measurement is usually performed with the use of uniform sampling. Next, on the basis of obtained measurement data, a relevant algorithm is applied to find the areas where the risk of occurring of large local surface irregularities is high. After conducting measurements in these areas the value of presumed coefficient is calculated. The value of this coefficient indicates if the measurements should be stopped or it is necessary to conduct the next series of measurements applying denser sampling. Such approach uses so-called kriging models quite often. The name of the method comes from the name South African engineer, Krige, who has the first researcher to apply this method to predict changes of the signal in geostatics in the sixties of the XX. century. At present, these models are more and more common to predict localization of subsequent sampling points.

In kriging models the signal is modified with the use of the set of base functions and their coefficients supplemented by a random component whose expected value is equal to zero and for whom the covariance between the input and output remains constant [5]. Apart from the application of kriging models, the localization of points, where the profile should be sampled can be conducted with the use of such optimization techniques as Tabu Search, Hybrid Search or coordinate search [6].

3 Concept of Adaptive Strategy of 3D Parameters or Rotary Elements Proposed by Authors

As it was shown in previous section, measurements of form deviations of 3D elements can be conducted with the use of various strategies, differing in a number and distribution of sampling points located on an investigated surface. Low number of sampling points or measured sections can lead to the situation that some surface irregularities are not detected by a measuring system. An application of higher density of sampling points, in turn, results in significant lengthening of measurement time, which is undesirable, if one takes into account requirements of modern manufacturing processes. This is the reason why authors have taken efforts aiming at development a novel, adaptive measurement strategy, different from those described in Sect. 2.3. Proposed adaptive strategy consists of two stages: a preliminary measurement and additional measurement (one of many – it depends on the results of the signal analysis). General concept of the strategy is shown in Fig. 1. During the preliminary measurement an investigated area is scanned along preselected trajectory. Authors propose to apply three strategies of preliminary scanning: along the helical trace, in preselected cross-sections and in preselected longitudinal sections.

The approach to the evaluation of form deviations of 3D parameters of rotary elements shown in Fig. 1 requires solving some specific problems. One of them is how we can decide what signal change should be considered as a large one.

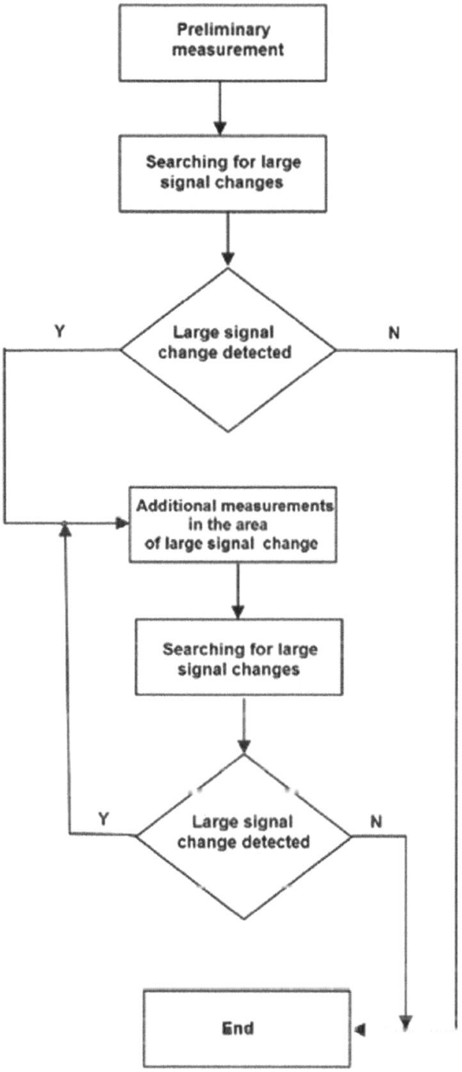

Fig. 1. Algorithm of the adaptive strategy proposed by authors

Authors propose to take such decision on the basis of the distance of sampling points from the reference feature. Considering approach applied in quality engineering we propose to use a value of experimental standard deviation s as an indicator in this case. Points that are located in a distance higher than 3s are the ones where additional measurements should be considered.

Thus, the procedure of searching points where additional measurements should be performed can be described as follows:

- performing preliminary measurement,
- calculation of the parameters of the reference feature,
- calculation of the value of the experimental standard deviation s,
- searching for points whose distance from the reference feature is higher than 3s.

Authors are going to apply cross-sections, longitudinal sections and helical trace strategies to measure such elements as: cylinders, spheres, cones and barrel-shaped workpieces. It will involve solving a number of theoretical problems, including development of the mathematical models of the strategy. Developed model should respect such problems as: correct calculation of a reference feature, filtering of measurement data and non-uniform sampling. On the basis of formulated equations computer procedures will be developed to allow qualitative (visual) and quantitative (with the use of suitable numerical parameters) evaluation of form deviations of measured workpiece.

4 Conclusions

Measurements of form deviations of 3D elements can be conducted with the use of various strategies, differing in a number and distribution of sampling points located on an investigated surface. Low number of sampling points or measured sections can lead to the situation that some surface irregularities are not detected by a measuring system. An application of higher density of sampling points, in turn, results in significant lengthening of measurement time, which is undesirable, if one takes into account requirements of modern manufacturing processes.

This is the reason why authors are conducting research work aiming at development of a novel, two-stages adaptive measurement strategy. Proposed adaptive strategy has not been discussed nor described in the scientific literature so far. Moreover, even leading manufacturers of modern metrological equipment do not offer similar options in software packages of their measuring instruments. Proposed strategy changes existing approach to measurements. It is assumed that it will permit achieving high measurement accuracy and reduction of measurement time when comparing it to strategies that are currently applied. However, it is noteworthy that the strategy will require development of mathematical models of measurement and evaluation of form deviations of rotary elements, which is under investigation at present time. After successful finish of the theoretical part of the work authors will verify the concept experimentally.

Acknowledgments. The paper has been elaborated within the framework of the research project entitled "Theoretical and experimental problems of integrated 3D measurements of elements' surfaces", reg. no.: 2015/19/B/ST8/02643, ID: 317012, financed by National Science Centre, Poland.

References

1. Taylor Hobson – training materials. https://www.taylorhobsonserviceusa.com/uploads/2/5/7/5/25756172/cylindricity.pdf. Accessed 30 Nov 2017
2. Adamczak, S., Stepien, K., Zmarzły, P.: An analysis of strategies of measurement of 3D rotary elements. In: Proceedings of the 28th DAAAM International Symposium on Intelligent Manufacturing and Automation, Zadar, Croatia, November 2017 (2017)
3. Capello, E., Semeraro, Q.: The harmonic fitting method for the assessment of the substitute geometry estimate error. Part I: 2D and 3D Theor. Int. J. Mach. Tools Manuf. **41**(8), 1071–1102 (2001)
4. Summerhays, K.D., et al.: Optimizing discrete point sample patterns and measurement data analysis on internal cylindrical surfaces with systematic form deviations. Precis. Eng. **26**, 105–121 (2001)
5. Pedone, P., Vicario, G., Romano, D.: Kriging-based sequential inspection plans for coordinate measuring machines. Appl. Stoch. Models Bus. Ind. **25**(2), 133–149 (2009)
6. Chen, J., Ren, Y., Zeng, G.: An improved multi-harmonic sine fitting algorithm based on Tabu Search. Measurement **59**, 258–267 (2015)

Re-engineering of Manufacturing Parts by Computed Tomography Data

Anil Akdogan[1(✉)], Ali Serdar Vanli[1], and Numan Durakbasa[2]

[1] Mechanical Engineering Department, Yildiz Technical University, 34349 Istanbul, Turkey
nomak@yildiz.edu.tr
[2] Industrial Metrology and Adaptronic Systems at the Institute of Production Engineering and Photonic Technologies, Vienna University of Technology, 1060 Vienna, Austria

Abstract. The number of computed tomography (CT) applications today is satisfactorily high and rapidly increasing in many environments. Many manufacturing industries use CT technologies. In order to able to reach very accurate internal dimensions even at complicated workpieces to achieve the internal structure of a manufactured object is one of the main advantages of the system. Its' non-destructive style, comparison capability with reference models and extremely high attribute imaging are the other issues. In this paper, CT systems, scanning capabilities and technological developments is examined in the frames of the latest technology and future CT technologies. It especially gives the latest information about inspectable materials and application examples of manufacturing industries. Because innovative design and development works of industrial products in manufacturing environments require valuable big data. The conducted experimental CT results were used for re-engineering applications for continuous improvement of the manufacturing process. The results were also valuable for quality control and sustainable quality assurance.

Keywords: Computed tomography · Re-engineering · Manufacturing · Casting · Quality control

1 Introduction

Until recently, CT imaging technology was available only for use in medical settings. In the last few years, it has found a new place in the world of production and has uncovered numerous opportunities that have never existed before. Computed tomography is now widely used in laboratory and online or in various industrial metrology environments for product quality control evaluation and re-verse engineering applications [1].

Industrial CT scanners use the same type of technology as CT scanners in hospitals. It takes multiple readings from different angles and converts CT grayscale images to pixel-based 3D point clouds. After the CT scanner creates the point cloud, an expert can create a comparison map from the CAD to the part, create a 3D model of the piece, or reverse the part according to its needs. For example, high-speed line scanning data

V. D. Majstorovic and N. Durakbasa (Eds.): IMEKOTC14 2019, LNME, pp. 114–121, 2019.
https://doi.org/10.1007/978-3-030-18177-2_12

can be examined for quality assurance, while online or lab scanning can instantly compare a product produced with the CAD program in a variety of ways [2].

Reverse engineering is one of the most common uses of 3D and CT scanning in industrial metrology applications. The process of taking measurements of an existing part or object and then creating a complete CAD copy is a well-known technique in reverse engineering. This technology is sometimes become extremely useful when you have a manufactured part or object, but no original prints or design data. The other applications of CT scanning in industrial metrology becomes as both non-contact and contact types. In non-contact measurement, the process of capturing the geometry of existing objects by using 3-D scanners like laser-emitting cameras. On the other hand in contact measurements the single point collection method is used. There is a wide range of solutions for contact measurement and two types of system solutions are often used with portable CMMs. Nowadays, the popular tools for this are articulated arms for small volumes and laser scanner and photogrammetric tools for larger volume objects.

3D imaging devices and long-range scanners serve a wide range of industries and provide the required level of testing and accuracy. The leading industry that currently uses this technology is the manufacturing industry. Such as metal casting, automotive, aerospace, energy, polymer and medical equipment manufacturing industries uses this technology widely, as well. There are many research papers showing the usage of CT technology in many manufacturing applications. The researchers examine manufacturing parts quality requirements by CT clearly. The CT scan reveals many kinds of porosities in manufactured parts. They were tested with CT scanner [3].

Typical CT scanners offer a wide range of measurement capabilities and features. The dimensions start with small desktop models that meet the toughest demands and have the smallest 3D detail of 0.25 μm. They can be equipped with different X-ray forces. Major browsers are 500 mm in diameter and 600 mm height of up to 50 kg, so that the work can be used for extremely fast CT data acquisition. 3-D analysis of the turbine blade can be scanned. The future of industrial scanning, especially when combined with 3D printing or other additive manufacturing methods, offers unlimited possibilities. The industrial scanning industry is planned to grow very rapidly in the next 5–10 years, as improvements in technology and performance and costs continue to fall [4].

2 Technical Principles

As known, typical CT scanners offer a wide range of measurement capabilities and features. The operator should match the appropriate X-ray source with high or low energy exposure, depending on the purpose of the scan with the work piece size and material. After that, the part is needed to place on the turning table between the digital detector panel and the X-ray radiation source. When the part rotates 360°, the X-ray source passes through the part. The variable intensity of the piece absorbs varying amounts of radiation. The remaining radiation goes to the detector panel that captures a 2D X-ray image. This process is repeated to capture hundreds of thousands of 2D X-ray images. Hundreds of thousands of captured 2D X-ray images are reconstructed mathematically to enhance the 3D model. This model is used for further internal and external analysis of the piece, whether it is defining internal faults or carrying out a dimensional analysis [5].

A typical CT has an X-ray source, an X-ray detector and kinematic system as hardware components. It has custom software for reconstruction and edge detection and/or size analyze. The measurement principle of computed tomography relies on the attenuation of X-rays by the measurement object which is dependent on the material and the thickness of the workpiece [6]. X-rays are spread on the workpiece. It is weakened due to absorption or scattering. The amount of weakening is determined by the path taken over by X-rays due to the material composition and density on the absorbent material. The measurement of weakening allows the presence of the material and the length of the various materials to be found [7].

The casting technique has a special importance in the manufacturing industry. Cast parts are usually produced with internal cavities which require dimensional CT measurements. In addition, besides lighter materials sometimes the formation of hollow parts also appears as a way to reduce mass and weight, particularly in industries such as automotive and aerospace. The resulting CT scan of a cast part is a series of pixel-based data describing the varying density of the piece with varying "grayscale" value. The density of a part absorbs radiation exposed to the scanned portion. Since cracks or hollows do not absorb any radiation, they will appear at higher intensity in the resulting image such as darker in grayscale. Areas with high density materials tend to absorb more radiation, so they appear lighter in grayscale. The images are analyzed in a similar way with any X-ray or radiographic test technique. A 3D model for X-ray CT scanning is reconfigured, which can be analyzed to name a few for internal faults, measurements and wall thickness. The results can be colored according to project requirements to facilitate the visualization of the analysis in the software [7].

3 Experimental Works for a Cast Part

No matter which casting method is used, the biggest defect of a cast part is porosity. High pressure die casting method technologies have come a long way in order to produce non-porous parts in many important fields from automotive to aero-space. All process parameters are carefully adjusted to process structural parts with less porosity and therefore high density and high mechanical properties. Today, foundries, quality control laboratories and researchers at universities use CT scanners for any quality control applications of many casting parts like engine blocks, cylinders, cylinder heads, pistons, joints and so on. Since reverse engineering supplies the creation of a digital dataset based on a physical representation by obtaining an idea through the CAD construction of the product, industrial CT has become one of the main tool for its' scientific applications. Especially, defect detection, error analysis, accessible non-geometric dimensional measurements, compilation or material properties are analyzed statistically. Today, the most important application of the CT is 3D digitization. First of all, the automotive industry and its suppliers show a strong interest in new possibilities. By using this new technology, it is possible to reduce the time to develop and market new products. Thus, companies can compete with serious advantages.

In this experimental work we examine a casting by CT scanner in order to get quality specifications. High pressure die casted automotive part needs to have high density so that high mechanical properties. The porosity observation and verification is conducted. The used CT scanner and its technical specifications are given in Fig. 1. Because the highly dynamic radioscopy technology provides detail detection and assists in visually determining the depth of casting flaws. Counting on detailed casting-flaw information is possible due to 3D reconstruction of the CT scan [1].

Attribute	Respective Value (standard 225 kV)
Maximum Sample Weight	60 [kg] (opt. 200 kg)
System Dimension	2200 x 1850 x 2750 [mm]
System Weight	~ 6,5 [t]
Sample Diameter max. [mm]	600 x 900 [mm]
Inspection Envelope CT (3D) Diameter x Height	~170 x 150 [mm]
Focus-Detector-Distance (FDD)	695 - 995 [mm]
X-ray Tube	160 - 450 [kV]
Detector Active Area	1024 x 1024 [px]

Fig. 1. YXLON MU2000-D CT scanner and technical data sheet [1].

4 Experimental Results

Firstly, after detective calibration of the CT scanner X-ray source is activated. The examination part is which is a leak proof automotive part named oil pump housing should be positioned automatically in the machine by the table. The table is rotated and confirmed that the part in completely visible. The scanning conditions are set at 225 kV voltages and 3.6 mA current. After scanning the data is analyzed at reconstruction program. Automatic or manual surface determination is advised at this stage. In addition, we have transparent (Fig. 2a) and render the solid. At the indicator section we determined the grayscale images and after alignment coordinates of the suspicious points were determined. You can cut, set planes, cylinders or lines or for instance intersect the selected geometries at that stage serving for your re-engineering works. We have best-fitted (Fig. 2b) the CAD data and the scanned data. Moreover, the comparison analysis between the actual and the designed tolerances were done. Figure 3 gives the colored determination results of the comparison. It gives the positive or negative deviations of the actual data scanned by CT and the nominal added in the software with the tolerance limitations.

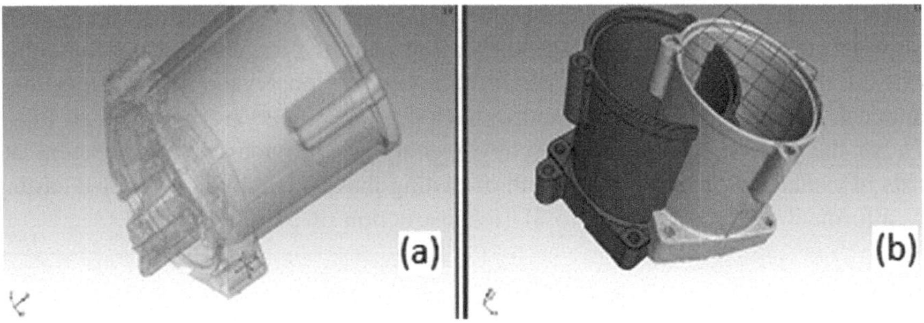

Fig. 2. Transparency (a) and Best-fit (b) figures

Fig. 3. Deviations of actual data from the designed

Finally, as one of the most important quality parameter of the cast part we have scanned the porosity of the object that might have. In conventional methods the determination of the porosity in cast parts sometimes might be troubled.

After removing small voids and indication of grayscale images porosity analyze has run. Figure 4 gives the porosity views of the selected section in the part.

Researchers can make many different parameter adjustments and/or filtration on the porosity analyze section. Specially, transparency works a lot to see the exact places of the hollow sections in the part. Their probability, radius, diameter and volume values were calculated by the software. The results are given in the Table 1 as the porosity report of the selected section of the part in Fig. 4.

When we view hollow sections in the part in 3D scanned with CT, we can see the deviation of the porosity in the internal structure of the part. A 3D view is given in Fig. 5. When we look at the images of the scanned parts, they are displayed in different colors by the computer according to their size. The coordinates of these regions can also be determined and examined for confirmation.

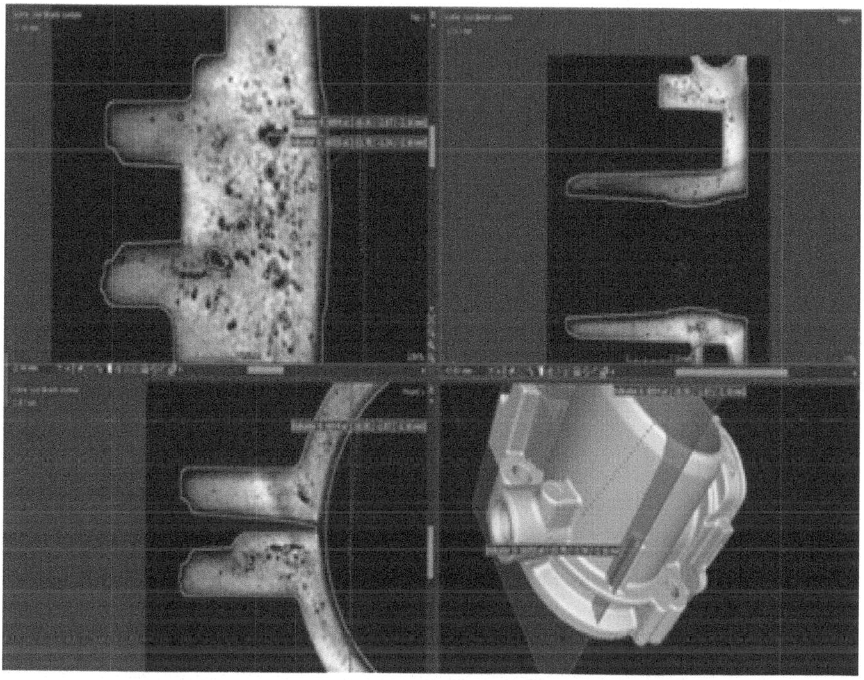

Fig. 4. Porosity in the selected section of the part

Table 1. Porosity report of the selected section of the part.

Defect no	Probability	Radius (mm)	Diameter (mm)	Volume (mm^3)
3304	1.12	1.85	3.70	1.36
943	2.06	1.74	3.47	0.55
174	3.64	1.55	3.09	1.01
2745	1.24	1.44	2.88	0.76
1508	1.67	1.21	2.42	0.51
1304	1.79	1.19	2.38	0.50
862	2.14	1.19	2.37	0.60
3976	1.00	1.13	2.26	0.34
2004	1.45	1.10	2.21	0.75
839	2.17	1.06	2.12	0.43
1768	1.54	1.05	2.10	0.17
3745	1.04	1.05	2.09	1.14
2153	1.40	1.03	2.06	0.51
232	3.40	1.01	2.02	0.69
425	2.81	1.00	2.01	0.32
3077	1.16	0.98	1.97	0.17

(*continued*)

Table 1. *(continued)*

Defect no	Probability	Radius (mm)	Diameter (mm)	Volume (mm^3)
1780	1.54	0.97	1.93	0.39
107	4.09	0.96	1.93	0.44
3622	1.06	0.96	1.92	0.16
1329	1.77	0.94	1.88	0.52
1646	1.60	0.94	1.87	0.26
142	3.88	0.93	1.87	0.39
488	2.68	0.92	1.85	0.58
1776	1.54	0.91	1.82	0.17
691	2.34	0.88	1.77	0.37
3323	1.12	0.88	1.75	0.40

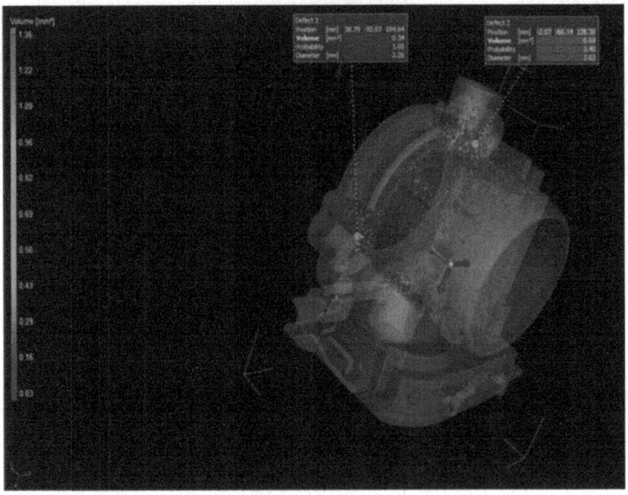

Fig. 5. 3D view of the porosity

5 Conclusions

Industrial scanning becomes more and more important day by day. New ways to produce higher quality products, reduce costs and ensure customer satisfaction are being researched in the market where is highly competitive as a result of technological developments. Reverse engineering is one of the most common uses of 3D and CT scanning in many industrial metrology applications. Many modern and mass productive foundries use these tools like the other industries. Nevertheless, the future of CT scanners provides unlimited possibilities, especially when combined with 3D printing or other additive manufacturing methods. CT is planned to grow rapidly in the coming years.

The interface produces data for use in areas such as machine learning, big data, the internet of objects, and machine-to-machine communication in intelligent foundries adapted to Industry 4.0. To ensure the acceptable inspection quality of components that are critical to safety, they comply with existing industry standards such as the requirements of the standards and specific car manufacturers.

References

1. Yxlon. https://www.yxlon.com/products/x-ray-and-ct-inspection-systems/yxlonmu2000-d
2. Drégelyi-Kiss, Á., Durakbasa, N.M.: Measurement error on the reconstruction step in case of industrial computed tomography. In: Durakbasa, N., Gencyilmaz, M. (eds.) Proceedings of the International Symposium for Production Research ISPR. Springer (2018)
3. De Chiffre, L., Carmignato, S., Kruth, J.-P., Schmitt, R., Weckenmann, A.: Industrial applications of computed tomography, vol. 2014, pp. 655 677 (2014)
4. Solberg, D.: Why use a CT scanner for an industrial job. Quality Magazine, 13 April 2016. LNCS. http://www.springer.com/lncs. Accessed 21 Nov 2016
5. ASTM E1672-12: Standard guide for computed tomography (CT) system selection, ASTM International, West Conshohocken, PA (2012)
6. Weckenmann, A., Krämer, P.: Computed tomography for application in manufacturing metrology. In: Key Engineering Materials, vol. 437, pp. 73–78 (2010)
7. Kruth, J.P., Bartscher, M., Carmignato, S., Schmitt, R., De Chiffre, L., Weckenmann, A.: Computed tomography for dimensional metrology. CIRP Ann. **60**(2), 821–842 (2011)

Manufacturing Automation for Magnesium Die Casting

Ali Serdar Vanli(✉) and Anil Akdogan

Mechanical Engineering Department, Yildiz Technical University,
34349 Istanbul, Turkey
svanli@yildiz.edu.tr

Abstract. High pressure die casting (HPDC) of magnesium alloy parts is mostly problematic. Because magnesium alloys oxidize at above 400 °C. They need to be protected by the proper gas atmospheres during manufacturing. Besides, other components of an integrated HPDC cell such as casting machine, melting-dosing unit, gas mixing system and die heating-cooling device must be carefully designed in order to obtain the desired high quality products. This work guides to design manufacturing automation of HPDC process in order to get succeeded results in product quality. It also details the manufacturing parameters like gate velocity and intensification pressure for better quality parts. The collected big data about manufacturing analyzed as computer aided quality parameters which were used for continuous improvement of the process. The obtained data from the manufacturing process was used to increase the product quality expectations like high density and satisfactory mechanical properties. At the end of the conducted manufacturing experiments at our industrial scale die casting cell the required quality values were obtained.

Keywords: High pressure die casting · Magnesium alloys · Big data · Product quality

1 Introduction

The first patent on die casting was made in 1905, in the United States. The first obvious use of the pressure die casting method is the production of gas masks during the World War I. In the first original version of the pressure die casting method, only zinc alloys were used, but later aluminum alloys and later all magnesium alloys found wide application in the pressure die casting industry.

Later, the production of light metals by pressure die casting has become a fully accepted process in the metal casting industry and has become available in all parts of mass production. The most important user of this technology is of course the automotive industry. Through the end of 1980s, while too much effort was being made to develop pressure die casting machines, today, alloy development, control of metal quality and process optimization have come to the fore [1]. HPDC can be defined as a casting method utilizing hydraulic energy source based on injection of molten metal under high speed and pressure into metal die. In die casting technology, the die consists of two halves and after the two die halves namely fixed and movable are closed and

V. D. Majstorovic and N. Durakbasa (Eds.): IMEKOTC14 2019, LNME, pp. 122–130, 2019.
https://doi.org/10.1007/978-3-030-18177-2_13

locked, the liquid metal is transferred to an injection chamber. This chamber may be cold or heated to metal temperature. The liquid metal in the chamber is filled into the metal die by means of a plunger. The air in the die is expelled from the ventilation duct systems on the die during filling. After the die cavity completely filled, pressure is applied until the part solidifies. The die is then opened and the casting part removed from the die. When the die is open, it is cleaned and lubricated and the previous operations are repeated [2]. Although production of magnesium alloys by die casting process is also a proven method for the production of a wide segment of casting parts from decorative or electronic components, these alloys oxidize at above 400 °C. They need to be protected by the proper gas atmospheres during manufacturing to handle the process difficulties [3].

HPDC is a very good casting method for manufacturing fully automatic, high efficiency and high capacity parts with a large scale of weight such as a few grams and a hundred kilograms. Although the basic principles of the equipment for mechanical and hydraulic parts have been used for a hundred years, the pressure die casting process is still in a very rapid development process. Manufacturers of pressure die casting machines are now producing fully automated casting lines ready for conversion of magnesium, aluminum, zinc and copper alloys into the desired form. This study primarily gives the details of HDPC with all its components and introduces the cold chamber HPDC mass production line of Magnesium alloy part manufacturing at the designed and established by the authors as an integrated die casting system at industrial scale at Yildiz Technical University, Die Casting Laboratory in Istanbul, Turkey.

2 HDPC Machines and Dies

The die-casting machines consist mainly of three main parts namely the body; which carries moving parts and die, keeping them in the most suitable position is the section that allows each to perform its function, the moving part; opens and closes mechanism of the die generally hydraulically and the injection system; which is the part of the machine sending the molten metal to the die cavity under high speed and pressure.

The basic function of a die casting machine is holding the two die halves locked in the exact axis, sending the molten metal to the die cavity under pressure, and opening the die to remove the spilled piece from the die. HPDC machines are divided into two main groups as hot chamber and cold chamber according to the injection systems. While hot chamber machines are for casting alloys with low melting temperatures, cold chamber machines are for casting alloys with high melting temperature [4]. Hot chamber machines have a furnace holding the molten metal at the casting temperature, and the injection process is usually provided by the piston-cylinder assembly, or the gooseneck, which is immersed in the molten metal. In the hot chamber machine, with the rising of the piston allows the molten metal to fill the cylinder. Cold chamber machines are for casting aluminum, magnesium and copper alloys. In cold chamber machines, the crucible holding the molten metal is separate from the machine. In cold chamber type pressure casting machines, the molten metal is transferred to the cold reservoir cylinder through the casting port. With the forward movement of the hydraulically operated piston, the casting port is closed and the liquid metal is injected under high pressure into the locked die [5, 6].

HPDC dies are made of tool steel, movable and fixed parts. The fixed die half is placed on the fixed plate on the injection system side where the molten metal is pressed. The movable die half ensures the finished casting part is removed when the die is opened. The movable part generally includes gating and nozzles that direct the molten metal to the die cavity. The movable die half also includes ejectors which allow the casting part to be removed from the die. Fixed and/or moving cores are frequently used in dies. If a fixed core is used, the axis of the core must be parallel to the opening axis of the die. The movable cores are usually arranged parallel to the parting surface, sometimes at an angle to the parting surface. Moving cores usually act in connection with a sliding system [7].

3 HDPC Process Parameters

There are many factors affecting the quality and mechanical properties of die casting. Some of these factors depend on part and die design. However, process parameters are one of the factors affecting the quality of parts and mechanical properties. Incorrect selection of parameters during the casting process cause faulty and poor mechanical properties. The process parameters are casting and die temperature, filling time, volumetric flow rate, nozzle and plunger velocity and injection pressure (specific pressure) constitute will be considered below.

3.1 Casting and Die Temperatures

The temperature of the casting should be kept as low as possible. The higher the temperature of the liquid metal the greater the solubility of gas in the metal.

These gases, which are dissolved in the metal during melting, cannot get out of the pieces in parallel with the reduced rate of dis-solution during solidification, which leads to the formation of gas porosity. Although the casting temperature varies according to the type of the alloy and the casting process, if the part is not very complex, it can be selected over 30–40 °C above the melting temperature. As a general rule, the die casting temperature ($t_{casting}$) is chosen 30–90 °C above the melting temperature ($t_{melting}$) [2]. In the die casting process, it is important that the die reaches a stable temperature distribution and remains at approximately this temperature for each injection. Immediately prior to injection, the recommended surface temperature is between 150 and 300 °C according to the type of the alloy. In the production of thin-walled parts; the heat energy loose due to radiation, convection, conduction and evaporation may be more than the heat energy supplied by the molten metal. In this case, the heating of the die from the outside becomes even more important. External heating of the die is usually provided by oil-operated heating/cooling units. The heating/cooling channels are present in both die halves and a preheated or cooled fluid passes through these channels. Except for some special cases (max. 300 °C), the heater/coolant temperature is limited around 240 °C [7, 8].

3.2 Filling Time and Flow Rate

The key factor in the design of the gating system, in determining the casting temperature and in selecting other parameters is the filling time required for the die cavity to fill. The filling time is also closely related to the occurrence of errors and the reduction of errors depends on the accuracy of the calculated filling times. In general, The North American Die Casting Association (NADCA)-developed Eq. 1 is used to calculate filling times [7].

$$T = 0.0346\left[\left(T_m - T_f + 2.5S\right)/\left(T_f - T_d\right)\right]T(s) \tag{1}$$

In the Eq. 1; t (s) indicates the required fill time for a successful casting in seconds. T_m (°C) contains the casting temperature of the molten metal; T_f (°C) minimum flow temperature; S (%) is the highest solidification rate that can occur without obstructing the feed; T_d (°C) indicates the die temperature and T (mm) mean wall thickness [8]. The value calculated in the application; depending on the wall thickness, alloy type, flow distance and die temperature, it is recommended to be between 10–100 ms. The total metal volume which is then entered into the die engraving is determined by calculating the volume v (m^3) of the die cavity and the volume flow rate Q (m^3/s) according to the filling time is found using Eq. 2 [2].

$$Q = v/t\left(m^3/s\right) \tag{2}$$

3.3 Nozzle Input and Plunger Velocity

After the volumetric flow rate is determined, a nozzle input rate V_g (m/s) is selected from the experience by utilizing the principle of constant flow through the flow. Accepted gate velocities according to different wall thickness are given in Table 1. According to the selected nozzle input speed and the volumetric flow rate, the gate cross section A_g (m^2) is calculated. The Eq. 3 is utilized for this. Since the volume flow rate and the gate velocity are known, the nozzle inlet section will be easily found. Finally, since a plunger diameter D_p (m) is selected and volumetric flow is known, the plunger velocity V_p (m/s) is calculated from the Eq. 4 [2].

Table 1. Gate velocities at hot and cold chamber HPDC [9]

Method and material	Cold chamber		Hot chamber	
	Al	Mg	Zn	Mg
Wall thickness	**20 m/s** (15–30 m/s)	**35 m/s** (25–35 m/s)	**30 m/s** (25–35 m/s)	**35 m/s** (25–35 m/s)
Wall thickness	**40 m/s** (35–45 m/s)	**45 m/s** (35–50 m/s)	**40 m/s** (35–45 m/s)	**45 m/s** (35–50 m/s)
Wall thickness	**60 m/s** (45–60 m/s)	**75 m/s** (50–90 m/s)	**50 m/s** (45–55 m/s)	**75 m/s** (50–90 m/s)

$$Q = A_g x\, V_g\,(m^3/s) \tag{3}$$

$$Q = A_p x\, V_p = A_g x\, V_g = \left(\pi x D_p^2 x V_p\right)/4\,(m^3/s) \tag{4}$$

3.4 Injection Pressure

After the determined volumetric flow rate and the plunger velocity as a result of the calculations, the appropriate injection pressure should be calculated. There are many charts and tables prepared for this purpose. One of them; is a graph showing the change of pressure applied to liquid metal with square of metal flow. These graphs, which are called P-Q^2 diagrams in short, have been optimized by optimizing the adjustments made on the nozzle inlet area, hydraulic pressure and plunger diameter, thus allowing the casting to be made under the most favorable conditions [4]. In addition, tables have been prepared, which are called monographs and show numerical connections to facilitate calculations [10]. In all calculations, the parameters required to fill the die to be laminar flow must be determined.

This is closely related to the pressure applied to the liquid metal and the volume flow of the metal. In cold and hot chamber-type pressure casting machines, the pressure applied during injection (in-die specific pressure) ranges from 100 to 1200 bar [8]. The recommended pressure values for different types of alloys are available in literature. The appropriate pressure according to the requirement of the part can be selected from such these tables by taking advantage of the experiences [9].

3.5 Calculation of Die Locking Force

As shown in Fig. 1, when determining the locking force required to be applied by the pressure casting machine, A_{proj} (cm^2) must be calculated for the total projection area including the gating and nozzle sections of the spilled part. When determining the total projection area, 1.75 times of the trace areas of the die engravings will be taken into account and gaps and overflow pockets will be taken into account (Eq. 5). Then, the die opening force F_e (kN) is calculated by taking into account by in-die specific pressure P_{sol} (bar) applied during the solidification (Eq. 6). The required locking force F_{lock} (kN) is obtained by taking into account the safety factor G_f and the die opening force F_e (Eq. 7). Safety factor for cold-chambered machines are between 1.1 and 1.3; for the hot chamber machines they are between 1.3 and 1.6. If the users need it in tons, use Eq. 8 [3, 8, 11].

$$A_{proj} = 1.75\, x\, A_g\,(cm^2) \tag{5}$$

$$F_e = \left(A_{proj} x\, P_{sol}\right)/100\,(kN) \tag{6}$$

$$F_{lock} = \left(G_f\, x\, F_e\right)(kN) \tag{7}$$

$$F_{ton} = (F_{lock}/g)(ton) \tag{8}$$

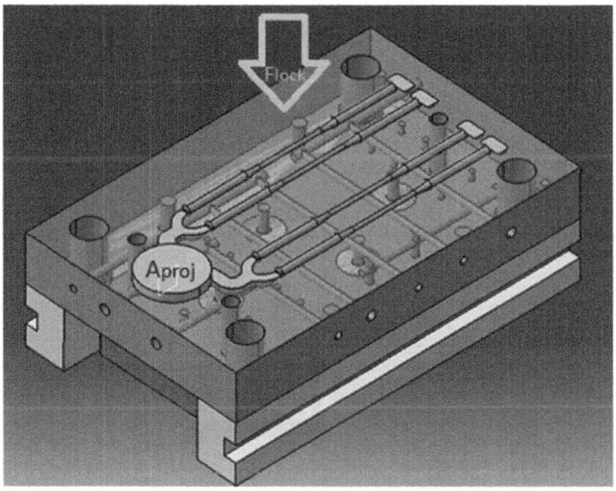

Fig. 1. Die locking force [8]

4 An Integrated Die Casting Process

After calculation of all process parameters given in Sect. 3 according to the given literature and empirical researches the system design is completed. Table 2 gives the manufacturing process characteristics of the cold-chamber HPDC machine in detail.

Table 2. Process characteristics of the cold-chamber HPDC machine

Characteristics	Unit	Specifications
Locking force	kN	1000
Size of platen	mm	525 × 525
Maximum casting force	kN	110
Injection stroke	mm	280
Plunger diameter	mm	30–50
Casting volume	cm^3	130–363
Casting area	cm^2	66–200
Specific injection pressure	MPa	50–150

The technical drawing of the integrated serial production line used to determine the process parameters is shown in Fig. 2. An integrated system has been established with a modified die casting machine, a melting furnace, a molten metal transfer system and a mold heating/cooling system. This system has a structure that allows continuous production. Many experimental works were conducted with the system for Magnesium alloy part manufacturing. Design of experimental works was conducted by Taguchi system design with orthogonal matrix. Table 3 gives the experimental factors and their levels used in L27 orthogonal array matrix.

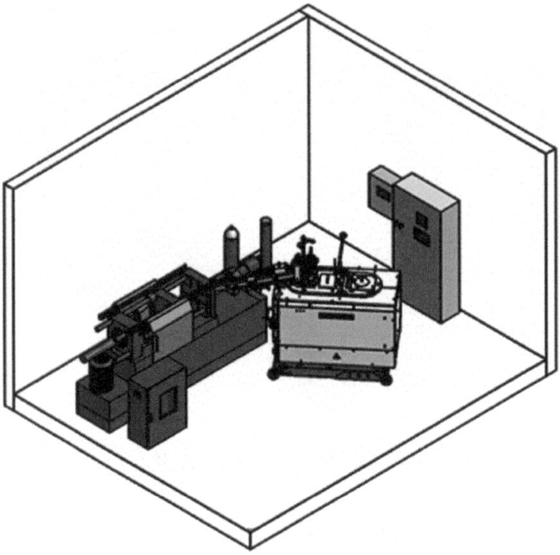

Fig. 2. An integrated die casting system for automation [13].

Table 3. Experimental factors and their levels

Levels	Bath temp.	Die temp. (°C)		Gas concen.	Intensif. pressure	Gate velocity
	°C	Cover side	Ejector side	%Vol.	MPa	$m.s^{-1}$
1	640	150	200	0.20	80	30
2	660	175	225	0.25	100	45
3	680	200	250	0.30	120	60

Algorithms and software tools for the analysis of big amounts of data and machine learning are available for many years. The collected big data by means of sensors at the each component of the system were analyzed to determine the optimum process parameters and the product quality. Within the vision of Industry 4.0 the standard for digital information exchange between devices had been determined and is OPC-UA (Open Platform Communications Unified Architecture). OPC-UA is a platform independent protocol and interface as well as a modeling language for digital information. Such smart foundries in a networked supply chain are a network of hundreds or thousands of cyber-physical systems, which are connected to a common Ethernet network exchanging data and information via OPC-UA [12]. The analyzed data were used for continuous improvement of the process. After reaching the optimum process parameters high quality standardized test samples were obtained.

As a result of the mechanical tests applied to the test samples obtained from the production line by the optimum process parameters, 160–175 MPa yield strength,

240–255 MPa tensile strength and 4–6% elongation values were obtained and it was determined. However, it is another important point that the sample production has been realized within the dimensional and form tolerances mentioned in the relevant standard. In addition, the total porosity values of 5% or less have been reached in all samples. The results show that all the process parameters used during the production allow the production of satisfactory quality products [13].

5 Conclusions

The high pressure die casting method offers important and economic advantages that other casting methods cannot provide. Of course, the die casting of molten metal is the perfect solution for converting the molten metal to the desired shape in the designed dimensions and in the shortest possible cycle time. Globally preferred die casting process; it is an ideal method for high volume production of magnesium alloy parts which are widely used nowadays. The installation high pressure die casting integrated mass production line was one of the first for Turkey supplies mass production. The obtained results from the system show that all the process parameters used during the production allow the high quality products. In this regard, the knowledge and experience required by our country and abroad industry has been provided.

References

1. Kaufmann, H., Uggowitzer, P.J.: Metallurgy and Processing of High-Integrity Light Metal Pressure Castings, 1st edn. Fachverlag Scihele & Schön GmbH, Berlin (2007)
2. Vanli, A.S.: Optimization of process parameters for product quality in high pressure die casting of magnesium alloys. Ph.D. thesis, Yildiz Technical University, Istanbul (2013)
3. Vanli, A.S., Akdogan, A., Durakbasa, M.N.: Integrated die casting manufacturing system for sustainable high quality of magnesium alloy products. In: International Conference on Production Research – Africa, Europe and the Middle East 4th International Conference on Quality and Innovation in Engineering and Management (QEIM) Proceedings, Cluj-Napoca, Romania (2016)
4. ASM Handbook: Casting, vol. 15, 4th edn. ASM International Handbook Committee, Ohio (1998)
5. Vanli, A.S., Akdogan, A.: Automation in HPDC with all its components I. Mach. Technol. (MakTek) J. **209**, 184–190 (2015)
6. Vanli, A.S., Akdogan, A.: Automation in HPDC with all its components II. Mach. Technol. (MakTek) J. **210**, 106–112 (2015)
7. ASM Specialty Handbook: Magnesium and Magnesium Alloys, 2nd edn. ASM International Handbook Committee, Ohio (1999)
8. Friedrich, H.E., Mordike, B.L.: Magnesium Technology: Metallurgy, Design Data, Applications, 1st edn. Springer Verlag, Berlin (2006)
9. Oskar Frech GmbH + Co. KG.: Basic Principles of Die and Process Design Education Manual, Schorndorf (2015)
10. Hydro Magnesium: Magnesium Die Casting, 1st edn. Hydro Media Inc., Brussels (2002)
11. Marinescu, I., Boothroyd, G.: Product Design for Manufacture and Assembly, 2nd edn. Marcel Decker Inc., New York (2002)

12. Vanli, A.S., Akdogan, A., Kerber, K., Ozbek, S., Durakbasa, M.N.: Smart die casting foundry according to industry 4.0. In: Acta Technica Napocensis, Applied Mathematics, Mechanics, and Engineering, vol. 61, no. 4, pp. 787–792 (2018)
13. Vanli, A.S., Akdogan, A., Sonmez, H.: Cold chamber high pressure die casting machine and process design for Mg alloy part manufacturing. In: TMMOB Chamber of Mechanical Engineers, Machinery Manufacturing Technologies Congress Proceedings (2013)

Multi-material Acceptance Testing for CT-Based Coordinate Measurement Systems

Fabricio Borges de Oliveira[1(✉)], Markus Bartscher[1], Ulrich Neuschaefer-Rube[1], Rainer Tutsch[2], and Jochen Hiller[3]

[1] Physikalisch-Technische Bundesanstalt, Bundesallee 100, 38116 Brunswick, Germany
{fabricio.borges,markus.bartscher, ulrich.neuschaefer-rube}@ptb.de

[2] Technische Universität Braunschweig, Schleinitzstraße 20, 38106 Brunswick, Germany
r.tutsch@tu-bs.de

[3] Technische Hochschule Deggendorf, Edlmairstraße 9, 94469 Deggendorf, Germany
jochen.hiller@iis.fraunhofer.de

Abstract. This contribution presents a multi-material acceptance test for computed tomography-based coordinate measurement systems (CT-based CMS). The multi-material test requirements and concepts – based on the international reference standard ISO 10360 – are presented. Also, a set of multi-material reference standards for the assessment of the probing error test (*P*-test) and length measurement error test (*E*-test) are presented. For the *P*-test, two half spheres made of different materials are assembled to a sphere. A multi-material hole cube standard is used for the assessment of the *E*-test. The hole cube consists of two symmetric half cubes made of different materials. For both, multi-material spheres and hole cube standards, the materials were selected to obtain two multi-material scenarios: (1) with high and (2) with at least a medium attenuation ratios. Thus, for the probing test, silicon nitride, aluminium oxide and lead free glass N-SF6 were used. The hole cubes, however, were made by paring aluminium, a special ceramic material called carbon fibre silicon carbide (Cesic) and titanium. Form error and size measurements were evaluated in the multi-material spheres as well as hole-based centre-to-centre distances were evaluated in the hole cubes. The proposed test and multi-material reference standards were successfully tested, as they appear to be suitable for evaluating the multi-material error characteristic of CT-based CMSs.

Keywords: Acceptance testing · Multi-material measurements · Computed tomography · CT-based CMS · Standardisation

1 Introduction

X-ray computed tomography (CT) is regarded as the third revolutionary development in coordinate metrology, after the presentation of the tactile- and optical-based coordinate measurement systems (CMSs) [1]. This status is explained by the high capability

V. D. Majstorovic and N. Durakbasa (Eds.): IMEKOTC14 2019, LNME, pp. 131–154, 2019.
https://doi.org/10.1007/978-3-030-18177-2_14

of CT to acquire the complete three-dimensional (3D) representation – of inner and outer structures – of objects, by allowing several kinds of measurements (e.g. dimensional measurements, wall thickness, actual/nominal comparison, et cetera) and non-destructive tests (e.g. defect analysis) with just one single scan, and by the fact that CT allows in principle the measurement of multi-material (MuMat) objects, even in their mounted state.

A typical CT measurement consists of several non-trivial processing steps each having a significant influence on the measurement result. A simplified workflow is presented in Fig. 1.

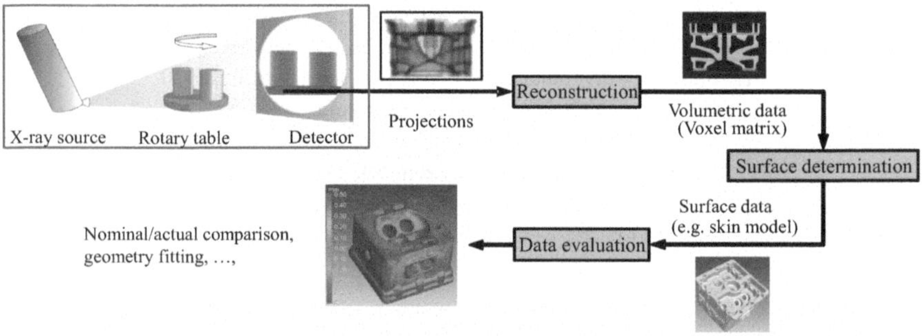

Fig. 1. Typical workflow of a CT measurement.

Since the first decade of the 21st century – when CT was applied as a CMS to a larger extent – significant effort has been dedicated to improving its measurement capabilities, enabling more precise and reliable measurements. Despite of this effort, CT measurements still have not yet reached the same level of reliability of the conventional CMSs technologies (i.e. tactile and optical). Traceability, for example, is still an important open issue for CT due to the great number and complexity of influencing factors impacting the measurement result, i.e. every step of the measurement workflow has a significant influence on the measurement result (cf. Fig. 1). Another important obstacle to a broader acceptance of CT as a CMS is the absence of international standards describing specifications and technical guidelines for the particular application of CT as a CMS, since standardisation contributes to create trust in a measurement technology.

In 2011, the Association of German Engineers (VDI/VDE) published the first national guideline for CT-based CMSs. On the international level, the standardisation activities have started in 2010, where the concepts of acceptance and reverification tests – being the scope of the ISO 10360 series of standards for testing CMS – are being adapted and extended to CT-based CMS [2].

Up to date, these on-going activities are focused on CT systems for the task of mono-material (MoMat) measurements, i.e. they do not deal with multi-material measurements.

Due to the increasing number of multi-material objects in industry, the measurement of multi-material workpieces has gained special interest. The demand of quality assurance in complex multi-material objects are, in many cases, of primary importance due to their functionality, e.g. car door latch systems. Therefore, industry, CT users and manufacturers are highly interested in a traceable and systematic way to verify the performance of CT-based CMS for the case of multi-material measurements.

CT-based CMSs made the task of non-destructive dimensional measurements in MuMat workpieces possible. However, a severe lack of knowledge regarding the error behaviour of the CMS when performing such complex measurements is to be identified. Up to date there exist very limited research on dimensional multi-material measurements, also the topic is still out of the scope of standardisation developments for CT-based CMSs.

Considering the given context, as a continuation of the published work in [3] and [4], this contribution presents a multi-material acceptance testing for CT-based CMS, aiming to:

- Integrate the existing knowledge to better understand the multi-material effects on CT measurements;
- Propose a traceable and systematic method of performance verification for CT-based CMSs applicable to multi-material measurements;
- Depict relevant multi-material effects on the measurements and encourage standardisation bodies to consider the effects observed;

2 Performance Verification and Multi-material Influencing Factors of CT-Based CMS

2.1 Acceptance Testing for CT-Based CMS

When a new CMS is bought, the buying decision is primarily based on the system specifications. Important numbers related to the system specifications are the so-called maximum permissible errors (MPEs). The manufacturer provides MPE statements, verified or falsified using test measurements, carried out under well-defined conditions. The test measurements provide information on the system performance for diverse metrological characteristics. The set of test measurements where the standardised characteristics are assessed and tested against the specified MPE limits are called acceptance and reverification tests.

The technical content of acceptance and reverification testing is usually identical. However, acceptance testing is performed normally when a new system is bought, delivered and installed to check if the CMS performs according to the manufacturer's specifications. Reverification tests are performed in regular time intervals to check if the system still performs within specification or after significant change or an unexpected event, e.g. system collision.

Besides the fact that acceptance testing enables technical-based decision-making, it also allows a systematic check and record of the CMS performance over time. It enables comparability between different CT-based CMSs or even between CMSs with different sensors (e.g. CT and tactile). Also, some aspects of traceability are covered in the scope of acceptance testing principles, however, to a limited extent. The uncertainty estimation of the acceptance test is based on the concepts of the test value uncertainty, described in general in the international standard ISO 23165. Furthermore, it is important to state that acceptance testing provides traceability to the metre only for the specified test characteristics. A CMS which has been proven by an acceptance test – to operate according to the manufacturer given specifications – cannot be declared to provide traceable measurements for tasks other than the ones which are tested by the acceptance test. Furthermore, the tests sample the performance only on an internationally agreed level. For example, the test of tactile CMSs is based on a sampling of 105 lengths, while a tactile CMS can measure an infinite number of lengths within its measurement volume.

From the legal and economic aspects, some warranty issues can be clarified by the test results.

The most relevant document of acceptance testing for CMS is the ISO 10360 series of international reference standards. In this normative document, the principle and procedures of acceptance and reverification testing for CMSs are described.

According to the ISO 10360, the acceptance testing is to perform an overall test of the entire performance of the CMS, where all dominant error behaviour of the CMS should be included. Thus, the CMS should be evaluated as an integrated system comprising every step of the measurement workflow. Another important principle is that the test measurements should reflect the standard use of the CMS. Therefore, real-life effects should be considered in the test design. However, the use of real-life workpieces is limited in the scope of acceptance testing, due to the complexity and variability of such objects, which can hinder the comparability between CMSs.

In the ISO 10360 standard the most common and well-stablished characteristics to be assessed are the local and global performance error characteristics.

The local performance is assessed as a probing error test (*P*-test) by means of measuring the size and the form of a (small) test sphere. The *P*-test assesses the ability of the system to precisely locate and measure the localised point coordinate on the surface of a structure under test within a small measurement volume of the entire system.

There are several metrological quantities under discussion for the realisation of the *P*-test in the ISO 10360 series for CT. These quantities are based on the developments of the ISO 10360-8 standard for CMSs with optical distance sensors. For form analysis, e.g. probing dispersion error and probing form error are used; and for analysis of size, probing error size all and probing size error. A detailed description of these four quantities is given following:

1. **Probing dispersion error** ($P_{\mathrm{Form.Sph.D95\%::CT}}$), the smallest possible width of all spherical shells that contain 95% of all data points.
2. **Probing form error** ($P_{\mathrm{Form.Sph.1x25::CT}}$), the error of indication within the range of the Gaussian radial distance determined by an unconstrained least-square fit of 25 representative points on a test sphere
3. **Probing size error All** ($P_{\mathrm{Size.Sph.All::CT}}$), the difference of the diameter of an unconstrained least-square fit of all points measured on a sphere and its calibrated diameter.

4. **Probing size error** ($P_{\mathrm{Size.Sph.1x25::CT}}$), the error of indication of the difference between the diameter of an unconstrained least-square fit of 25 representative points on a test sphere and its calibrated diameter.

– Remark: The above – in brackets – stated notation for the *P*-characteristics is deduced from the current evolution of standards in ISO 10360 (especially ISO 10360-8). It is in part described in [2], but is currently under development. There is no guarantee that the quantities and the notation used here will be fully implemented in a first future ISO 10360 standard for CT. The same statement needs to be made for the test itself. The final ISO 10360 *P*–test for CT-based CMSs may differ from the above draft statements.

For probing form error and probing size error, the geometrical element (sphere) is created based on least-square fit of 25 representative points on a test sphere. These representative points are calculated based on patch operators each assessing the localised surface within a limited spatial region of the test sphere, an exemplary patch creation workflow can be seen in Fig. 2.

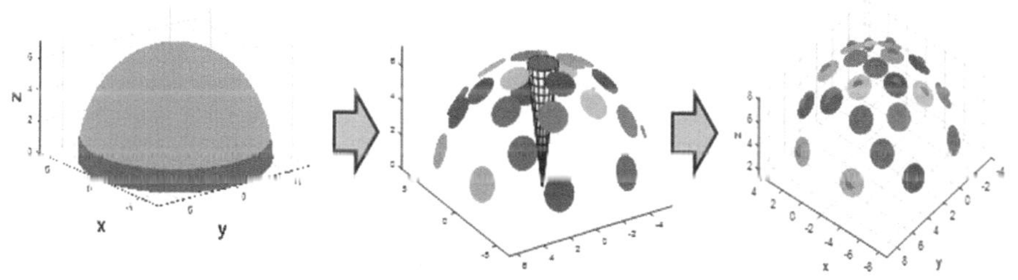

Fig. 2. Exemplary patch operator creation process.

Global performance describes the 3D error behaviour of a CMS in the entire measurement volume, and it is assessed as a length measurement error test (*E* test) by means of measuring (long) length reference standards. Examples of length reference standards used to assess the *E*-test are shown in Fig. 3.

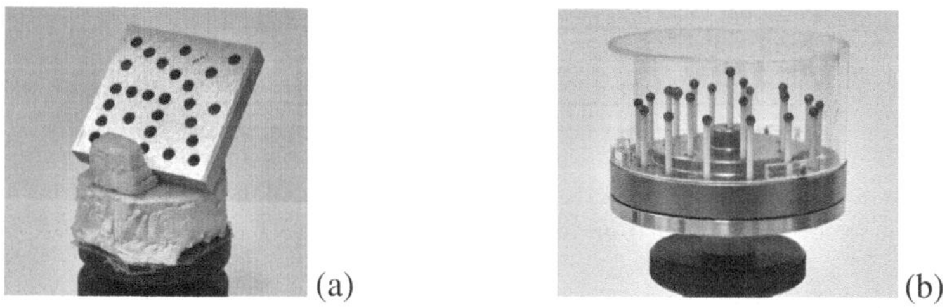

Fig. 3. Example of length reference standards (source: [5]) (a) PTB aluminium hole plate; (b) multi-sphere standard.

Length measurements can be evaluated as bi- and/or unidirectional measurements as well as averaged measurement of the distance between sphere or circle centres, see Fig. 4. The difference between unidirectional and bidirectional measurements is the probing direction when creating a point-to-point distance. A bidirectional length is obtained when both arrows creating a length are pointing to opposite directions as shown by the arrows in Fig. 4 (right). When both arrows creating a length are pointing to the same direction, this characterises a unidirectional measurement. Additionally, when the length is calculated by the centre of two fitted geometrical elements, e.g. circles, the unidirectional centre-to-centre distance is created. The last type of length measurement does not reveal potential local errors of the systems, e.g. the surface determination influence and beam hardening influence on CT measurements are included to a limited extent in the centre-to-centre measurements. The limitation of the centre-to-centre approach is due to the massive data averaging when creating the elements.

The basic *E*-characteristics under discussion to be assessed are:

1. **Bidirectional length measurement error** ($E_{\mathrm{Bi::CT}}$), error of indication when bidirectionally measuring a calibrated test length; according to Fig. 4 this can be an inner or an outer test length
2. **Length measurement error average** ($E_{\mathrm{Avg::CT}}$), error of indication when measuring a calibrated test length where the error is deduced from the distance between two averaged representative points each created from multiple measurement points at two respective geometrical elements.

Important progress has been made for standardisation development. However, it is only focused on mono-material measurements. Thus, an important deficit in standardisation developments on testing multi-material characteristics of the system is identified.

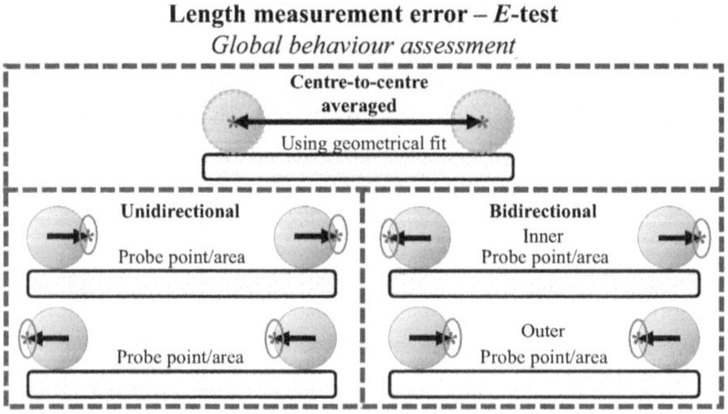

Fig. 4. Assessment of the global performance characteristic of a CMS by means of length measurements. (Source: Adapted from [6])

2.2 Influencing Factors of Multi-material CT Measurements

Current standard CT systems use the absorption contrast to measure the geometry of workpieces. Thus, the basic quantity for CT measurements is the local attenuation of X-rays by the object material, represented – after the CT scan – in a 3D voxel matrix (i.e. volumetric pixel matrix). This matrix is obtained by the reconstruction of a set of two-dimensional (2D) X-ray images – called projections or radiographies – where the sum of the attenuation of the beam path is detected and recorded by an X-ray detector.

The physical phenomena that explain the attenuation of X-rays by the workpiece material are: Photoelectric absorption, Compton or incoherent scattering, Rayleigh or Thompson or coherent scattering and pair productions. However, for the level of energy used for most of industrial CT-based CMS (i.e. up to 600 kV), pair production is physically not possible. More details on the radiation-matter interaction, see [7]. The attenuation depends on the material and X-ray energy, see Fig. 5 showing the X-ray absorption of aluminium (Al) and iron (Fe) as examples.

In Fig. 5 is possible to see that the photons – of a polychromatic spectrum – with low energy are easier attenuated than the more energetic ones. The high energy photons, which have enough energy to go through the material, increase the effective beam energy, i.e. cause a so-called harder beam. This phenomenon is the origin of the beam hardening artefact.

From a certain energetic level, incoherent scattering becomes the predominant physical phenomenon of X-ray attenuation, cf. attenuation curve of Al from ~50 keV and of Fe from ~100 in Fig. 5. It becomes clear that scattering radiation is a physical propriety of materials which also depends on the X-ray energy. Incoherent scattered radiation – as the name suggests – are X-ray photons that have deviated or scatter from its incident path. This is the origin of the scattering artefacts as current reconstruction algorithms usually do not treat scattered radiation.

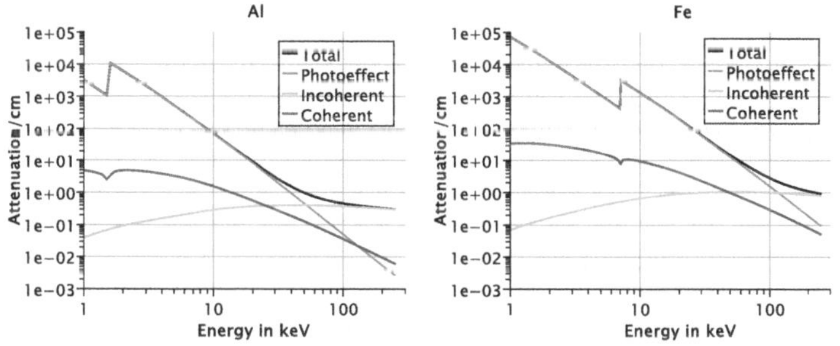

Fig. 5. Attenuation curves for Al and Fe for energies up to 250 kV

The standard reconstruction algorithms for CT-based CMS consider that the X-rays are linearly attenuated according to Beer's law for all X-ray paths and angles. In reality, physical effects which modify the X-ray spectrum while traveling in the material, such as beam hardening and scattered radiation, are sources of differences with the expected

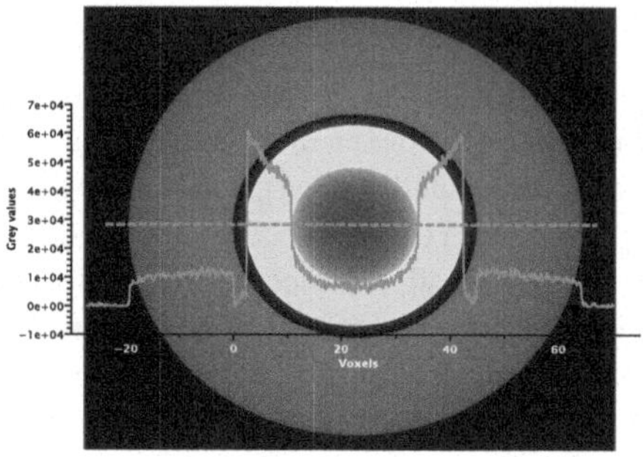

Fig. 6. Beam hardening artefact effect, full red line represents the profile in grey values along the dashed red line

– by the reconstruction algorithms – spectrum. These divergences lead to artefacts in the reconstructed image, see Fig. 6.

Beam hardening mainly appears in the image as material inhomogeneity, see profile in Fig. 6. Derived from beam hardening effects, cupping and strike artefacts are more extreme cases of beam hardening. An example of a moderate beam hardening artefact scenario and cupping artefacts is shown in Fig. 6.

Scattering artefacts can contribute to an unwanted background signal to the projections, being difficult to recognise in the image or to separate from background noise. In Fig. 6, scattering artefact is presumably visible in the inner region (consisting of air) of the inner cylinder.

2.3 Relevant Effects in the Measurements/Surface

The influence of scattering and beam hardening artefacts on dimensional evaluations remains – to a certain extent – unclear and not fully predictable. Some authors have studied the influence of these artefacts on dimensional measurements. Lifton et al. [8] studied experimentally the influence of scatter and beam hardening on dimensional CT measurements. The results showed that the gradient-based surface determination algorithms are robust enough to the influence of these artefacts for the measurement scenarios present in the contribution. Lifton and Carmignato [9] studied the influence of scatter and beam hardening using simulation data. It was found that beam hardening artefacts influence significantly the measurement results even when applying the state-of-the-art gradient-based surface determination algorithm. Likewise, Maier et al. [10] proposed a simulation-based approach to calculate a correction term that compensates the contribution of artefacts on dimensional evaluations. Beam hardening, scattering, off focal radiation, partial volume and cone-beam artefacts were considered in the paper. Bartscher et al. [5] showed the importance to consider beam hardening artefacts on performance verification tests.

The influence of multi-material effects on the dimensional measurements is even less researched than the effects of beam hardening and scattering artefacts. Reiter et al. [11] have shown – using a multi-material reference standard – that beam hardening artefact correction methods designed for medical applications can have a positive impact in dimensional measurements as well. Jansson [12] identified inconsistencies in measurements of a multi-material reference object and proposed a dual-energy-based correction approach. The results of the approach appear to improve the measurement results.

Also, limitations of the state-of-the-art evaluation software for CT impair the measurement of multi-material objects. The main problem is that it is not possible to obtain a surface determination optimised for both materials in the same volume at the same time, when there is a substantial difference in the attenuation ratio, see Fig. 7. For materials with higher attenuation ratios, local-based surface determination algorithms are able to overcome the problem, enabling multi-material measurements [13].

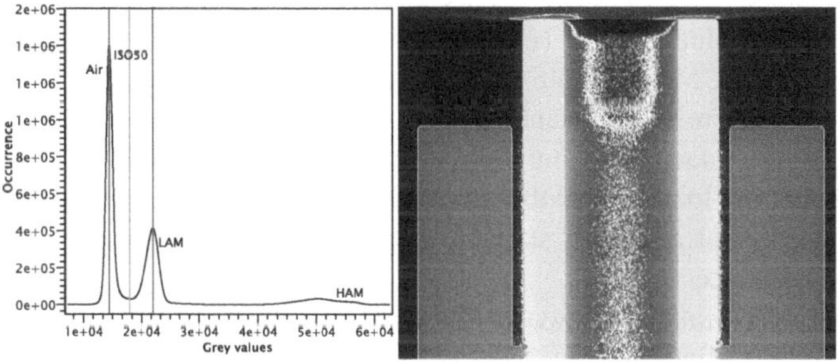

Fig. 7. Surface determination in a multi-material assembly (same object as Fig. 6): (left) optimised only to low absorption material (LAM), (right) surface in the volume data, the inner cylinder is not measurable (white contour represents the estimated surface)

3 Multi-material Acceptance Testing for CT-Based CMS

In Sect. 1, three main issues were identified representing problems of multi-material measurements in the industrial scenario: (a) the lack of traceable and systematic methods for the assessment of multi-material CT measurements; (b) the absence of standards and technical guidelines for multi-material measurements; and (c) the not fully understood multi-material induced effects on the measurement, hindering reliability and measurement uncertainty analyses of multi-material CT measurements.

The concept of acceptance and reverification testing offers a potential, systematic approach for testing the performance of CT-based CMSs, where dominant error behaviour is assessed. Thus, in this contribution a multi-material test design based on the concepts of the standard mono-material acceptance testing (i.e. ISO 10360) for evaluating CMS is proposed.

In the next sections test design requirements, testing concepts as well as the practical implementation of the multi-material acceptance testing for CT-based CMS are presented.

3.1 Test Design Requirements

The establishment of the test design requirements based on the identified needs is the first step of the development of the multi-material acceptance testing. The requirements for the multi-material test are:

- The multi-material acceptance testing shall follow the main concepts described in the ISO 10360 series of standards for acceptance testing of CMSs;
- The multi-material test should be complementary to the mono-material test, as it covers multi-material effects not covered by the mono-material test. Considering that the multi-material performance of a CT-based CMS is not always relevant (e.g. for mono-material measurements), and the acceptance test should be as efficient as possible, unnecessary effort and costs should be avoided;
- The test shall evaluate global (*E*-test) and local (*P*-test) performance characteristics of the CT-based CMS for multi-material measurements by means of measurements of multi-material reference standards;
- Design and manufacture of multi-material (with two materials or more) reference standards for the local and global performance assessment;
- The multi-material reference standard shall reveal potential multi-material effects and error influences;
- The test should feature different attenuation coefficient scenarios, where high and medium values of attenuation ratios are present;
- The proposed test should feature sufficient low test value uncertainty enabling multi-material-related specification statements to be verified;
- The multi-material reference standards should include accessible features with simple regular geometries (i.e. planes, circles, cylinders, spheres) allowing the tactile-based calibration;
- The design should allow absolute-based data analyses (i.e. comparison with the reference values), as well as relative-based analyses;

_P_-test Specific Requirements

- The multi-material *P*-test should use the existing concepts of the standard mono-material *P*-test, the measurement of size and form of a test sphere;
- The size of the test sphere should represent a small portion of the measurement volume (e.g. 20% of the longest measurement volume diagonal) comprised by two or more materials;
- The geometrical and surface proprieties of the multi-material test sphere(s) shall have a negligible influence on the measurement result by CT;
- The mechanical design and manufacture induced errors of the components of the multi-material *P*-test shall not influence significantly the test characteristics.

_E_-test Specific Requirements

- The multi-material *E*-test should be evaluated as length measurements using length reference standard(s) comprised by two or more materials;
- The size of the standard should allow mid-range magnification, assuming that the multi-material effects are generic in nature;

- The length multi-material reference standard shall feature short and long lengths to be measured in different spatial directions;
- Multi- inter-material and in-material measurands as well as mono-material measurands should be performed in the test (for definition of in- and inter-material measurands see Fig. 9);
- The measurement strategy shall also allow uni- and bidirectional length measurands, centre-to-centre averaged, pointwise and patch-based length measurands;
- The length standard should feature different material ratios along the standard; ideally a single CT scan is able to provide different material ratios in one shot;
- Fixing elements (e.g. screws and nuts) should not influence significantly the measurement, however they shall guarantee the mechanical stability of the standard;
- The *E*-test reference standard shall feature sufficiently low geometrical errors and surface roughness having a negligible influence on the measurement results; the above statement includes assembly induced effects.

3.2 Multi-material Acceptance Test Concepts

The multi-material acceptance testing is designed to assess the performance of CMSs with X-ray CT-based measurement principle featuring at least one rotary axis, an X-ray source and an X-ray detector. Medical imaging CT, native non-destructive CT as well as non-X-ray CT systems, e.g. Neutron CT, THz-CT, are currently excluded from the scope of this test.

The proposed test tries to reach maximum comparability with the principles and concepts described in the ISO 10360 series of standard for testing CMSs. In the scope of the ISO 10360, the development of a document focused on mono material CT measurements is already in progress, cf. Sect. 2.1. The proposed multi-material test is intended to be supplementary to this future ISO mono-material test, and it is not intended to substitute it.

The multi-material performance verification comprises the assessment of local (*P*) and global (*E*) multi-material error behaviour characteristics of CT-based CMS, to be performed in multi-material reference standards. A multi-material reference standard is defined, in this work, as a technical workpiece where two or more materials (excluding air) are used to perform dimensional or geometrical measurements; or two or more materials which significantly influence the measurement by substantial attenuation of the radiation for the used X-ray spectrum.

The CMS is to be specified for a medium/large range of attenuation coefficients, test scenarios using combination of materials with, at least, high and medium attenuation coefficient ratios are to be tested. For the high attenuation ratio scenario, a minimum attenuation ratio number – for a defined energy – of 0.8 is allowed. For the medium/low attenuation ratio scenario, the ratio should be below 0.7, see Table 1. Thus, at least three materials are used to create the test. Additionally, the mono-material case should be included in the test for verification purposes. In Table 1, attenuation coefficient ratios being close to 1 shows that the materials have similar X-ray attenuation, whereas attenuation coefficient ratios being close to 0, indicates that the materials have different X-ray attenuation.

Table 1. Definition of the attenuation coefficient scenarios.

High attenuation ratios	$0 < \mu_2 \leq 0.2\ \mu_1$
Transition high-medium	$0.2\ \mu_1 < \mu_2 \leq 0.3\ \mu_1$
Medium attenuation ratios	$0.3\ \mu_1 < \mu_2 \leq 0.7\ \mu_1$
Transition medium-low	$0.7\ \mu_1 < \mu_2 \leq 0.8\ \mu_1$
Low attenuation ratios	$0.8\ \mu_1 < \mu_2 \leq 1\ \mu_1$

An important factor to be considered in the multi-material acceptance testing is the intrinsic uncertainty of the test. The concepts described in the ISO/TS 23165:2006 are adapted to this proposal for the estimation of the test value uncertainty.

Besides this, CT simulation-based studies supported the development of this multi-material test. More details are given in the specific test approach description.

Conceptual Approach of the Multi-material *P*-test

The main objective of the multi-material *P*-test is to assess the three-dimensional error behaviour within a small portion of the entire measurement volume for multi-material measurement scenarios.

In general, the *P*-test value is determined by the errors related to the sensor of the CMS – e.g. noise, digitalisation error, geometry-dependent interactions with the reference standard and it is – for the case of CT – largely influenced by the material-dependent interaction with the X-ray radiation.

Targeting maximum comparability with the standard mono-material test, the multi-material *P*-test adopts also a test sphere for the test measurements. In contrast to the standard case, the proposed multi-material reference standard for performing *P*-test consists of a compound test sphere consisting of two symmetric half spheres (HS) made of different materials, see Fig. 8.

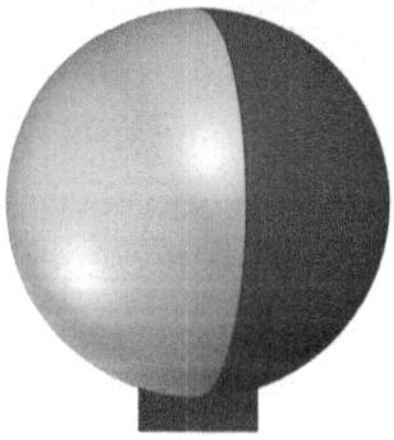

Fig. 8. Design of the multi-material test sphere used for the proposed *P*-test.

At least two multi-material spheres – featuring materials with high and medium/low attenuation ratios – are to be measured during the test. The precise values and ranges of the attenuation coefficients and respective performance values (MPEs) are part of the manufacturer specification.

Size (i.e. diameter) and form error are to be measured in such multi-material spheres. The results, to be compared with reference values, are to be used to state the multi-material influence on the probing error test for the high and medium/low attenuation ratio scenarios.

Conceptual Approach of the Multi-material *E*-test

The main concept of the multi-material length measurement error test is the evaluation of the three-dimensional error behaviour of the CMS represented by a large portion of the entire measurement volume. In CT based-CMS, the value of the length measurement errors can be determined largely by the sensor errors and by the attenuation proprieties of the material(s) to be penetrated by X-rays. Therefore, the multi-material *E*-test attempts to quantify the multi-material influence on length measurements.

The creation of the multi-material *E*-test design was mainly based on the principles of the standard mono-material acceptance testing described in the ISO 10360 series of standards, providing maximum compatibility with the mono-material test and the list of requirements listed in Sect. 3.1.

The principle of measuring a long length reference standard based on the ISO 10360-2 is also adopted in the multi-material *E*-test. However, for the proposed *E*-test, the test measurements are to be performed in a multi-material length reference standard.

The data evaluation is based on length (i.e. distance) measurements between geometric elements (e.g. cylinder, spheres, etc.) to be assessed using different measurands: Centre-centre measurements; and inner and outer bidirectional length measurements performed in short, middle and long lengths. The results of the CT length measurements will be compared with calibrated reference measurements. Besides this, different multi-material measurement scenarios are present, where different material ratios are obtained along the designed scheme. Inter- as well as in-material measurements can also be evaluated.

The test design attempts to facilitate the recognition of multi-material effects by allowing different measurands concerning material interface. Three measurement types are classified: mono-material measurements, inter-material measurements and In-material measurements, see Fig. 9.

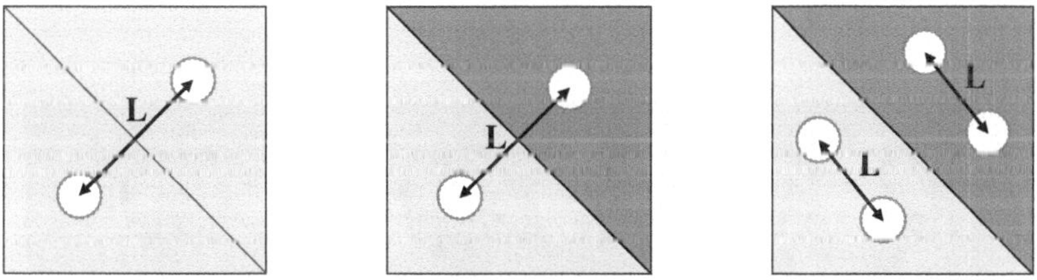

Mono-material measurements (equal materials) | Inter-material measurements (multi-material) | In material measurements (multi-material)

Fig. 9. Multi-material measurement scenarios based on the material interface: mono-material measurement, multi-material inter-material and multi-material in-material measurements.

3.3 Implementation of the Multi-material Acceptance Testing

This section addresses the challenge of practically implement the multi-material *P*- and *E*-test for CT-based CMS.

Implementation of the *P*-test

For the implementation of the multi-material *P*-test, two half spheres (HS) made of different materials are used to create a complete sphere. The HS have (nominally) the same geometry and size – 9/16 inches (approximately 14.3 mm) in diameter. The construction of the half spheres was carried out using grinding process of a complete sphere to its half, and the grinded surface was subsequently polished and then glued to a second HS using epoxy resin-based glue creating a multi-material sphere.

Silicon nitrite (Si_3N_4), aluminium oxide (Al_2O_3) and lead-free glass (N-SF6) were selected based on X-ray proprieties, availability, costs, surface and geometrical quality.

With these materials, a medium range of attenuation coefficient was achieved. Table 2 presents the attenuation coefficient ratios (μ_2/μ_1), at 150 kV X-ray voltage with no physical filter on the tube. Thus, in the case studied here, N–SF6 absorption is larger than Si_3N_4 by a factor of ~2.5.

Table 2. Attenuation ratios between the materials for the *P*-test at 150 kV, no filter.

	Al_2O_3 & Si_3N_4	Si_3N_4 & N-SF6	Al_2O_3 & N-SF6	Si_3N_4 & Si_3N_4	Al_2O_3 & Al_2O_3	N-SF6 & N-SF6
Attenuation ratio (μ_2/μ_1)	0.9	0.4	0.5	1	1	1

The three materials were mounted in pairs – on a carbon fibre reinforced polymer (CFRP) shaft – resulting in three MuMat compound spheres and – as a reference – three MoMat compound spheres, see Fig. 10.

Fig. 10. Set of multi- and mono-material spheres used for the assessment of the *P*-test. From left to right: three MuMat-spheres Al_2O_3 & Si_3N_4, Si_3N_4 & N-SF6 and Al_2O_3 & N-SF6; and three MoMat-spheres Si_3N_4 & Si_3N_4, Al_2O_3 & Al_2O_3 and N-SF6 & N-SF6.

Mounting-related effects, e.g. different half spheres diameters, different cut position of the HS, HS positional error, et cetera, are present in any realistic scenario of such multi-material sphere. These mounting effects might affect the geometry as well as the size of the multi-material sphere and consequently the measurement result, if no data handling is applied. The mounting-related effects of the HS should have a negligible influence on the measurement results, enabling separation of effects and consequently, statement about the performance of the CMS.

Simulation tools were used – using the software aRTist from BAM, Berlin – to verify the mounting-related effects. The preliminary results have shown a non-negligible influence on the results, unless data handling is applied, e.g. translation and rescaling of a HS relative to another HS. Therefore, the data evaluation was performed in two half spheres separately.

The multi- and mono-material spheres were all calibrated using a tactile CMS. The calibration strategy is based on measuring the half spheres separately (but in the glued compound state). Two spheres are calculated from the probed points: half sphere material 1 (HS1), half sphere material 2 (HS2). An area of approximately 120° opening angle near the pole (i.e. opposite side of the CFRP shaft) is covered by tactile points using a single point probing strategy. A total of 64 points evenly distributed on each half sphere – excluding the gap/glue area and its vicinity – were acquired. By this means, the mounting-related effects were excluded from the analyses. Diameter and form deviations of each half sphere were determined. The form deviation of all half spheres was found to be below 0.5 μm. The expanded measurement uncertainty U ($k = 2$) of each single point was of the order of 1 μm or less. These numbers are comparable to the specified form error of complete spheres. Thus, it shows that cutting or grinding full spheres to create half spheres made of the given materials does not cause a significant degradation of the form. A good quality sphere is required in the test enabling separation of effects. Therefore, the extent of the form errors must be smaller than the typical voxel sizes, ensuring that the impact of half spheres as reference standards, does not impair the conducted tests.

The data analysis of the multi-material probing error test is based on form error measurements as well as the deviations of the sphere diameter measured by CT from the tactile reference measurements. Also, comparison between the measurement scenarios, i.e. mono-material scenario, large and small absorption difference, is part of the data analyses. Due to the mounting-related effects, the multi-material P-test data for size and form are evaluated on each HS separately enabling a fair test scenario for the probing test in a multi-material assembled sphere.

Two measurands are considered in the multi-material P-test: diameter and form error of the multi-material half spheres.

Considering the difference between the nominal and the actual geometry of the multi-material spheres, misalignments deriving from the imperfect assembly process, and the limitation of the current surface determination algorithms regarding multi-material scenarios, a new multi-step surface determination workflow for the multi-material P-test is proposed. A simplified data analysis workflow is following.

1. CT scans of the multi-material spheres → CT volume;
2. Surface determination optimized for the high absorption material (HAM) → HAM CT surface;
3. Extraction of the volume based on the HAM → HAM volume including surface data;

4. Surface determination optimized for the low absorption material (LAM) → LAM CT surface;
5. Extraction of the volume based on the LAM → LAM volume including surface data
6. Assessment of four *P* characteristics separately in the two HS volumes.

Implementation of the *E*-test

For the implementation of the multi-material *E*-test, a new reference standard is presented. The multi-material hole cube (MuMat-HC) design is presented in Fig. 11.

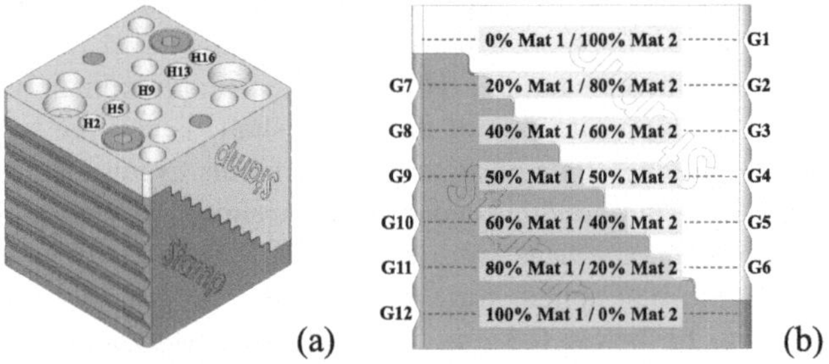

Fig. 11. Multi-material hole cube design for the length measurement error test (*E*-test); (a) Isometric view of the cube; (b) Different material ratios along the cube

The design consists of two symmetric half cubes made of different materials. The fixture of the assembly is guaranteed by fitting pins, screws and nuts made of the polymer polyether ether ketone (PEEK), due to its low X-ray absorption and relatively good mechanical stability. The MuMat-HC has a size of 30 mm × 30 mm × 30 mm featuring 17 holes and 12 "V"-shaped grooves. The design also features a step-like "cut" shape enabling different and well-defined multi-material ratios along the standard's height, see Fig. 11b. The measurands are mainly based on distances between the holes. Hole-based measurements allow the evaluation of different probing strategies, i.e. centre-centre, point- or patch-based uni- and bidirectional measurements in multi-material as well as mono-material scenarios, cf. Fig. 4, G1 and G12 in Fig. 11b. Also, the multi-material influence on the creation of the fitting element based on the mono-, in- and inter material measurements are included in the MuMat-HC design, cf. Fig. 9.

Medium/large range of attenuation coefficients was covered by the combination of a composite material – carbon fibre reinforced silicon carbide (Cesic), aluminium and titanium. These materials were used due to their large applicability in industry, adequate mechanical proprieties, machinability and adequate X-ray attenuation coefficients.

Table 3 presents the attenuation coefficient ratios ($\mu 2/\mu 1$), at 195 kV X-ray tube voltage and 1 mm Cu filter on the tube.

Six cubes are created, three mono- and three multi-material, see Fig. 12. All the cubes were manufactured by electro discharge machining. The sufficient electrical conductivity of all three materials, even of the composite material, has shown significant advantages over other materials considered for the design. Since all three

Table 3. Attenuation ratio between the materials at 195 kV, 1 mm Cu.

	Al & Cesic	Cesic & Ti	Al & Ti	Al & Al	Cesic & Cesic	Ti & Ti
Attenuation ratio (μ_2/μ_1)	0.92	0.39	0.36	1	1	1

Fig. 12. Multi-material hole cubes from left to right: MuMat-HCs Al & Cesic, Cesic & Ti and Al & Ti; and the MoMat-HCs Al & Al, Cesic & Cesic and Ti & Ti

materials are manufactured using the same manufacturing technology, thus, similar manufacturing quality is expected.

Reference measurements – using a tactile-based CMS – were carried out in all hole cubes. The strategy is based on the measurement in scanning mode of 7 circumferential lines at seven different heights (indicated by the groove position) to be measured inside of the holes. This measurement approach allows the evaluation of centre-centre, point- or patch-based uni- and bidirectional measurements.

The evaluation of the multi-material influence on the *E*-test was based on the deviation from tactile CMS reference measurements.

To ensure comparable analyses, the same measurement strategy as those used in the reference measurements were also applied to the CT data. The workflow of the analysis is described below:

1. Reference measurement of the MuMat-HC using a tactile CMS → Tactile CMS points
2. Calculation of reference lengths tactile CMS data (for e.g. centre-to-centre, bidirectional patch-based lengths) → tactile CMS reference length results
3. CT scans and surface determination of the MuMat-HC (surface determination for multi-material cubes are performed in two steps: optimised for the high absorption material (HAM) and optimised for the low absorption material (LAM) → CT surface (in two volumes)
4. Load and fit the tactile CMS points into the CT surface(s) (CT data aligned) → CT points (two point clouds; one for HAM and one for LAM)
5. Length measurements of the CT data (for e.g. centre-to-centre, bidirectional patch-based lengths) → CT results
6. Difference between CT results and tactile CMS reference results is the length measurement error

Geometrical elements present only in one material are used for the alignment procedure of the MuMat-HC, as in the version of the data analysis software in use (VG Studio Max 3.2), a common and optimal surface determination for different materials in the same scene is not possible, cf. Fig. 7.

Centre-to-centre length measurements – as distance defined by the two least-squares short cylinder centre points at a specific height – were evaluated in this paper.

4 Experimental Application

The experimental application of the proposed multi-material acceptance test was carried out in the metrological CT system Nikon MCT225 of the *Physikalisch-Technische Bundesanstalt* (PTB), Braunschweig, Germany. The main objective is to assess the multi-material acceptance testing concepts, test operability, reference standards as well as identify the potential improvements not considered in the previous phases.

The CT system features an X-ray source with a reflection target with a maximum acceleration voltage of 225 kV and maximum power of 225 W. In this X-ray source, tungsten is used as anode material and a beryllium window of 0.1 mm thickness placed in the exit of the X-rays from the source. The detector unit is a flat panel PerkinElmer 1620 AN3CS detector with caesium iodine scintillator material with size of 400 mm × 400 mm with a quadratic pixel size of 200 μm. An aluminium foil of 0.75 mm thickness is placed in front of the detector for protection. All the measurements were carried out using circular trajectory and continuous scanning mode. The air temperature measured close to the measurement object, inside of the CT cabin, was recorded for all measurements to be in the range of (20,0 ± 1,0) °C. The data acquisition as well as CT data reconstruction were done based on the implemented manufacturer solution, i.e. X-Inspect and CT Pro 3D version 3.1.9 from Nikon Metrology, Tring, UK. Adaptive local surface determination implemented in the commercial data analysis software for CT VG Studio Max version 3.0.1, Heidelberg, Germany was used for all the scans.

_P_-test Measurement Setup

The multi-material test spheres are CT scanned to verify the multi-material local performance of the system. The complete test comprises of three multi-material test spheres, but also three mono-material test spheres. The first is to verify the performance of the system when measuring objects with high and medium attenuation coefficient ratios and the second serves as standard and verification step to the multi-material spheres (also to prove that the half spheres approach is not significantly impaired by the cut/glue, consequently appropriate for the proposed test). Diameter and form error of the multi-material spheres are assessed based on the 4 metrological characteristics presented in Sect. 2.1. Avoiding the half spheres mounting-related errors, the data evaluation is carried out on each half sphere separately.

The multi-material spheres were all positioned with the glue/gap parallel to the flat panel detector's centre column. All scans were performed with the same magnification of 10 times which leads to a voxel size of (20 μm)^3 and 1700 projections. The scanning parameters, which are reported in Table 4, were selected for each assembly in such a way as to yield a similar noise level in the multi-material measurements but also to minimise beam hardening effects.

Table 4. CT scanning parameters used for each mounted sphere.

Parameter	Unit	Si_3N_4 & Al_2O_3	Si_3N_4 & N-SF6	Al_2O_3 & N-SF6	Si_3N_4 & Si_3N_4	Al_2O_3 & Al_2O_3	N-SF6 & N-SF6
Voltage	kV	200	220	220	200	200	220
Current	μA	46	55	55	46	46	55
Power	W	9.2	12.1	12.1	9.2	9.2	12.1
Cu filter thickness	mm	0,25	1	1	0,25	0,25	1
Exposure time	ms	2000	2829	2829	2000	2000	2829
Scan time	min	56	80	80	56	56	80
Beam hardening correction		None	Soft	Soft	None	None	Soft

Additionally, to verify the influence of beam hardening on multi-material *P*-test measurements, datasets of three spheres (Al_2O_3 & Al_2O_3, N-SF6 & N-SF6 and Al_2O_3 & N-SF6) with no beam hardening correction and applying soft beam hardening correction[1] were evaluated. It is worth to remark that no new CT scans were performed. The datasets for the two MoMat-spheres and one MuMat-sphere were reconstructed again using different beam hardening correction settings.

E-test Measurement Setup

The MuMat-HC standards are scanned to verify the multi-material influence on the length measurement error, to check the multi-material *E*-test concepts and to verify the suitability of the multi-material hole cube standards. The complete set consists of 3 MoMat-HCs and 3 MuMat-HCs.

For the CT scans, the cubes were positioned in the CT system with the hole axes parallel to the rotary axis. Thus, the X-ray penetration length per material at each groove (height) is kept constant, due to the cube's design. All scans were performed with the same magnification of 6.4 times which leads to a voxel size of $(32\ \mu m)^3$ and 3000 projections. The CT scanning parameters – shown in Table 5 – were selected different for each hole cube to reduce beam hardening artefacts and to keep the noise level similar for all scans. In order to remove residual scaling errors, and thus improving the accuracy of the analyses, a multi-sphere-based scale correction (using the standard visible in Fig. 3b) was applied to all measurements.

Averaged centre-to-centre length measurements were evaluated for all hole distances per height (i.e. different material ratios) in the cubes. In order to provide a better view of the material influence depending on the material ratios, the average, the range and the standard deviation of length measurement errors per groove are also evaluated.

[1] Nikon Metrology CT PRO 3D version 3.1.9 standard beam hardening correction based on a polynomial function of order 2 (soft) was carried out during the reconstruction of the projections.

Table 5. CT scanning parameters used for the hole cube standards, highlighted in blue the setting of new lower spectrum energy scan.

Parameter	Unit	Al & Cesic	Cesic & Ti	Al & Ti	Al & Ti	Al & Al	Cesic & Cesic	Ti & Ti
Voltage	kV	200	225	225	200	200	200	225
Current	μA	75	110	110	30	75	75	110
Power	W	15	25	25	6	15	15	25
Filter thickness	mm	1 Cu	1 Ag	1 Ag	None	1 Cu	1 Cu	1 Ag
Exposure time	ms	2829	4000	4000	2000	2829	2829	4000
Scan time	min	141	200	200	100	141	141	200
Beam hardening correction		none	soft	soft	none	none	none	soft

An additional scan of the cube featuring the smallest attenuation ratio (Al & Ti) to verify the influence of the X-ray spectrum on the multi-material length measurements was carried out and evaluated. The spectrum energy used for this scan (200 kV, no filter) was significantly lower compared to the standard high energy spectrum (225 kV, 1 mm Ag) used for the other scans, see Table 5. The results between the two scans – for centre-to-centre lengths – were compared.

5 Results

The results of the multi-material acceptance testing for assessment of local and global error behaviour are presented.

***P*-test Results**

The results of all MoMat- and MuMat-spheres for form and size are presented in Fig. 13. Bars with similar colour – e.g. full red and dashed red – represent two HSs of the same mounted sphere, i.e. same scan. A simplified notation of the *P*–test characteristics: PF25 = $P_{Form.Sph.1x25::CT}$, PF95 = $P_{Form.Sph.D95\%::CT}$, PS25 = $P_{Size.Sph.1x25::CT}$, and PSall = $P_{Size.Sph.All::CT}$ was used. The size and form errors of all the spheres were below one voxel size. However, the multi-material effect appears to have a strong influence on the *P*-test measurements. The LAM half sphere in low attenuation ratio assemblies (i.e. Si_3N_4 & N-SF6 and Al_2O_3 & N-SF6) suffer a strong degradation of the form measurements, when using 95% of measured points. This effect was verified by the measurements of the MoMat-spheres and with the sphere with high attenuation ratio (i.e. Si_3N_4 & Al_2O_3). The form deviation is more than 3 times worse in the MuMat-sphere measurement than the MoMat-spheres and the high attenuation ratio sphere measurements. For form measurements using 25 patch-based representative

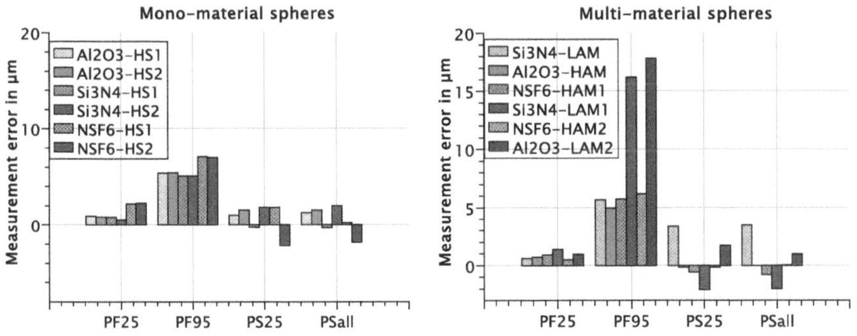

Fig. 13. *P*-test measurements (left) mono- and (right) multi-material spheres.

points, no significant effect was observed, presumably due to the data averaging. For size, the multi-material effect appears not to have a significant influence on the measurements, once the results were significantly smaller than the voxel size.

Despite the multi-material effect observed, a good agreement between the MuMat- and the MoMat-spheres was observed, since the measurements of the MoMat-spheres and the sphere with high attenuation ratio presented similar results. This indicates that the glue and gap between the HSs do not influence significantly the measurements.

Additionally, the influence of beam hardening artefacts on the multi-material *P*-test was evaluated. The results from two MoMat-spheres and one MuMat-sphere is presented in Fig. 14. The use of beam hardening correction for the *P*-test has shown a negative impact on the measurements for the measurement set-up presented in this paper. It is possible to observe that for the tested MuMat-sphere (N-SF6 & Al_2O_3) the form error seems to be worsened with the used beam hardening correction. This is an indication that the noise of the scan was increased with the beam hardening correction. Also, an overcorrection in the MoMat-sphere with material featuring low attenuation coefficient (Al_2O_3 & Al_2O_3) was observed, as the measurement errors of the measurement of size were higher than the scan with no beam hardening correction.

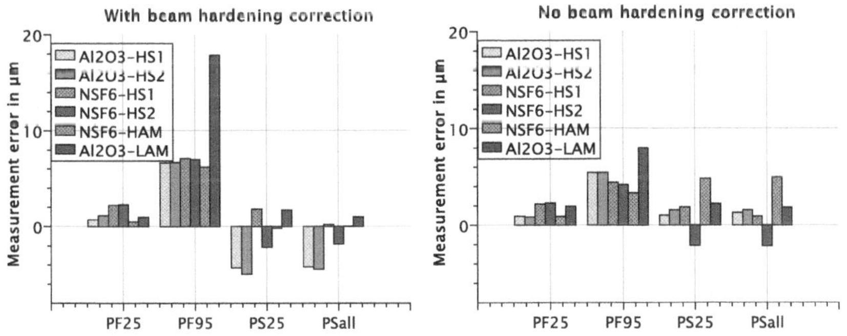

Fig. 14. Influence of the beam hardening correction on the *P*-test.

E-test Results

The measurements of the multi-material length measurement error test are presented in Fig. 15. In the plots, red rectangles, black bars and ranges represent the average, standard deviation and maximum/minimum range of all lengths per groove, respectively.

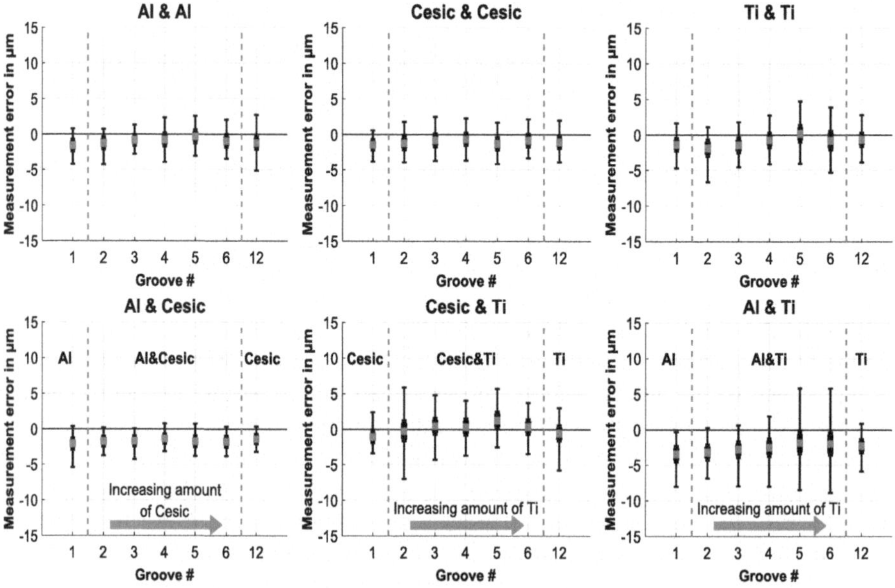

Fig. 15. Centre-to-centre length measurements of the hole cube reference standards.

Although, the measurement error for all cubes for centre-to-centre lengths are below half of the voxel size, a multi-material influence on the centre-to-centre length measurement was observed. An increase of the standard deviation and total range for the cubes with low attenuation ratios (i.e. Al & Ti and Cesic & Ti) was observed in the grooves containing more HAM (i.e. Ti). The multi-material effect was also confirmed by the mono-material cubes, where the results were all within ±7 µm, comparable to the results obtained with the cube with high attenuation ratio (i.e. Al & Cesic), while for the MuMat-HCs with low attenuation ratios the results were within ±10 µm.

To verify the dependence of the multi-material effect on the spectrum energy, an additional CT scan of the Al & Ti cube with low spectrum energy was performed, evaluated and compared with the high energy scan.

The results of the two energy spectra are presented in Fig. 16. From the results it is possible to observe that the low energy spectrum delivers results with much higher measurement errors, and the measurement errors increase in the grooves with a high content of Ti (i.e. grooves 4, 5 and 6 in Fig. 16). This indicates that the multi-material effect strongly depends on the spectrum energy.

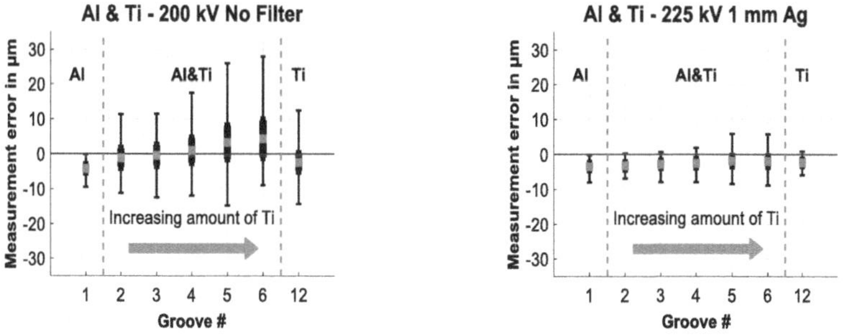

Fig. 16. Centre-to-centre measurements of two scans: (right) low energy spectrum and (left) high energy spectrum.

6 Conclusion and Outlook

A multi-material acceptance test for CT-based CMS was presented in this paper. Test concepts and the practical implementation of the multi-material *P*- and *E*-test for local and global error behaviour assessment were presented. The test concepts and designed reference standards were successfully tested based on test measurements on multi-material spheres and multi-material hole cube standards. Strong multi-material influence in both *P*- (for form error and size) and *E*-test (for centre-to-centre) measurements was observed. It is expected that the multi-material effect is clearer visible for bidirectional length measurements, to be tested in future work. Additionally, the influence of a beam hardening correction method on the *P*-test measurements was carried out. The beam hardening correction seems to have a negative impact on the *P*- measurements, since an increase of the noise and an overcorrection of the data was observed. Furthermore, the dependence of the multi-material influence on the spectrum energy was also tested by comparing two scans with different energy spectra. The results showed a strong multi-material influence in the low energy spectrum.

The set of multi-material reference standards has shown to be suitable for testing the multi-material characteristics of CT-based CMSs.

However, the estimation of the test value uncertainty of the test is a topic for future development of the multi-material acceptance testing. Furthermore, a sensibility study of the multi-material acceptance test proposed here – based on the test value uncertainty estimation – is a further topic of future work.

References

1. De Chiffre, L., Carmignato, S., Kruth, J.-P., Schmitt, R., Weckenmann, A.: Industrial applications of computed tomography. CIRP Ann. **63**, 655–677 (2014). https://doi.org/10.1016/j.cirp.2014.05.011
2. Bartscher, M., Sato, O., Härtig, F., Neuschaefer-Rube, U.: Current state of standardization in the field of dimensional computed tomography. Measur. Sci. Technol. **25**, 064013 (2014). https://doi.org/10.1088/0957-0233/25/6/064013

3. Borges de Oliveira, F., Stolfi, A., Bartscher, M., Neugebauer, M.: Creating a multi-material probing error test for the acceptance testing of dimensional computed tomography systems. In: Proceedings of iCT, Leuven, Belgium, p. 8 (2017)
4. Borges de Oliveira, F., Bartscher, M., Neuschaefer-Rube, U., Tutsch, R., Hiller, J.: Creating a multi-material length measurement error test for the acceptance testing of dimensional computed tomography systems. In: Proceedings of iCT, Leuven, Belgium, p. 12 (2017)
5. Bartscher, M., Illemann, J., Neuschaefer-Rube, U.: ISO test survey on material influence in dimensional computed tomography. Case Stud. Nondestr. Test. Eval. **6**, 79–92 (2016). https://doi.org/10.1016/j.csndt.2016.04.001
6. Carmignato, S., Dewulf, W., Leach, R.: Industrial X-Ray Computed Tomography. Springer, Heidelberg (2018)
7. Buzug, T.M.: Computed Tomography. Springer, Heidelberg (2008)
8. Lifton, J.J., Malcolm, A.A., McBride, J.W.: An experimental study on the influence of scatter and beam hardening in x-ray CT for dimensional metrology. Meas. Sci. Technol. **27**, 015007 (2016). https://doi.org/10.1088/0957-0233/27/1/015007
9. Lifton, J.J., Carmignato, S.: Simulating the influence of scatter and beam hardening in dimensional computed tomography. Measur. Sci. Technol. **28**, 104001 (2017). https://doi.org/10.1088/1361-6501/aa80b2
10. Maier, J., Leinweber, C., Sawall, S., Stoschus, H., Ballach, F., Müller, T., Hammer, M., Christoph, R., Kachelrieß, M.: Simulation-based metal artifact reduction for computed tomography of multi-material components. In: Proceedings of iCT, Wels, Austria (2016)
11. Reiter, M., Borges de Oliveira, F., Bartscher, M., Gusenbauer, C., Kastner, J.: Case study of empirical beam hardening correction methods for dimensional x-ray computed tomography using a dedicated multi-material reference standard. J. Nondestr. Eval. **38** (2019). https://doi.org/10.1007/s10921-018-0548-3
12. Jansson, A., Pejryd, L.: A dual-energy approach for improvement of the measurement consistency in computed tomography. Measur. Sci. Technol. **27**, 115013 (2016). https://doi.org/10.1088/0957-0233/27/11/115013
13. Borges de Oliveira, F., Stolfi, A., Bartscher, M., De Chiffre, L., Neuschaefer-Rube, U.: Experimental investigation of surface determination process on multi-material components for dimensional computed tomography. Case Stud. Nondestr. Test. Eval. **6**, 93–103 (2016). https://doi.org/10.1016/j.csndt.2016.04.003

Adaptive Calibration for Articulated Arm Coordinate Measuring Machine

Ryoshu Furutani(✉)

School of Engineering, Department of Advanced Machinery Engineering, Tokyo Denki University, Tokyo, Japan
ryo@cck.dendai.ac.jp

Abstract. The calibration of the coordinate measuring machines is performed based on the mathematical model of those machine. The feature of the artifact is measured by CMM, the kinematic parameters are determined by comparing the calculated coordinates and the calibrated coordinates. However, it is assumed that the CMM is rigid and stable in any orientations and locations. The actual CMM is not sufficiently rigid and stable. Especially the articulated arm CMM is not. In order to consider the CMMs' deformation and improve the accuracy of those, it is proposed that the model equation, the kinematic parameters and the estimators are replaced with the neural network. The calibration method using the neural network is called the adaptive calibration. The neural network is trained based on the relationship between the inputs and the coordinates successfully. In this article, the adaptive calibration is applied to the articulated arm coordinate measuring machine.

Keywords: Adaptive calibration · Kinematic calibration · Kinematic parameters · Coordinate measuring machine · Neural network · Artifact

1 Introduction

The research of the calibration and test method of the coordinate measuring machines were mainly focused on the Cartesian type coordinate measuring machines [1–6]. On the other hand, the coordinate measuring machine which has 6 or more rotational joints, is the articulated arm coordinate measuring machine. It is abbreviated to the arm CMM in this paper. Because the accuracy of the arm CMMs are improved and they are widely used in the manufacturing site, the calibration method and test method of the arm CMMs were studied [7–10].

The Cartesian CMMs has the Cartesian coordinate system as nature. However the arm CMM is not based on the Cartesian coordinate system, so the geometric errors of the arm CMMs are more complicated than those of the Cartesian CMMs.

In this article, the calibration process using the artifact is discussed about the arm CMMs. The kinematic models are shown and it is explained how to determine the kinematic parameters of those.

In order to improve the accuracy of the CMMs, the model equation of the CMMs gets much more complicated. So it is proposed that the model equation, the kinematic

V. D. Majstorovic and N. Durakbasa (Eds.): IMEKOTC14 2019, LNME, pp. 155–161, 2019.
https://doi.org/10.1007/978-3-030-18177-2_15

parameters and the estimator of the kinematic parameters are replaced with the neural network estimator of kinematic parameters in this article.

2 Usual Calibration and Measurement of CMM

The calibration process of the CMMs is shown in Fig. 1(a). The actual CMM, e.g. Arm CMM, measures the coordinates of the feature of the artifact and the sensor information is detected by the actual CMM. The coordinates are temporally calculated using the model of the CMM and the estimated kinematic parameters.

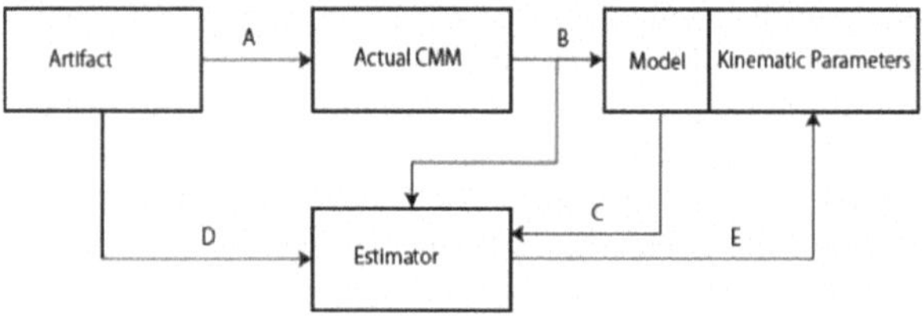

A: feature of artifact,
B: sensor information
C: calculated coordinates
D: calibrated coordinates
E: estimated kinematic parameters

(a) Diagram of estimation of kinematic parameters using artefact

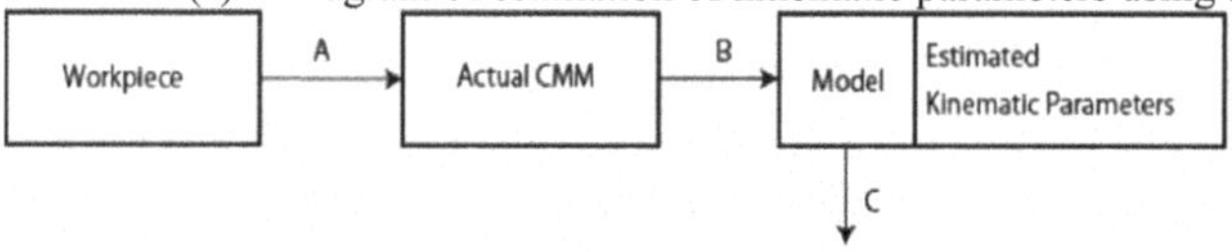

A: feature of workpiece
B: sensor information
C: calculated coordinates

(b) Diagram of measurement using estimated kinematic parameters

Fig. 1. Usual calibration and measurement

In the estimator, the temporally calculated coordinates are compared with the calibrated coordinates and the kinematic parameters are estimated as minimizing the difference between the calibrated coordinates and the calculated coordinates of the artifact. This estimated kinematic parameters are used with the model of the CMM as shown in Fig. 1(b).

The measurement process is shown in Fig. 1(b). The coordinate measuring machines are used to measure the coordinates of the workpiece. In measurement process, the actual CMM measures the coordinates of the feature of the workpiece and the sensor information is detected by the actual CMM. The coordinates are calculated using the model of the CMM and the estimated kinematic parameters.

It is necessary that the model and the kinematic parameters are clearly separated and defined in the measurement system based on the usual calibration using the artifact as shown in Fig. 1.

3 Articulated Arm Coordinate Measuring Machine

The arm CMM as shown in Fig. 2 has 7 rotational joints. The relationship between joint i + 1 and joint i is described by DH notation as shown in Fig. 3. The stylus coordinates are expressed and calculated based on DH notation. The model equation of the arm CMM with 7 rotational joints is expressed as shown in Eqs. (1) and (2).

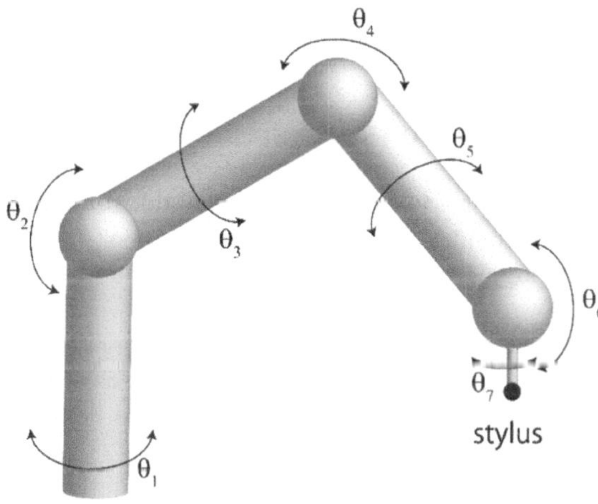

Fig. 2. Articulated arm coordinate measuring machine

$$ {}^{i-1}A_i = \begin{pmatrix} \cos\theta_i & -\sin\theta_i\cos\alpha_i & \sin\theta_i\sin\alpha_i & a_i\cos\theta_i \\ \sin\theta_i & \cos\theta_i\cos\alpha_i & -\cos\theta_i\sin\alpha_i & a_i\sin\theta_i \\ 0 & \sin\alpha_i & \cos\alpha_i & d_i \\ 0 & 0 & 0 & 1 \end{pmatrix} \quad (1) $$

$$ T = {}^0A_1\,{}^1A_2\,{}^2A_3\,{}^3A_4\,{}^4A_5\,{}^5A_6\,{}^6A_7 \quad (2) $$

The kinematic parameters are described as (θ_i, d_i, a_i, α_i) in DH notation. It is assumed that the arms of the arm CMM is not deformed. In order to improve the accuracy of the arm CMM, the measuring posture of that is taken into consideration. The arm of the arm CMM is assumed as cantilever, and each arm has weight and has the deformation by the weight itself as shown in Fig. 4 [10]. As the effect of weight is varied according to the orientation of the arm, Eqs. (1) and (2) become varied and much complicated.

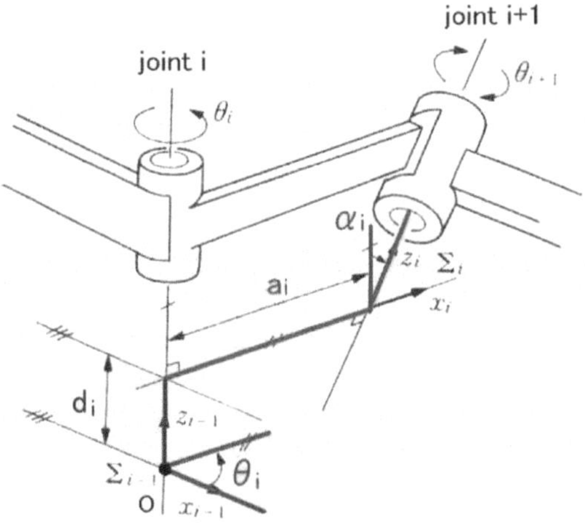

Fig. 3. DH notation

4 Adaptive Calibration and Measurement

As the behavior of the arm CMM is complicated, it is difficult to express the model of the CMM as the physical model and mathematical equation.

In this paper, it is proposed that the complicated equations are replaced with the trained neural network.

In the calibration, the model equations, the kinematic parameters and the estimators as shown in Fig. 1(a) are replaced with the neural network as shown in Fig. 5(a). The calibration method using the neural network is feasible and flexible.

So it is called the adaptive calibration.

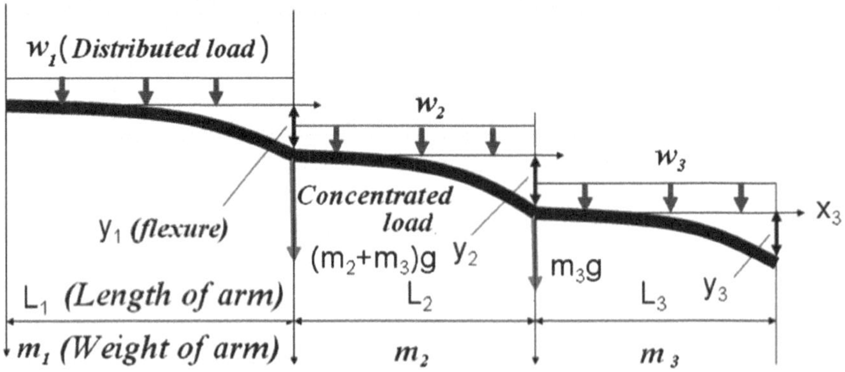

Fig. 4. Deformation model of the arm with weight

The neural network studies the relationship between the sensor information and the calibrated coordinates of the artifact in training.

In measurement, the model equation and the kinematic parameters as shown in Fig. 1(b) are also replace with the neural network as shown in Fig. 5(b). In Fig. 5, the neural network is trained only in the adaptive calibration and it is not trained in the measurement.

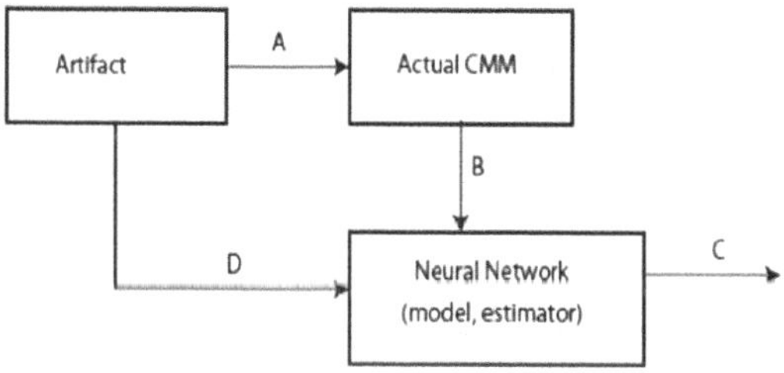

(a) Kinematic parameters are estimated adaptively using artifact

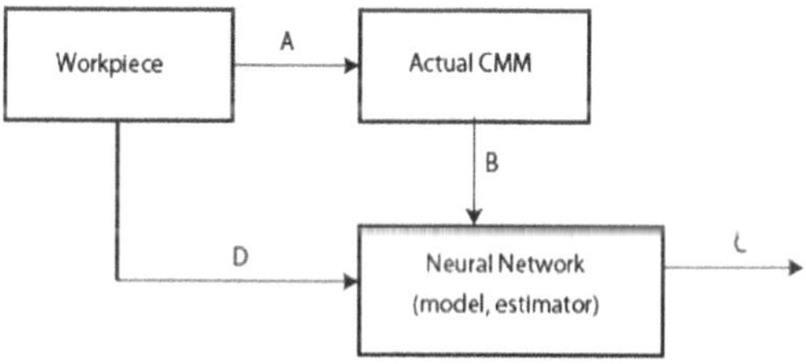

(b) Coordinates are calculated using adaptively estimated kinematic parameters.

Fig. 5. Adaptive calibration and measurement

5 Experiment

Figure 6 shows the artifact which has 8 poles with the reference points of which the coordinates are calibrated. The height of reference points are not identical, and all reference points are not on the identical plane. The reference points on poles are measured repeatedly by the arm CMM, and the coordinates of the reference points and all angles of joints are recorded.

In this case, the neural network consists of the input layer, output layer and 5 hidden layers. The input layer has 7 nodes corresponding to the angles of joints and the output layer has 3 nodes corresponding to the calibrated coordinates of the artifact. The hidden layers with 32, 64, 128, 64 and 32 nodes is tested and the other hidden layers with the other patterns of nodes are tested.

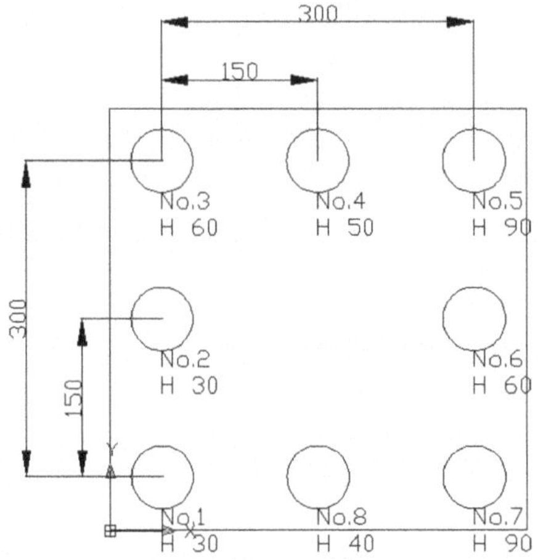

Fig. 6. Artifact has 8 poles with reference points

At first, the neural network get a good result on the angles for training, but get a bad result on the other angles.

After that, the dropout technique is introduced. Finally, the neural network get a good result less than 10 μm on the angles for training and the other angles.

6 Conclusion

In the case of the articulated arm coordinate measuring machine, it is proposed that the equation of physical model and the kinematic parameters are replaced with the neural network. The calibration process is replaced with training process of the neural network with dropout.

At this moment, the small number of experiments are performed. It is necessary to check the robustness of the neural network calibration for the different articulated arm coordinate measuring machines.

References

1. Schewenke, H., Knapp, W., Haitjema, H., Weckenmann, A., Schmitt, R., Delbressine, F.: Geometric error measurement and compensation of machines – an update. CIRP Ann. Manuf. Technol. **57**, 660–675 (2008)
2. Ohnishi, T., Takase, S., Takamasu, K.: Study of improvement methods of CMM (Coordinate measuring machine) in workshop environment – correction of squareness error. J. JSPE **73** (7), 818–822 (2007)
3. Agaplou, J.S., Du, H.: Assuring the day to –day accuracy of coordinate measuring machines – a comparison of tools and procedures. J. Manuf. Process **9**(2), 109–120 (2007)
4. Curran, E., Phelan, P.: Quick check error verification of coordinate measuring machines. J. Mater. Process. Technol. **5648**, 1207–1212 (2004)

5. Bringmann, B., Kung, A.: A measuring artefact for true 3D machine testing and calibration. CIRP Ann. Manuf. Technol. **54**(1), 471–474 (2005)
6. Asano, Y., Furutani, R., Ozaki, M.: Verification of interim check method of CMM. Int. J. Autom. Technol. **5**(2), 115–119 (2011). Au5-2-4609S
7. Santolaria, J., Yague, J.A., Jimenez, R., Aguilar, J.J.: Calibration-based thermal error model for articulated arm coordinate measuring machines. J. Int. Soc. Precis. Eng. Nanotechnol. **33**, 476–485 (2009)
8. Takamasu, K., Furutani, R., Shimojima, K., Ozono, S.: Artifact calibration of co-ordinate measuring machine. J. JSPE **69**(6), 851–855 (2003)
9. Shimojima, K., Furutani, R., Ozono, S., Hiraki, M., Araki, K.: The estimation method of uncertainty of coordinate measuring machine(1st Report). J. JSPE **70**(9), 841–845 (2004)
10. Hamana, H., Tominaga, M., Ozaki, M., Furutani, R.: Calibration of articulated arm coordinate measuring machine considering measuring posture. Int. J. Autom. Technol. **5**(2), 109–114 (2011). Au5-2-4610

Training in the Aeronautic Industry for Geometrical Quality Control and Large Scale Metrology

Luis Rocha[1(✉)], Paul Bills[2], Michael Marxer[3], and Enrico Savio[4]

[1] CATIM – Technological Centre for the Metal Working Industry, Porto, Portugal
luis.rocha@catim.pt

[2] Centre for Precision Technologies, University of Huddersfield, Huddersfield HD1 3DH, UK
p.j.bills@hud.ac.uk

[3] NTB Interstate University of Applied Science, Buchs, Switzerland
michael.marxer@ntb.ch

[4] DII – Dipartimento di Ingegneria Industriale, Universitàa di Padova, Padua, Italy
enrico.savio@unipd.it

Abstract. Advanced manufacturing of aircraft, and the large parts needed in this field, is based on multiple suppliers located in different countries and continents, intensively using automation, data exchange, advanced manufacturing technologies embedded in the digital era. Several national and international initiatives (e.g. Industrie 4.0) are oriented to support this effort and development. New opportunities for advanced manufacturing of aeronautic products are based on measuring technologies for Geometrical Quality Control and Large Scale Metrology. In the Aeronautic Industry they are an essential tool for the implementation of the initiatives in modern product engineering and process control. To operate, programme and manage the most advanced measuring systems, highly competent and skilled personnel is required.

The authors will describe new developments in the framework of the international project "WINGS+", addressing the training needs of the Aeronautic industry, and suggest innovative training solutions with focus on the competent use of relevant measuring systems.

Keywords: Large Scale Metrology · Training · Aeronautic industry

1 Introduction

Globalization has moved to low-wage countries in other continents many labour-intensive manufacturing activities, therefore the current focus in the EU is on the creation of high-added value products. The Aeronautic industry is one important example of high-value production and the needs for high quality training is a must for the aeronautic industry sector and sub-sectors [1]. The Industrial Advisory Group of the Factories of the Future PPP (public-private partnership) and EFFRA (European

V. D. Majstorovic and N. Durakbasa (Eds.): IMEKOTC14 2019, LNME, pp. 162–169, 2019.
https://doi.org/10.1007/978-3-030-18177-2_16

Factories of the Future Research Association) have identified key technologies and enablers for this required transition [2]. These include innovative manufacturing technologies (e.g. 3D printing of complex shaped parts), manufacturing with new and advanced materials (e.g. composite and lightweight materials) and the use of measuring technologies to support manufacturing, including assembly operations [3]. A key factor for competitiveness and quality assurance is the competent use of innovative equipment for the 3D measurement and digitisation of large parts, as needed for advanced product/process engineering and quality control. It is worth noticing that almost 1% of EU GDP is spent each year in this sector (metrology) by a diversity of organizations in society, industry and official organizations [4].

2 Aeronautic Industry, the Demands in Coordinate Metrology Training

Coordinate Metrology is a topic very relevant to a number of sectors of industry throughout the EU, especially in several industrial sub-sectors, the aeronautical industry being one of them. At present, within the European Union training market there is still a dearth of providers of manufacturer-independent vocational education as well as training relating to newly available measuring technologies and related standards.

There are current initiatives in coordinate metrology training which have developed and co-existed over a period of years. Two such initiatives are AUKOM [5] which was begun on a German national project in 1998–2001 and is largely based around single manufacturer systems, and the CM Train (Coordinate Metrology Training) association [6], in this case a result from an EU-funded project EUKOM (European Training for Coordinate Metrology) [7] with the main objective to harmonise the coordinate measuring machines (CMM) training across Europe. However, none of the initiatives include currently the measuring technologies most relevant for the Aeronautic industry, such as articulated arm CMMs and laser trackers.

New valued learning opportunities for lifelong education on innovative measuring technologies (including: articulated arms, laser tracking, fringe and projection systems) as well as enriched, updated content from already existing training activities is needed in a number of languages.

The analysis of student feedback to previous courses ran by CM Train, over the last ten years, covering around 900 participants from industry and universities, show a number of trends, specifically requesting changes to learning material including:

- More interactivity in learning material using media such as videos, animations, comprehension quizzes, etc.
- Fast update of learning material with newest technology
- Access to the learning product through different platforms (mobile devices)
- Having the possibility to choose a more tailored and efficient learning path (taking pre-knowledge into account)

3 Project Public Target

The main target group for the new EU-funded project as outlined in this paper are industry employees, with a special focus in aeronautic manufacturing, in particular those working in small and medium enterprises and those that are newcomers to 3D measuring technologies (including articulated arm CMMs and laser trackers) in the supply chain of the aeronautic industry. Other target groups that will benefit from the project are students and teachers at VET institutions in mechanical and mechatronics subjects having limited access to advanced measuring equipment, as happens in most practical cases due to lack of funds and/or unavailability of adequate teaching staff.

Practical work using measuring equipment is vital for successful VET in Geometrical Quality Control. This needs state-of-the-art infrastructure and well trained tutors with daily experience, but of course both resources are expensive. Blended learning can efficiently support VET e.g. in Coordinate Metrology (a subset of technologies for Geometrical Quality Control) as demonstrated by the results of the EUKOM project [7] and has been shown to be very effective for learning the latest innovations in the field. Follow-up experience also revealed a strong demand of potential learners to use new technologies for the distribution of learning content for an enriched, motivating and high-availability learning experience.

4 Learning Material Structuring

In a way to take into account the pre-knowledge of each individual learner, the structure of the learning modules follows a two-level approach. On the basis of this two-level approach further expert modules are developed to meet the specific requirements of potential users. In these two levels, the learning content is spread across ten sections (see Table 1).

These sections are structured in a way that the natural workflow of metrologists is taken into account [8], and this is consistent with the integrated framework of reference that has been proposed for the qualification of personnel on different types of

Table 1. Ten-sections structure

Section	Section title
1	Identification of measurement requirements
2	Inspection planning
3	Equipment selection
4	Workpiece preparation
5	Measuring system preparation
6	Measuring process execution
7	Evaluation strategy definition
8	Measurement uncertainty
9	Documentation
10	Infrastructure and environment

Coordinate Measuring Systems [9]. Each of these ten sections contains several chapters with a number of learning units that deliver the content to the learner in logical fractions.

4.1 Section 1: Identification of Measurement Requirements

This section serves as the interface to the product specification. With information out of the technical drawing and other information sources, the requirements for planning and performing the measurement tasks are gained. The learning content covers the identification of measurement requirements and other important fundamental aspects required at the measurement planning stage.

4.2 Section 2: Inspection Planning

Based on the measurement requirements, the points for the measurement strategy are discussed. These steps cover points as e.g. datums and datum systems, coordinate systems, probing strategy as well as the influence of modifiers found in the specification such as e.g. indications for filter settings.

4.3 Section 3: Equipment Selection

The selection of appropriate equipment is detailed. It covers aspects such as measuring system architecture and measuring system components to be able to choose suitable equipment for performing the measuring procedure. Indicators for performance specification of measuring systems and their application is also included here.

4.4 Section 4: Workpiece Preparation

Based on the findings in the measurement requirements, important points for workpiece preparation are pointed out in this section.

In this part, fixturing requirements and constraints are demonstrated, in addition details of special procedures for surface treatment are covered that may be needed in situations such as when using certain optical sensors for the measurement of glossy workpieces.

4.5 Section 5: Measuring System Preparation

Section 5 details the correct procedure for the preparation of measuring systems covering qualification procedures with a particular emphasis on verification. Different probing systems and their needs and procedures together with suitable measurement artefacts are discussed in this part.

4.6 Section 6: Measuring Process Execution

Section 6 covers the important points for measuring process execution such as acquisition of measuring points and measuring conditions based on the measuring requirements.

4.7 Section 7: Evaluation Strategy Definition

In this part, the association criteria, data processing as well as data analyses and evaluation and data representation are covered.

Beside these topics, this section covers reverse engineering needs and possibilities in particular.

4.8 Section 8: Measurement Uncertainty

The main focus of section 8 is the application of decision rules. These are the rules for verifying conformity or nonconformity with a given tolerance, for a characteristic of a workpiece. It is further explained that metrologists have to be aware of the role of measurement uncertainty and its estimation.

As a result in this section, decision rules, questions about tolerance and uncertainty, error sources as well as statistical parameters are covered.

4.9 Section 9: Documentation

Different forms of measurement result documentation with respect to the target group are covered. The required content of a measurement report is explained as well as the way, how to generate them are discussed.

4.10 Section 10: Infrastructure and Environment

The last section is oriented to the need for infrastructure maintenance as well as verification and performance checking and its procedures are explained.

5 Learning Material Production

Since the development of the learning content is made by several international partners, new forms of collaboration, sharing of documents and ideas, were used in an efficient exchange process, allowing a structured distribution of learning modules across chapters, sections and levels. The solution came from dividing the large amount of learning material into small parts by defining learning units, starting in English, and after the review process, create the other several language versions (see Fig. 1).

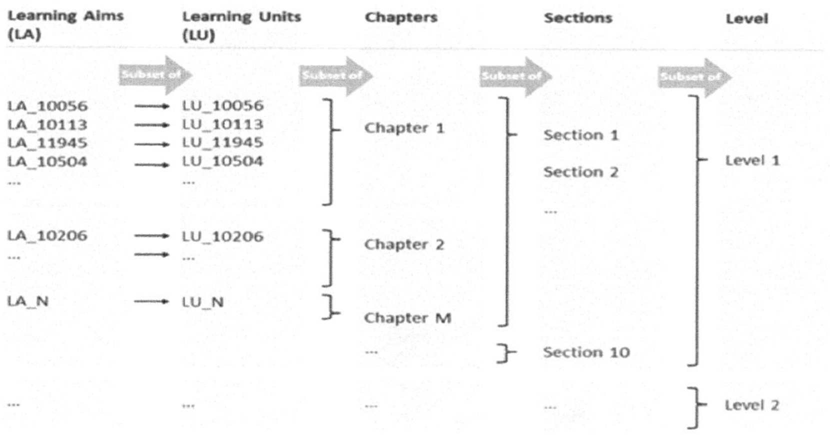

Fig. 1. Schema for the structure of learning contents, their relation between learning units, chapters, sections and levels [8]

The learning modules were developed using an open source software tool (eXeLearning) [10], defining first a technical template that was then utilised by the different partners, following the same design rules. For each learning module, the didactical template for further production was defined to have a logic path for the learner and for the tutor (see Figs. 2 and 3).

Learning content structure	Short objective
Learning aims	Inform learner which are the aims for the present learning module
Introduction	Gives an overview about the relation to other learning units. Also offer motivation to the learner
Explanation	The learning content (knowledge) is delivered to the learner.
Examples	Practical examples is show to demonstrate the application of the learning content
Exercises	Some examples illustrate the delivered learning content and let the learner consolidate the theory.
Comprehension questions	Questions to the learner for validating the learning process with Immediate feedback
Workshop	Some ideas and approaches to take on future workshops (visible for tutor)
Examination	Examination questions for integrate in the Learning management System to use in future exams (visible for tutors)

Fig. 2. Schema for the structure of learning contents, their relation between learning units, chapters, sections and levels [8]

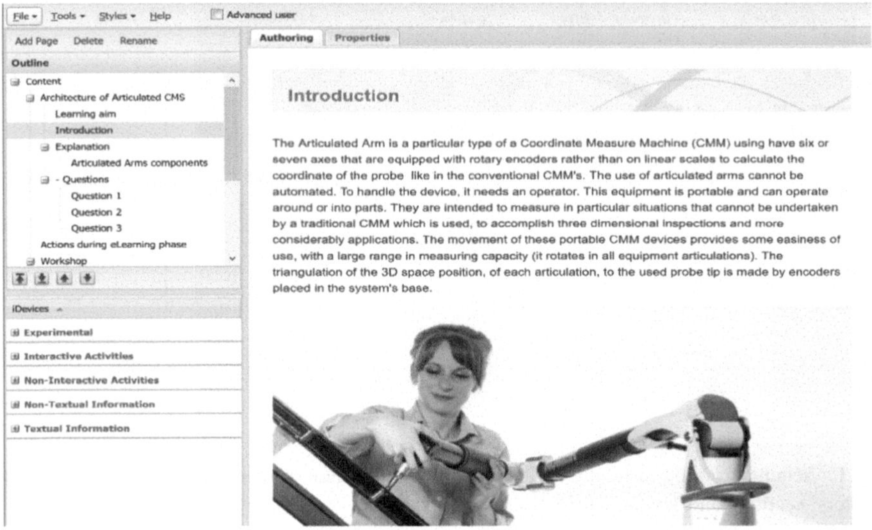

Fig. 3. Editor screen on the eXeLearning tool, for learning content module production

6 Conclusions

The chosen approach to develop learning material oriented to the learning aims formulations, guide us to an effective procedure that supported the flow of information of learning contents. Based on the common understanding of learning aims, it was possible to install the development of learning material in a parallel procedure that enable higher development speed and quality than in conventional procedures used. The international collaboration in an expert group consisting of teachers, designers and metrologists for structuring and developing a learning system, in this case for coordinate metrology, showed that the use of shared online documents that allow real time cooperation in development, is a potential form to work together and share ideas.

Acknowledgements. The authors wish to acknowledge the support of the European Union in the Programme Erasmus + through the project WINGS+: Geometrical Quality Control and Large Scale Metrology in the Aeronautic Industry, agreement no 2016-1-PT01-KA202-022964. The authors also wish to acknowledge the support of the members of CM Train association [6] and gratefully appreciate the feedback of the manufacturing companies, and university students that have participated in the field tests, and training editions, for their important contributions.

The present paper reflects the perspectives, information and experience of the authors, and the Commission cannot be held responsible for any use which may be made of the information contained therein.

References

1. van der Heiden, P., Pohl, C., Mansor, S.B., van Genderen, J.: The role of education and training in absorptive capacity of international technology transfer in the aerospace sector. Prog. Aerosp. Sci. **76**, 42–54 (2015)
2. Factories of the Future 2020 Roadmap - Consultation document, Brussels (2012). ISBN 978-92-79-31238-0
3. Bauer, J., Bas, G., Durakbasa, N., Kopacek, P.: Development trends in automation and metrology. IFAC-PapersOnLine **48**(24), 168–172 (2015)
4. Kunzmann, H., et al.: Productive metrology—adding value to manufacture. Keynote paper. Ann. CIRP **54**(2), 155–168 (2005)
5. AUKOM homepage. http://www.aukom.info. Accessed 19 Feb 2019
6. CMTrain homepage. http://www.cmt-train.org. Accessed 20 Feb 2019
7. Keferstein, C.P., Marxer, M., Weckenmann, A., Beetz, S.: Ein europäisch abgestimmtes, auf eLearning basierendes, neues Ausbildungskonzept für die Koordinatenmesstechnik. Innovative Entwicklungen im Fokus des Anwenders. Tagung Braunschweig 15 und 16, pp. 109–118. VDI Verlag, Düsseldorf (VDI-Berichte 1914)
8. Marxer, M., Rocha, L., Anwer, N., Savio, E.: New development and distribution concepts for education in coordinate metrology. Proc. CIRP **75**, 320–324 (2018). https://doi.org/10.1016/j.procir.2018.04.063
9. Savio, E., Menoncin, M., Marxer, M., Anwer, N., Hausotte, T., Bills, P., Liam, B.: An integrated framework of reference for the qualification of personnel in coordinate metrology. In: 18th International Conference of EUSPEN. Bilbao, Spain, 3rd–7th June 2019 (2019)
10. eXelearning homepage. https://exelearning.org. Accessed 20 Feb 2019

New Improved Method of Setting the Jaw's Coordinate System

Srdjan Živković[1(✉)], Nemanja Majstorović[2], Branislav Glišić[3], and Davorin Kramar[4]

[1] Military Technical Institute, Coordinate Metrology Lab, Belgrade, Serbia
srdjan.vti@gmail.com
[2] Faculty of Dentistry, Clinic for Jaw Orthopedics, University of Belgrade, Belgrade, Serbia
[3] Faculty of Mechanical Engineering, Department for Production Engineering, University of Belgrade, Belgrade, Serbia
[4] Faculty of Mechanical Engineering, Quality Assurance Lab, University of Ljubljana, Ljubljana, Slovenia

Abstract. Precise and reproducible alignment of the coordinate system plays a key role in the processes of monitoring geometric parameters. In cases of long-term processes, which are executed in multiple phases, errors in setting the coordinate system can lead to wrong conclusions and mismanagement of these processes. Orthodontic therapy, ie teeth leveling, lasts for one year with controls every month. The geometric parameters (teeth positions) are monitored by the dental arch equations. During each control, the current state is compared with previous control. *Conditio sine qua non* is precisely and repeatably setting of the jaw's coordinate system. The ABO method (American Board of Orthodontics; digital model orientation) does not provide repeatability in successive controls. In this study, a new method of setting the coordinate system of the jaw is presented. To ensure as user-friendly procedure as possible a simple algorithm is listed. A case study on digital dental model analyses for dental arch curve mathematical definition is presented at the end of the study. This is an example of the application of engineering methods in non-engineering areas.

Keywords: Coordinate system · Digital dental model · Dental arch curve

1 Introduction

Digital *3D* models (i.e. digital models achieved through plaster casts scan) are increasingly represented in everyday orthodontic practices. Their numerous advantages make them find their use in every third dental office in developed countries [1, 2] as a new and advanced technology [3–5]. These models are used not only in diagnostics and orthodontic treatment stages planning, but also in following of the stages for assessment analysis of each of the tooth position, that is, their alignment during first 6-8 months of treatment. During this process, using and defining of *3D* coordinate system proves to be of a very important. From an orthodontic perspective in general and in relation to teeth alignment particularly, *3D* models are used for:

V. D. Majstorovic and N. Durakbasa (Eds.): IMEKOTC14 2019, LNME, pp. 170–184, 2019.
https://doi.org/10.1007/978-3-030-18177-2_17

- Following and analyzing of overcrowding/spacing in a patient's jaw, carrying out of the spatial analysis, application of Bolton's ratio,
- Dental arch form defining and modeling, PAR (Peer Assessment Rating) index defining,
- Following of teeth alignment to define validity, accuracy, reproducibility and reliability of measuring on *3D* digital models in comparison to conventional ones; and
- Analysis and synthesis of teeth motions during orthodontic treatment, particularly through the phase of their alignment when these motions may have both translational and rotational properties within the space.

The analysis leads us to conclude, *3D* coordinate system is extremely important in all areas of its application. Today, it is based on the ABO digital model requirements [6].

2 Digital Model Orientation and Coordinate System Definition

ABO (American Board of Orthodontics) digital model requirements represent a commonly accepted *de facto* world standard for digital models of jaws [6]. ABO defines requirements for digital models (scan resolution and accuracy, measuring units, file formats and mesh topology) and the procedure for adjusting the coordinate system. In accordance with the ABO guidelines, digital models are generated in PLY (Polygon File Format), STL (Stereo lithography) or OBJ (Object) file formats, from which further analyses are performed. In addition, the scanners used for generating 3D digital models need to have resolution of 0.10 mm or better, and scan accuracy must be at 0.20 mm or better [9].

The curve defines dental arch in normal occlusion, meaning it is in X-Y (occlusal) plane in orthodontic coordinate system of the jaw digital model, called Global Coordinate System of the Jaw (GCSJ) [7]. One of the basic characteristics of the procedure for determining GCSJ is that it provides repeatability of measuring (scanning) one jaw model on different scanners, which is extremely significant for 3D digital models.

ABO procedure for digital model orientation, in essence, is very similar to the procedures used by engineers in coordinate metrology. This is quite logical, because this is completely an engineering problem. The jaw coordinate system adjusting procedure is a typical "3–2–1" procedure. Numbers "3–2–1" in the procedure name define degrees of freedom that are fixed in each of three steps. To define the coordinate system exactly, it is necessary to fix all 6 degrees of freedom (3 rotations and 3 translations) [10].

ABO procedure has two major drawbacks:

1. It is not possible to provide a satisfactory repeatability of Y axis orientation in successive controls of a patient jaw.
2. The origin position is not precisely defined. During treatment incisors are the most shifted teeth; their intake for reference is completely wrong. ABO procedure uses the term "approximately" for origin position. "Approximately" it is not good enough for accurate measurement, especially in the case of dental arch creation and their comparison in successive controls.

The diagram in Fig. 1 clearly indicates the imprecision of the ABO definition [6] (quote): "*The world origin (0, 0, 0) is located on the mid-sagittal (Y-Z) and occlusal (X-Y) planes at a point that lies approximately half-way between the most anterior and most posterior teeth.*"

Table 1. Positions of the coordinate system origin (X0Y0Z0) defined according to the ABO requirements in successive controls in relation to the fixed point.

Controls	X0 [ABO]	Y0 [ABO]	Z0 [ABO]
Before therapy	0,8286	21,5629	0,4045
1	−0,0358	20,9945	0,1270
2	−0,3614	20,2563	−0,2387
3	0,8734	19,8558	0,2720
4	−0,5786	19,7980	−0,4000
5	−0,5958	19,4712	−0,2746
6	0,3788	19,2639	0,4682
7	−0,4993	19,4481	−0,1644
8	−0,2119	19,2798	−0,2621
9	−0,3348	18,8855	−0,4098
10	−0,4782	18,9921	−0,1198

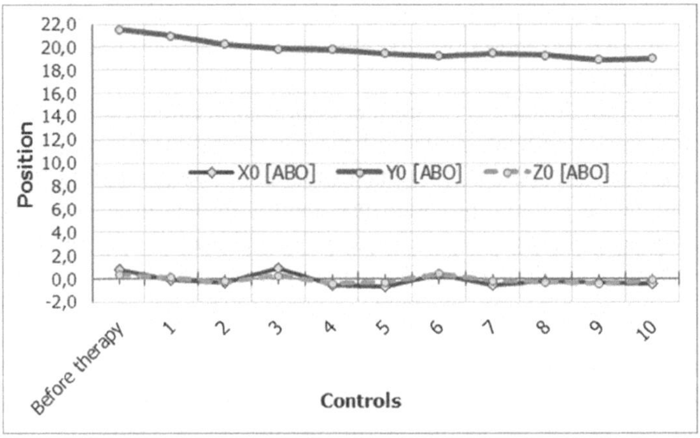

Fig. 1. Plotted results from Table 1.

If ABO definition is followed, the coordinate system moves from the initial state (before the start of therapy) in each of the successive controls (a total of 10 controls). How is it possible to compare the dental arc equations in all stages of therapy? Of course it is not possible! It is necessary the coordinate system of the jaw to be always in a fixed place. It is obvious, dentists do not understand the principles [10] of coordinate metrology.

Additionally, ABO instruction does not explain how to find maximum point on the molars (molars cusp). At the beginning, Z-axis direction is not known and very often dentist indicates these points imprecisely. Setting of XY plane is crucial for whole process. Errors that are made in the beginning may not be corrected and became systematic. Because of that comparison of dental arch is not possible. Incorrect assumptions lead to wrong conclusions.

This paper explains the improvements of ABO requirements and elimination of ABO jaw coordinate system settings procedure weakness. Described method [8] opens new possibilities of digital dental modeling in orthodontic treatment and eliminates a subjective factor of the dentist.

3 Precise Alignment of Jaw Coordinate System: Proposed Method

Jaws plaster models are initially scanned in an arbitrary coordinate system, Fig. 2. Reference vector that determine the maximum points on the tips of molars is unknown. First, it is necessary to determine reference vector. After defining vector, it is possible to determine the maximum points on the molars and premolars. Occlusal (horizontal) plane will coincide with the world *XY* plane; following steps (1 → 5) precisely define XY plane.

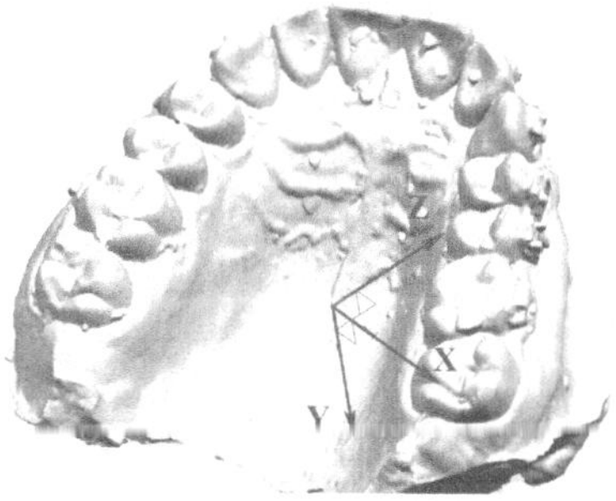

Fig. 2. Scanning coordinate system; arbitrary orientation and position

1st Step: Mark (*Paint Faced Body*) the area where they expect maximum points on the molars and premolars: eight zones on the left and the eight zone on the right side of the jaw, Fig. 3.

This is the only step that is highly dependent on the dentist. In the term of necessary knowledge, this step does not require any prior knowledge of mathematics and IT. This step only requires visual recognition molars and premolars peaks.

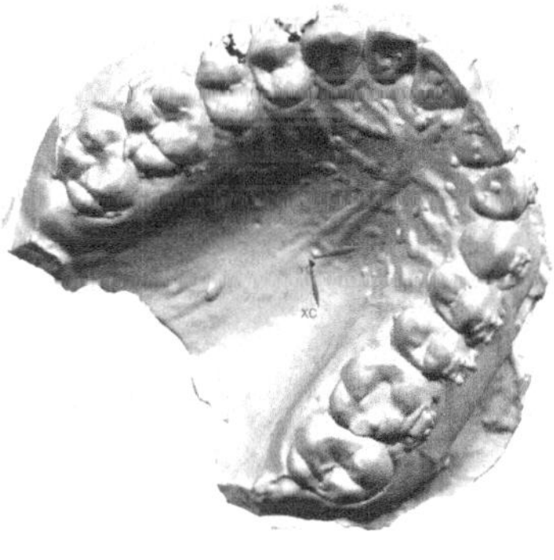

Fig. 3. Marked molars and premolars

2nd Step: Fitting a spheres thru sixteen marked zones (Selecting all of the same color). For each of the marked zone computer system give to user: radius, maximum and average error of fitted spheres. Automatic points rejection is enabled if the maximum errors are greater than scanning accuracy, defined by ABO requirements (0.2 mm).

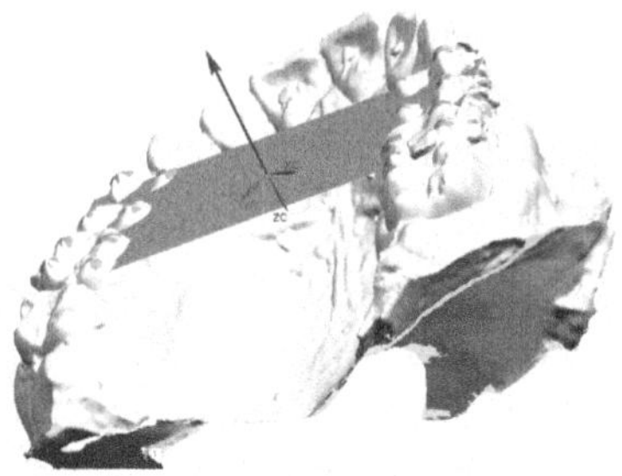

Fig. 4. Fitted auxiliary plane

3rd Step: Fitting a plane, using least squares method, thru centers of sixteen spheres, Fig. 3. This is an auxiliary plane, which defines the user main direction (vector) for molar and premolar cusp. Main direction is marked with blue arrow on the Fig. 4.

4th Step: Projection of 16 points in the spheres centers on faced body (jaw digital model). Figure 5 shows these points marks using small red crosses. These 16 points are, for sure, molars and premolars cups tips.

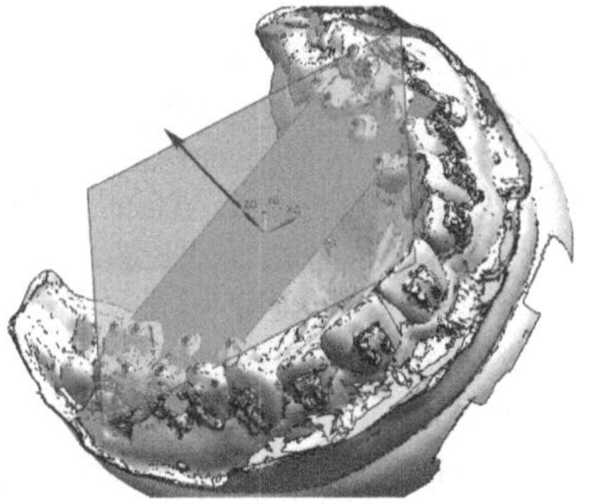

Fig. 5. Fitted XY (occlusal) plane

5th Step: Fitting a plane, using least square method, thru 16 points generated in 4th step. This is a XY (occlusal) plane of the jaw, determinate precisely. Figure 5 shows XY plane painted in red; light blue plane is auxiliary plane, created in 3rd step. A plane containing the mid-palatal raphe will be considered as the mid-sagittal plane and will coincide with the jaw *Y-Z* plane; following steps (6 → 8) precisely define *YZ* plane.

6th Step: Marking 4 points on the left side and 4 points on the right side of the jaw. Figure 6 shows these 8 points: P_{L1}, P_{L2}, P_{L3} and P_{L4} on the left side; P_{R1}, P_{R2}, P_{R3} and P_{R4} on the right side. ABO procedure use only 4 points; two on the left (P_{L1} & P_{L4}) and two on the right (P_{R1} & P_{R4}). Using 8 points, instead 4 points, will ensure more precise determination of Y-axis direction.

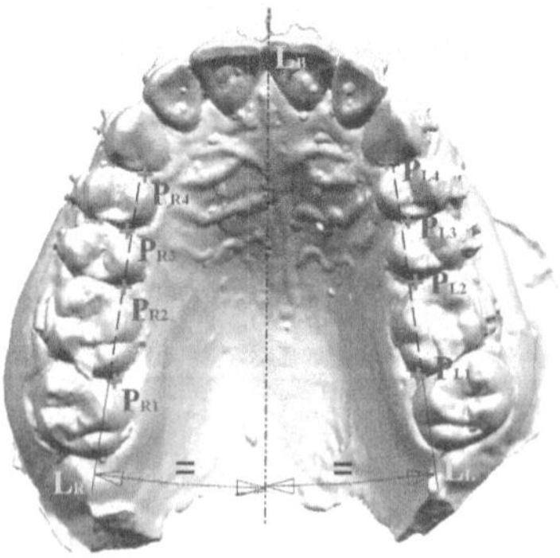

Fig. 6. Y axis orientation (mid-sagittal), bisector of two auxiliary lines

7th Step: Two auxiliary lines are created; on the left side line L_L using least square method thru following points: P_{L1}, P_{L2}, P_{L3} & P_{L4}; on the right side line L_R using least square method thru following points: P_{R1}, P_{R2}, P_{R3} & P_{R4}. Lines L_L and L_R, are shown on Fig. 6.

8th Step: Direction of Y-axis is defined as bisector of lines L_L and L_R, created in 7th step. Bisector line L_B is shown on Fig. 6. *YZ* plane is perpendicular to *XY* plane, created in steps 1 → 5.

Origin *(0, 0, 0)* is defined in the 9th step; this is the final step.

9th Step: The origin is determined by the position of the molars, which cannot be displaced (or very minimally). The incisors are the most displaced teeth during orthodontic treatment and should not be taken as a reference for coordinate system origin. First, it necessary to create two auxiliary points on the seventh tooth (second molar) on the left and on the right side. Points P_{R7} & P_{L7}, shown on Fig. 7, are created on the same manner as explained in steps 1 → 5.

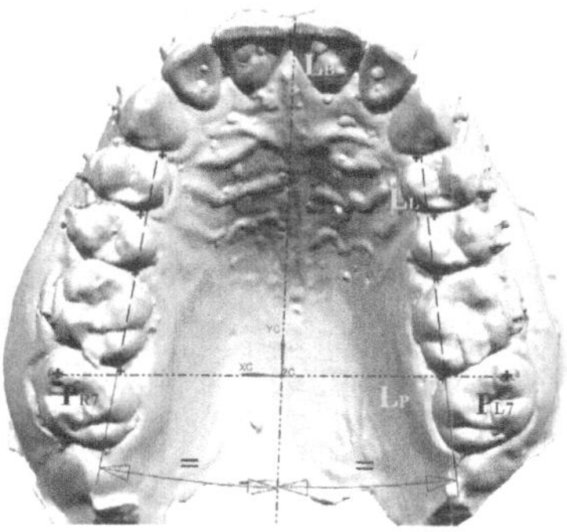

Fig. 7. Origin (0, 0, 0) definition

Points P_{R7} & P_{L7} define the line L_P. Intersection of two lines: L_B (bisector) and L_P (defined by points P_{R7} & P_{L7}) precisely define the digital model coordinate origin *(0, 0, 0)* of the jaw. Origin *(0, 0, 0)* position is shown in Fig. 7. Due to imperfection, X axis does not pass through points P_{R7} & P_{L7} as expected, Fig. 7. The human body is never perfectly symmetrical.

By applying the steps 1 → 9 coordinate system jaw is unambiguously determined. The coordinate system is very precisely defined, and easily reproducible. Implementation of procedure not requires additional knowledge of mathematics and computer science. Few simple steps are manual and requires careful work. All other steps can be automated, Fig. 8:

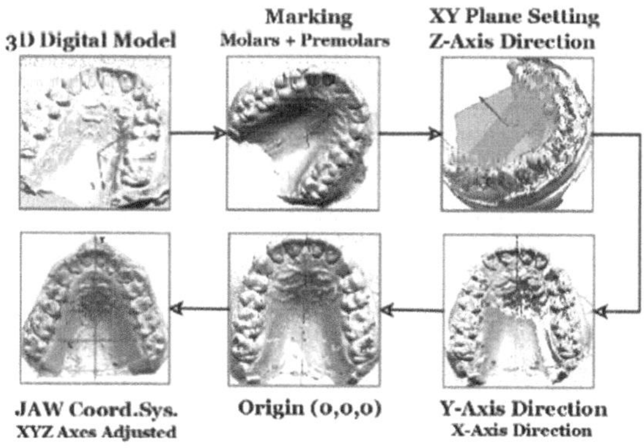

Fig. 8. Improved ABO method for setting the jaw's coordinate system

1. *XY* Plane: (manual) Paint Facet Body → (automated) Fitting Sphere → Sphere Center Points → Fitting Plane → Points Projection on Facet Body → Fitting Plane → *XY* Plane; Z = 0.
2. Y axis direction: (manual) 4 + 4 Points on Facet body → (automated) Fitting Lines 1 + 1 (Left + Right) → Bisector → Y axis direction; X=0.
3. Origin (*0, 0, 0*): (manual) Paint Molars 7th Left & Right Cusp → (automated) Fitting Sphere → Sphere Center Points → 2 Points Projected on XY Plane → Line → Intersection with Y axis → X = 0 & Y = 0. Coordinate system origin (0, 0, 0) is defined! X axis: Perpendicular to Y-axis.

ABO procedure [6] provides repeatability when only one unchanged jaw model is scanned, using different scanners. This is not the case during the monitoring of the entire orthodontic therapy. To make comparisons of two digital models of the jaw, it is essential that the coordinate systems are always in the same position and oriented in the same manner. The procedure defined by the ABO does not provide repeatability of the orientation of the coordinate system in successive controls during orthodontics treatment.

If we want to compare dental arch curves, as has been done in this paper, adjusting the coordinate system is the most important step in the entire procedure. The largest impact on the measurement error has adjustment of the coordinate system [10]. Mistakes made in the initial steps inevitably lead to wrong conclusions.

4 Determination of Dental Arch Curves

4.1 History of Dental Arch Equitation

The beginnings of defining the dental arch dated from the early 20th century. In his book [11] published 1902nd, G.V. Black described the upper teeth as being arranged in a semi-ellipse. Hawley 1905th proposed that the ideal dental arch was based on an equilateral triangle with a base representing the inter-condylar width. Edward, developing its own system 1907th, in a detailed discussion of the "line of occlusion" maintained that this line represented a parabolic curve [12]. Leon Williams [13] developed typical form theory, form of tooth depends of form of the bony face frame – the ovoid, square and tapering forms. William arrived at this classification after extensive anthropological study and was able to interest a manufacturer. This method is probably still the way in which most dentists select anterior artificial teeth. In contrast, in an investigation based on the skulls of apes and humans, Hellman (1919) was unable to find any relation between the size of the teeth and the form of the dental arches and he rejected the theories of arch predetermination based on measurements of specific teeth. Other investigators (Stanton 1922; Izard 1927) found a variability in arch form within a human sample, represented by an ellipse in 75%, a parabola in 20% and a square shape in 5% of the cases [14]. McConnail and Scher (1949) suggested that the ideal curve would correspond to a catenary curve that is formed when a fine chain is suspended at both ends. Scott (1957) as well as Pepe (1975) claimed that a catenary curve best represented a good average fit for arch form [15]. Lu (1966) claimed that the dental arch could be described by a polynomial equation of the 4th degree [16]. Brader

(1972) used mathematical model of arch form based on trifocal ellipse [17]. BeGole (1980) used the cubic spline function to model the form of the dental arches [18]. On the basis of digitized data, splines were fitted to the dental arches. Sampson (1981) described dental arch forms by means of an algorithm. The technique consisted of fitting arcs of conic sections to the data points representing the dental arches [19]. The conics were fit to the data using Sampson's refinement of Bookstein's algorithm. Braun [20] said that the human arch form could be portrayed by a complex mathematical formula, known as beta function. The model was defined by depth and width of the dental arch at second molar region.

4.2 Dental Arch for Orthodontic Treatment Analyses

Ideal occlusion is a hypothetical state, an ideal situation. McDonald and Ireland [21] defined ideal occlusions as a condition in which maxilla and mandible have their skeletal bases of correct size relative to one another, and the teeth are in correct relationship in the three spatial planes at rest.

There are a large number of studies considering appropriate mathematical equations to define the dental arch. Some of these studies have examined hundreds of patients. The shape of the dental arch depends on gender, age, ethnic origin [22], even geographic location [23]. However, several hundred patients are still too small sample for the population of a couple of million. It is very difficult and unreliable to try and draw general conclusions because the critics argue that "the sample is too small". Large sample studies cannot generate results regarding an individual.

A review of the literature shows, however, several assumptions are made: There must be an algebraic or geometric formula which explains ideal arch form; All ideal arches are the same shape and differ only in size; Every ideal arch is considered to be symmetrical. In existing literature extensive mathematical analysis of dental arch shape in persons with normal occlusion in the course of orthodontic treatment are very rare. None of the most cited relevant paper [16–20] did not explain adjustment of the jaw coordinate system in their analysis.

ABO recommendations for adjusting the coordinate system are relatively new, were published in 2013. Last release (June 2016) has been supplemented by rules for jaw 3D printing [12]. The chapter relates to the adjustment of the jaw coordinate system remained the same as in the first release.

The core of the problem is that during treatment the position of teeth in the jaw changes, which leads to a change in the position of coordinate system of the jaw, based on which the function of dental arch is modelled. The consequence of this is that the shapes of dental arch for two conditions – stages of treatment – can be compared with a relatively low precision. As a result, here we established a global coordinate system of the jaw (GCSJ) following the improved method of ABO. By eliminating this negative impact, so that dental arches can be absolutely accurately compared at each individual stage of orthodontic treatment.

This research represents a part of the overall model for monitoring and analysis of orthodontic treatment, from the aspect of teeth alignment, using 3D digital models [1]. The shape of the dental arch and deviation from ideal arch form is the most important information to the dentist during therapy using arch wire.

4.3 Dental Arch Curve Fitting

For fitting curves which define the dental arch we used CAD/CAM/CAE (Computer Aided Design, Computer Aided Manufacturing, Computer Aided Engineering) Siemens PLM NX12. NX12 is used for general purpose [24]; it is primarily intended for applications in mechanical engineering. Its flexibility and openness of the architecture allows it to be easily applied in very different areas. Its advantage over specialized software packages in the area of orthodontics is its dedication to sophisticated modules for complex spatial forms (free-form, sculptured surfaces). The fact is that modern medicine and dentistry cannot be imagined without the involvement of mechanical and electrical engineers. This research is an example of the interdisciplinary engineering modelling and its applications in orthodontics.

All splines created with Siemens NX12 are NURBS (Non Uniform Rational B-Splines) [24]. The literature providing a mathematical foundation is widespread and easily accessible. Splines are widely used, not only in various engineering areas. Their utilization in the CAD/CAM/CAE systems is simple and intuitive. There is no modern CAD/CAM/CAE system that does not include the ability to create splines. In this section, the terms "B-Spline" and "Spline" are used interchangeably. There are three creation methods for splines: thru points, by poles and fit.

Therefore, the aim of this paper is to define dental arch function as a polynomial, from 3rd to 8th degree, and to perform its thorough analysis, following all the stages of orthodontic treatment.

The curve generated by proposed mathematical functions is adjusted in accordance with marks on the teeth believed to precisely define a dental arch. With the aim of discovering the optimal mathematical function to describe dental arch curves, a number of authors have tested various mathematical functions (models) to best serve this purpose.

The main characteristic of examined studies is that dental arch shape and features, are examined within the context of functional characteristics of the jaw and the position of the teeth in it, which is the primary task of clinical orthodontics. In this study, splines were fitted through the points of digital models, which are located at the tips of the teeth (from the right to left side of jaw): 7–6_1–6_2–5–4–3–2–1–1–2–3–4–5–6_1–6_2–7; small blue crosses marked in Fig. 9. A total of 16 points are used for all the interpolations. As shown in Fig. 9, interpolated curve does not pass through the points on the digital model. Some points are closer to the curve, others are away from it. Siemens NX12 marks the most distant point from the fitted curve. In the example in Fig. 9, the point with maximum deviation is marked with a small red circle on Fig. 9. Siemens NX12 shows the value of the maximum deviation; this value is 1.656 mm on the second tooth on the left side. This is a clear indicator for the dentist which tooth to pay more attention to. In addition, Siemens NX12 gives the value of the average deviation of all points on the fitted curve. The user has information on the average deviation, maximum deviation value and the location of maximum deviation.

Another important aspect is the degree of the fitted spline. Every spline has a degree – a mathematical concept referring to the degree of the polynomial that defines the curve. A higher degree curve is stiffer in the sense that its poles have to be moved a long way to produce any appreciable change in the shape of the curve. Lower degree curves are more pliable, and tend to follow their poles much more closely.

There are a large number of studies considering appropriate mathematical equations to define the dental arch, as previously mentioned. It is difficult to generalize their findings or make final conclusions for lot a number of reasons: different objectives, different study samples with different criteria, and different methodology. In order to ideally describe the dental arch curve, a polynomial of a higher degree is to be used. This curve should possess significant flexibility, so it can be adjusted to suit any dental arch size, and has to include jaw asymmetries, if any. Starting from the fact that a dental arch is an imaginary curve, the descriptive information, i.e. function, represents a set of discrete points. For that reason, a mathematically obtained curve should be adjusted to different persons, individually.

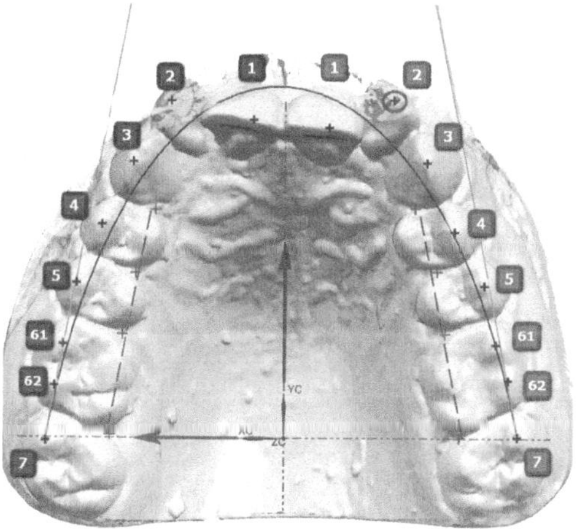

Fig. 9. Fitted dental arch curve (cubic spline)

Due to the previously stated reasons, the authors of this study have started from the premise that each individual is a subject unto itself. Personal orthodontics finds the appropriate equation of dental arch for each person separately. For some person 3rd degrees equations will be most appropriate, for other 6th degrees and so on. The shape of the jaw bone is the most appropriate template for the equation of the dental arch. Figure 10 adequately illustrates the problem of selecting the degree of fitted curve of dental arch. The figure shows the interpolated splines from 3rd to 8th degrees. It is hard to say which of the displayed curves is the most appropriate. Selection of the curve's degree is often very subjective. Is it possible to define an objective mathematical criterion for selection? Is there a mathematical criterion that orthodontic treatment conducted in the desired direction and to the desired objective? The authors have developed a method that gives a positive answer to the previous question.

Due to the computer speed and implemented algorithms, several spline fittings can be instantly executed. The user (dentist) interpolates splines of 3rd, 4th, 5th, 6th, 7th and 8th degrees in all stages of orthodontic treatment. This activity can be carried out automatically and does not require any additional knowledge or training. The convergence of the maximum and average deviation of the fitted splines are the criteria for selecting appropriate curve's degree. The curves whose degree is not suitable will not converge but diverge. A key requirement is the constancy the position of the coordinate system. If this condition is not fulfilled, the comparisons are not possible. As already pointed out, adjustment of the coordinate system is a critical step. Errors made in the initial step are leading to the wrong conclusions.

Zero or a very small value (smaller than the accuracy of the scanner [9]) of average and maximum deviation of the fitted spline for dental arch is a clear indicator that the treatment is reaching its completion.

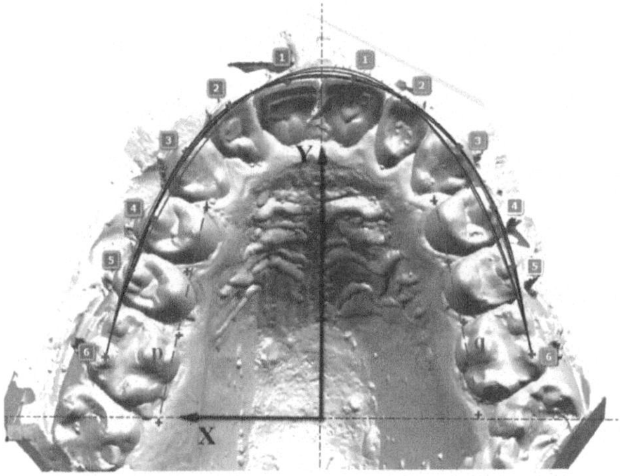

Fig. 10. Dental arch – fitted splines; 3rd, 4th, 5th, 6th, 7th and 8th degrees

4.4 Implementation of Developed Method

The developed method was applied on a patient undergoing orthodontic treatment. In each of the stages of the treatment an impression was taken on the basis of which a plaster model was created. All plaster models were scanned using optical scanners [9] in STL format (Stereo Lithography). The obtained digital models were transferred onto engineering software [24] where they served as a basis for the application of previously described method.

The initial state (without the fitted braces) was marked as "*Before therapy*". Each of the ten controls was performed in an interval of one month of wearing braces. Splines (3rd, 4th, 5th, 6th, 7th and 8th degrees) are fitted from the initial state in all 10 successive controls ($K_1, K_2, K_3, \ldots K_{10}$). All splines were fitted through 16 points, shown in Fig. 10. Points for interpolation are marked with red crosses in Fig. 3. Each interpolated spline at its beginning has a slope of line "L_R" and at its end has a slope of line "L_L", shown in Fig. 5. For each fitted spline CAD/CAM gives to the user the following information set: value of Average Fitting Error, value of Maximum Fitting Error and indicates with asterisk which points have the maximum value of fitting error. The values obtained for all 11 digital models are presented in Table 2. Column "*Deg.*" indicates the degrees of fitted splines. Column "*E*" is divided into two rows marked "*M*" and "*A*" for each degree of interpolated splines. "*M*" stands for Maximum Fitting Error and "*A*" stands for Average Fitting Error. Columns "Before Therapy, K_1, K_2, K_3, K_4, K_5, K_6, K_7, K_8, K_9, K_{10}" indicate all of the stages of the orthodontics treatment.

Table 2. Values of maximum and average splines fitting errors

Deg.	E.	Befo.	K_1	K_2	K_3	K_4	K_5	K_6	K_7	K_8	K_9	K_{10}
3	*M*	2.642	2.078	1.727	0.820	0.927	1.679	1.188	1.862	2.529	2.175	1.895
	A	1.244	0.882	0.934	0.285	0.372	0.547	0.446	0.910	1.076	1.199	0.891
4	*M*	2.016	1.879	1.831	0.681	0.935	1.232	1.508	2.205	2.475	2.486	2.186
	A	0.976	0.828	0.721	0.322	0.436	0.612	0.673	1.075	1.236	1.314	1.099
5	*M*	1.717	1.627	1.337	0.830	0.707	0.397	0.692	0.613	0.804	0.495	0.271
	A	0.811	0.726	0.737	0.253	0.209	0.199	0.243	0.185	0.227	0.242	0.106
6	*M*	1.615	1.446	0.858	0.566	0.376	0.444	0.404	0.312	0.531	0.435	0.247
	A	0.702	0.623	0.346	0.231	0.132	0.207	0.199	0.104	0.168	0.202	0.097
7	*M*	1.510	1.308	0.627	0.364	0.173	0.262	0.252	0.226	0.538	0.406	0.219
	A	0.651	0.622	0.260	0.099	0.074	0.112	0.096	0.086	0.172	0.195	0.091
8	*M*	1.373	1.225	0.483	0.360	0.079	0.253	0.208	0.255	0.544	0.436	0.206
	A	0.711	0.630	0.213	0.103	0.042	0.115	0.091	0.077	0.180	0.180	0.099

Maximum and average deviations values from Table 2 are shown in the diagrams, Fig. 11. The horizontal axis represents the stage of orthodontic treatment. Vertical axis shows the deviation in mm. Each degree of fitted splines is displayed in a different color and marked with different graphic symbol.

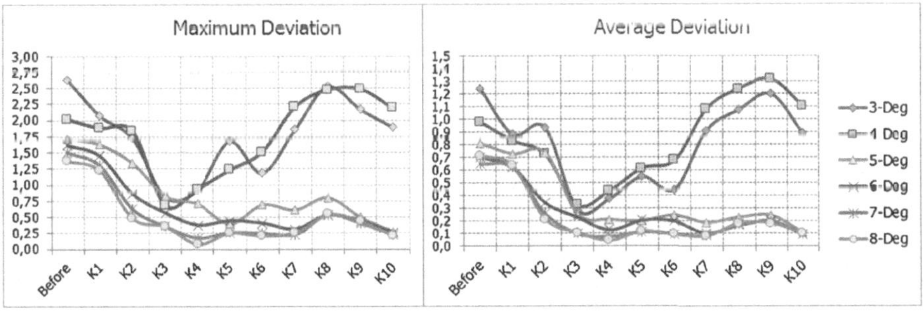

Fig. 11. Dental arch curves: maximum & average fitting error deviation diagram

What is noticeable is that the values of the maximum and average deviation of fitted splines of 3rd and 4th degrees diverge. A clear conclusion is that the third and fourth degrees of curves are not appropriate for this patient. Higher degree splines (6th, 7th, 8th) define a good dental arch of the patient. Table 2 shows that the best result has the 7th degrees spline (10th control – K_{10}): maximum deviation is 0.219 mm at the left incisor; average deviation is 0.091 mm. The obtained value of the average deviation is smaller than the accuracy of the used scanner.

5 Conclusion

This study describes a newly proposed method for determination of relative coordinate system in 3D digital dental model. The method is based on [6], but allows positional repeatability and comparison of different dental models of successive treatment with increased accuracy and repeatability of the measurement. The deviations were calculated with the comparison of different dental 3D scanned models of successive treatment.

In proposed method the precise alignment of jaw coordinate system consists of 9 steps. As the result, the coordinate reference point and orientation with 3 perpendicular orthodontic planes of *3D* digital dental model is defined. In this way, the coordinate system is very precisely defined, and easily reproducible. The implementation of procedure does not require additional knowledge of mathematics and computer science. To ensure as user-friendly procedure as possible a simple algorithm is listed.

Furthermore, based on developed method, analysis of dental arch curve is given. The method for dental arch curve construction and mathematical equation definition is proposed. The curve is defined with a NURBS, from 3rd to 8th degree, used for the analysis of all stages of successive orthodontic treatment. Authors claim that the appropriate equation of dental arch for each person individually should be found by personal orthodontics. The convergence of the maximum and average deviation of the fitted splines are the criteria for selecting appropriate curve's degree.

The proposed method has been tested in a case study of a patient undergoing orthodontic treatment. In each stage of the treatment an impression was taken on the basis of which a plaster model was created. All plaster models were scanned using optical scanners. The obtained digital models where used as a basis for the application of previously described method. From the results of the case study is noticeable, that the values of the maximum and average deviation of fitted splines of 3rd and 4th degrees diverge, so they are not appropriate for this case of patient. The higher degree splines (6th, 7th, 8th) define a good dental arch of the patient. The best fitting of the curve is obtained with the use of 7th degrees spline. The obtained value of the average deviation is smaller than the accuracy of the used scanner. Such a very small value of average and maximum deviation of the fitted spline for dental arch curve is also a clear indicator that the treatment is reaching its completion. As such, described method opens new possibilities of using digital dental modeling in orthodontic treatment and eliminates a subjective factor. Future research in this subject will be using on testing this approach to newly clinical study.

References

1. Kihara, T., Tanimoto, K.K., Michida, M., Yoshimi, Y., Nagasaki, T., Murayama, T., Tanne, K., Nikawae, H.: Construction of orthodontic setup models on a computer. Am. J. Orthod. Dentofac. Orthop. **141**, 806–813 (2012). https://doi.org/10.1016/j.ajodo.2011.10.027
2. Harrell, W.: 3D diagnosis and treatment planning in orthodontics. Semin. Orthod. **15**, 35–41 (2009). https://doi.org/10.1053/j.sodo.2008.09.004
3. Hernández-Soler, J., Enciso, R., Cisneros, G.: The virtual patient specific-model and the virtual dental model. Semin. Orthod. **17**, 46–48 (2011). https://doi.org/10.1053/j.sodo.2010.08.009
4. Bidra, A., Taylor, T., Agar, J.: Computer-aided technology for fabricating complete dentures – systematic review of historical background, current status, and future perspectives. J. Prosthet. Dent. **109**, 361–366 (2013). https://doi.org/10.1016/S0022-3913(13)60318-2
5. Hurt, A.: Digital technology in the orthodontic laboratory. Am. J. Orthod. Dentofac. Orthop. **141**, 245–257 (2012). https://doi.org/10.1016/j.ajodo.2011.06.045
6. Digital Model and 3D Printing Requirements. The American Board of Orthodontics (ABO), Updated 30 July 2016. https://www.americanboardortho.com/media/1157/abo-digital-model-requirements.pdf. Accessed 11 Nov 2018
7. Majstorović, N., Živković, S., Glišić, B.: The advanced model definition and analysis of orthodontic parameters on 3D digital models. J. Serb. Arch. Med. **1–2**, 49–57 (2017). https://doi.org/10.2298/SARH151207011M
8. Majstorović, N., Živković, S., Glišić, B.: Dental arch monitoring by splines fitting error during orthodontic treatment using 3D digital models. Vojn. Pregl. (2017). https://doi.org/10.2298/VSP161212067M
9. Majstorović, N., Čerče, L., Kramar, D., Soković, M., Glišić, B., Majstorović, V., Živković, S.: Examination of scanner precision by analysing orthodontic parameters. Balkan J. Dent. Med. **21**(1), 132–143 (2017). https://doi.org/10.1515/bjdm-2017-0005
10. Živković, S.: Coordinate Metrology in Manufacturing of the Complex Spatial Forms with Applications to the Aerodynamic Surfaces, vol. LI, no. 2. Military Technical Institute, Belgrade (2014). ISBN 978-86-81123-68-3
11. Black, G.V.: Descriptive Anatomy of the Human Teeth. The S. S. White Dental Manufacturing Co., Philadelphia (1902). Digitized: 2010, http://www.archive.org/details/descriptiveanato1902blac. https://ia600304.us.archive.org/6/items/descriptiveanato1902blac/descriptiveanato1902blac.pdf. Accessed 20 Jan 2017
12. Angle, E.H.: Treatment of Malocclusion of the Teeth: Angle's System, 7th ed. S. S. White dental manufacturing Co., Philadelphia (1907). MLA Citation. https://babel.hathitrust.org/cgi/pt?id=uc1.b5242461;view=1up;seq=7. Accessed 20 Jan 2017
13. Clapp, G.W.: James Leon Williams D.D.S., L.D.S., F.A.C.D., F.R.A.I. J. Dent. Res. **12**(6), 876–885 (1932). https://doi.org/10.1177/00220345320120060101
14. Rudge, S.J.: Dental arch analysis: arch form - a review of the literature. Eur. J. Orthod. **3**, 279–284 (1981). No: 0141-5387/81/00940279/S02.00
15. Pepe, S.H.: Polynomial and catenary curve fits to human dental arches. J. Dent. Res. **54**(6), 1124–1132 (1975). https://doi.org/10.1177/00220345750540060501
16. Lu, K.H.: Analysis of dental arc symmetry. J. Dent. Res. **43**, 780–790 (1964). http://ci.nii.ac.jp/naid/10016647144/en/. Accessed 11 Nov 2018
17. Brader, A.C.: Dental arch form related with intraoral forces: PR=C. Am. J. Orthod. Dentofac. Orthop. **61**(6), 541–561 (1972). https://doi.org/10.1016/0002-9416(72)90106-6
18. BeGole, E.A.: Application of the cubic spline function in the description of dental arch form. J. Dent. Res. **59**(9), 1549–1556 (1980). https://doi.org/10.1177/00220345800590092901

19. Sampson, P.D.: Dental arch shape: a statistical analysis using conic sections. Am. J. Orthod. **59**(5), 535–548 (1981). https://doi.org/10.1016/S0002-9416(81)90464-4
20. Braun, S., Hnat, W.P., Fender, D.E., Legan, H.L.: The form of the human dental arch. Angle Orthod. **68**(1), 29–36 (1998)
21. McDonald, F., Ireland, A.J.: Diagnosis of the Orthodontic Patient. Oxford University Press, New York (1998). ISBN 9780192628893
22. Williams, A.M.: Comparing occlusal arch form and basal bone arch form using CBCT in black, white and Mexican American mandibles. M.Sc thesis, Graduate Faculty of Saint Louis University, USA (2013). https://www.slu.edu/Documents/cade/thesis/Williams%20Thesis.pdf. Accessed 11 Nov 2018
23. Schaefer, K., Lauc, T., Mitteroecker, P., Gunz, P., Bookstein, F.L.: Dental arch asymmetry in an isolated adriatic community. Am. J. Phys. Anthropol. **129**, 132–142 (2006). https://doi.org/10.1002/ajpa.20224
24. Siemens PLM NX12 documentation. https://docs.plm.automation.siemens.com/tdoc/nx/12/nx_help/#uid:index. Accessed 11 Nov 2018

Determination of the Optimal Regression Model for the Measurement Quality Characteristics of the Micro Cutting Stone-Based Materials

Miloš Pjevic(✉), Slavenko Stojadinovic, Ljubodrag Tanović, Mihajlo Popović, Goran Mladenović, and Radovan Puzović

Faculty of Mechanical Engineering, Production Engineering Department, University of Belgrade, Kraljice Marije 16, 11120 Belgrade, Serbia
mpjevic@mas.bg.ac.rs

Abstract. Micro cutting represents one modern approach of processing the materials in order to achieve parts that feature high surface quality with low intensity of residual cracks. In the most cases, these parts are small sizes. Mechanisms that occur during micro cutting are not similar to those that occur during macro cutting. Even in the case of micro cutting brittle materials, the micro cutting mechanism becomes much more complex. The possibility of processing brittle materials in a ductile mode allows reduction of the intensity of the residual cracks within the material, since there is no initialization of the cracks within the material in this regime. However, although in this mode part is only plastically deformed, the intensity values of the components of the cutting force can vary considerably, especially if the processing is carried out on stone-based materials that are highly heterogeneous. In order to establish an adequate dependence of the cutting force components of the processing regimes, it is necessary to apply optimal regression model on the experimental results that will optimally cover all existing conditions, which is also the topic of this paper. The material over which micro-cutting experiments were carried out was marble *Plavi tok*. Experiments were carried out with two different tools whose tip radius value were R0.2 and R0.15 mm, and the value of the micro-cutting speed was v_s = 25 m/s.

Keywords: Micro cutting · Brittle · Ductile mode · Surface quality characteristics · Regression model

1 Introduction

Stone-based materials are widely used in construction industry, especially in decoration and for lining the aesthetic surfaces. Granite and marble are the most used materials, which are required with high quality surface finish. The measures and shapes accuracies are placed in the second plan.

It is required that stone-based materials are examined before they are recommended for any specific case of use, due to their unique properties.

V. D. Majstorovic and N. Durakbasa (Eds.): IMEKOTC14 2019, LNME, pp. 185–200, 2019.
https://doi.org/10.1007/978-3-030-18177-2_18

Marble is one of the most used stone-based materials with increasing usage in the industry. Under marble, in addition to the right marble and all other carbonate rocks, also atypical rocks can be included. These atypical rocks have ability to be easily and well polished. By its nature, the marble is a very brittle material with a heterogeneous structure. In dependence of the type of minerals that enter the composition of marble, its hardness can vary considerably. In previous studies, marble, like the most other brittle materials, can be cut in both the brittle fracturing mode and the ductile mode [1, 2]. However, until now, a small number of studies have been published on this topic.

The main goal of the micro cutting of materials, brittle by its nature, where are included stone-based materials, is to provide a smooth surface with no traces of material destruction. The grinding and polishing techniques are the most commonly used, considering the preservation of the structure and high shine finish surface. It is well known that the tool wear intensity, as a consequence of marble abrasiveness, is related to the friction and cutting speed and cutting forces [3].

The technological basis for the cutting stone-based brittle materials requires a thorough understanding the main mechanisms of micro cutting. So far, most of the mechanisms are followed by two approaches. The first approach reduces the interaction of the abrasive (diamond tool) and the workpiece to the idealized sequence of successive indentation. A large number of researchers in the field of forming a model of deformation and destruction of brittle material identified the work of the diamond grain in the grinding wheel by indentation, the so-called "Indentation fracture" approach [4–8]. When it comes to indentation, the research goes in two directions: according to the mode of action of force during indentation (static and dynamic) and according to the shape of the indenter (Brinell, Vickers, Knoop Rockwell), and analysis of the shape of cracks and destruction with the aim of finding critical load and penetration depth in which cracks and destruction of material occurs.

Another approach used by researchers is so-called "Machining approach" with a single diamond grain [9–19], as well as turning, milling and grinding operations [20–22]. This approach involves measuring the cutting force (normal and tangential component) and microscopic observation of the machined surface.

It is established that during the interaction between the diamond grain and the brittle material three zones are visualized: elastic, plastic and cutting zones. Cutting is achieved by plastic deformation as well as by brittle fracturing/destruction of the materials itself.

Both approaches provide a significant insight into the mechanisms of micro cutting. Specific energy is the fundamental parameter derived from the measurement of the cutting force. As its experimental value of force relatively small, it is necessary to use precise methods to present experimental data.

In common to all researches is that the destruction of the material during indentation is a result of the presence of cracks within material. This conditioned that the performed investigations deal with the finding of critical loads and the measured depth at the moment when initialization of the cracks appears. In support of this, it is necessary to develop an optimal regression model for experimental - analytical determination of critical loads as a dependent variables and depth as an independent variable. Thus, the moment when the initialization of the cracks appears, within the limits of the accuracy (error) of the optimal regression model, can be predicted. On this way, based on the critical load, the input parameters where the loads arise are determined.

Whether it is about macro or micro cutting, experimental data needs to be processed in a certain way. In order to achieve certain dependencies, regression analysis is one way of evaluating experimental data. Its application can be versatile. In some researches, the relations among brittleness concepts for rock cutting efficiency were established using regression analysis [23]. On the other hand, some researchers have conducted experimental investigations in order to determine nominal value of the cutting force using varies types of regression models such as polynomial [24]. Other types of the regression model can be also found among researches with purpose to determine nominal values of the cutting force, such as linear and exponential [25]. Until today, there have not been established optimal regression model for every type of data. Because of that, it is necessary to conduct detailed analysis to determine optimal regression model for specific case. This is particularly significant for determining the nominal value of the cutting force during micro cutting heterogeneous materials, such as marble.

In this paper, a nonlinear regression model based on two parameters is presented. As it is known for determining the dependence, the regression model gives the best results only when it is presented with the best curve, which follows the distribution of experimental data, and while the same curve (parameters of the curve) represents the least possible error. Therefore, the first parameter represents R^2 as the degree of goodness of the prediction of the curve of the dependence and the Normalized Mean Square Error (NMSE) as a standard error that takes into account the deviation of the data relates to the selected curve. The minimum value of this error is obtained by optimizing the equation parameters of the curve. In order to establish optimal regression model for micro cutting force, the experimental data had to be acquired by micro cutting experiment conducted on brittle and heterogeneous material, marble *Plavi tok*.

In this paper, mechanism of the micro cutting and the experimental setup were presented in order to planed experiments be successfully performed.

2 Material and Method

Transition from the macro to the micro cutting level includes, among influential factors, and value of the tool tip radius [26–29], as well as micro cutting speed [13]. The most important cause is that tool tip radius has a significantly lower value of the cutting depth. With this relation between cutting depth and the tool tip radius, the cutting edge can no longer be regarded as ideal sharp (Fig. 1). Such a change leads to the fact that the mechanism of chip formation (material removal) is diametrically different from the one set by Merchant [30, 31]. In other words, the chip formation process will not be based on the shear principle along the shear plane, but it can be divided into two groups. These groups are in the function of the material itself.

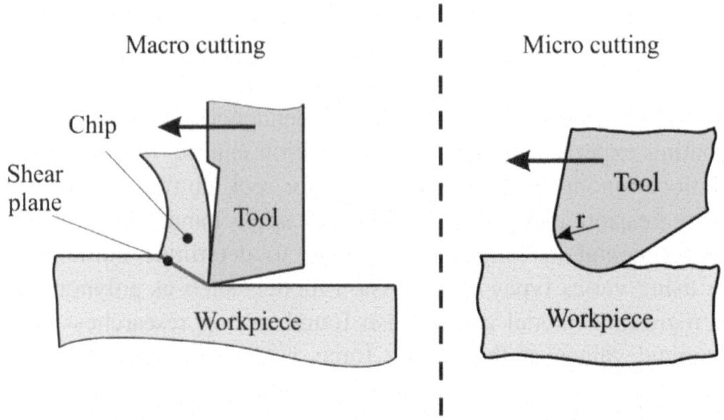

Fig. 1. Comparison between the tool/workpiece interaction in macro and micro cutting.

In the first case, which corresponds to the micro cutting of materials that have distinct plastic properties, such as aluminum, steel, etc., at cutting depth that do not exceed the critical value, the cutting process will only be based on reversible elastic and plastic deformations [32]. Crossing over the critical value, the formation of a chip in the form of extrusion of material occurs (Fig. 2a). This is different in compared to the process of the formation of the chip presented by Merchant.

The second case is referred to the micro cutting of brittle materials, where various types of ceramics, glass, as well as stone-based materials can be classified. During these materials micro cutting, separation of material will not be present at depths that are lower than critical value. In fact, the material will be elastically and plastically deformed.

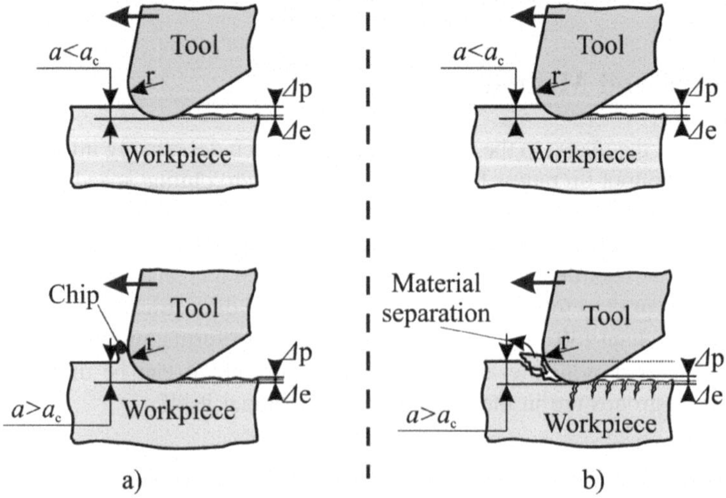

Fig. 2. Micro cutting mechanisms for (a) materials with plastic and (b) brittle properties.

This type of cutting occurs in the plastic deformation regime (ductile mode). After micro cutting depth is increased over the critical value, initialization and uncontrolled growth of the cracks within the material occurs, which leads to the separation of the material (brittle fracturing mode), Fig. 2b. Since stone-based materials belong to the group of materials with extremely heterogeneous structure [33], the appearance of ductile or brittle fracturing mode is in the function of the properties of the mineral that is currently in the contact with the tool. The variability of the minerals hardness themselves can affect the continuity of the cutting in the ductile mode, as well as the longevity of the tool itself.

The variability of the hardness of the mineral, not only affects the above-mentioned sustainability of the ductile mode, but also influences the intensity of the components of the force appearing in the cutting zone, which fall into the quality indicators of the micro cutting. The variability in the intensity of the components of the cutting force, in addition to the hardness of the minerals, is also in function of the crack intensity occurring in the brittle fracturing mode.

3 Experimental Setup

Micro cutting experiments were carried out at the ILR HMC-500 processing center. The tool used during the experiment consists of a rotating disc with defined diameter and a diamond tip of a precisely defined geometry attached to it (Fig. 3). The values of

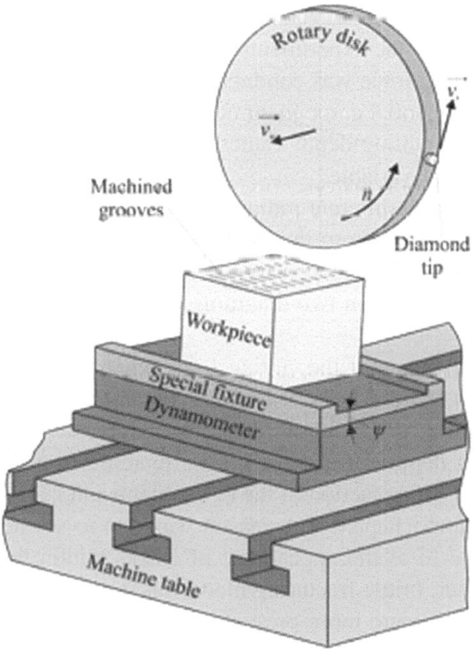

Fig. 3. Schematic representation of the experimental setup [14].

tool tip radius used during the experiments itself were R0.2 and R0.15 mm. The speed of the micro cutting was not varied, and it was $v_s = 25$ m/s. This was within the recommendation limits for cutting speeds that are used for stone-based materials, such as, marble *Plavi tok*.

The workpiece of marble *Plavi tok* was pre-cut on the 50 × 50 × 50 mm dimensions. All workpiece surfaces have been grinded and polished with a goal to eliminate as far as possible the irregularities in the form of cracks caused by the previous processing, or during the formation of the material itself. During the passage of the tool through the material, the formation of arched shape grooves was made, which contained a phase formed in the ductile mode, as well as formed in the brittle fracturing mode.

During the experiments, the measurement of the normal F_n and tangential F_t components of the cutting force was done using a dynamometer, on which a special fixture was placed. In order to ensure the variability of the maximum depth of the groove, the surface of the workpiece was locked at an angle ψ in relation to the direction of movement of the tool. This was performed with the help of a special fixture.

4 A Regression Analysis Model

As known, a regression analysis model represents sufficiently accurate correlation between the dependent and the independent variable of measurement quality characteristics (MQCs). Depending on the relationship of the variables of the MQCs, regression can be linear and nonlinear.

As already said, during the experiments, measurement of the normal and tangential components of the cutting force was conducted. Therefore, in this model of regression analysis, normal and tangential component of the cutting force are dependent variables of the MQCs, while the cutting depth values achieved by the tools of the radii R0.15 and R0.2 are independent variable.

Usage of two tools with different radius in the experiment generates two groups of data as independent variable. From the other side, two components of force are other two groups of data for the dependent variables. Thus, there are four types of data for regression, they will be shown in two diagrams, for the of tool radius R0.15 and R0.2 separately.

Values of the components of the cutting force when R0.15 mm tool was used are presented on the Fig. 4. As can be seen from the obtained experimental data, when the cutting depths were lower, dispersion of the components intensities was lower as well. Increase in the cutting depth correlates with the increase in a dispersion of the force components intensities. This reaction of the components intensity of the cutting force is explained in the previous chapter. The reason was due to variability of the hardness between minerals, as well as the appearance of cracks within the material. When the cutting depths are higher, brittle fracturing mode was dominant, in which initializations and growth of the cracks were more present.

The main goal of the conducted regression analysis was determination of the optimal regression model for the MQC. An optimal solution is provided with the help of a double analysis:

- Coefficient of determination (R^2),
- Normalized Mean Square Error (NMSQ).

The first parameter includes the degree of goodness of the selected regression model (curve), known as the goodness of fit. It shows how many points fall on the regression line. The R^2 value was calculated from the total sum of squares, more precisely; it is the sum of the squared deviations of the original data from the mean.

The second parameter covers the minimize NMSE (Normalized Mean Square Error) as an estimator of overall deviations between predicted and measured values. The main characteristic of this parameter is minimized of parameters of regression model (curve). NMSE generally showed the most striking differences among regression models. If a model has a very low NMSE, then it is well performing both of MQCs.

On the other hand, high NMSE values do not necessarily mean that a model is completely wrong. For this reason, the analysis was done according to the first parameters.

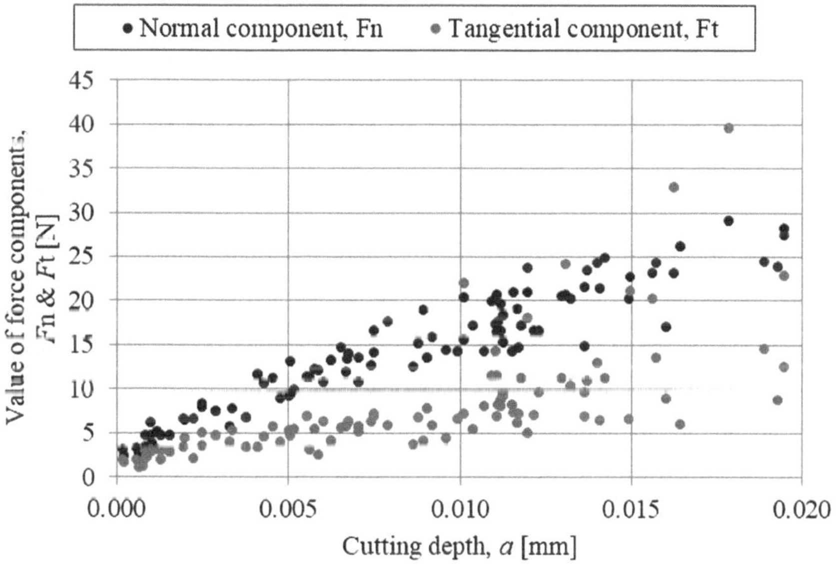

Fig. 4. Experimental values of the normal F_n and tangential F_t cutting force components during micro cutting marble *Plavi tok* with tool tip radius R0.15 mm.

An increase in the tool tip radius from R0.15 to R0.2 mm, led to increase of cracks growth within material, which further led to higher dispersion of the measured values of the components intensities of the cutting force (Fig. 5). Since the tool with tool tip radius R0.2 had much larger contact area than the tool with tip radius R0.15, increase in the intensities value were expected.

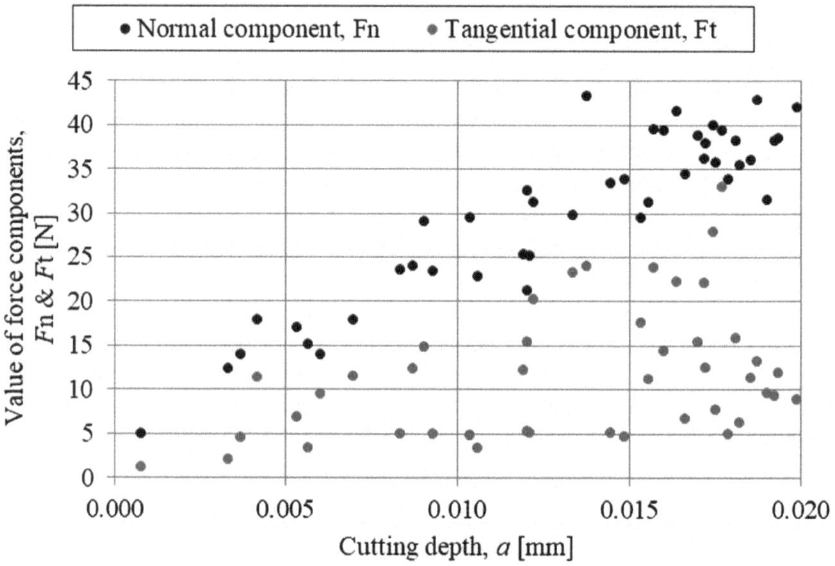

Fig. 5. Experimental values of the normal F_n and tangential F_t cutting force components during micro cutting marble *Plavi tok* with tool tip radius R0.2 mm.

Higher dispersion of the measured results, can case the inadequate estimation of the nominal value of the forces intensities, which are quality indicators of the micro cutting. This represents one of the main reasons for determination of the optimal regression model, which will, determine the regression curve of the components of the cutting force, in the most correct way when all the results are covered. In this paper, three regression analyses were carried out: polynomial, exponential and power.

4.1 Polynomial Regression Model

The first used regression model was polynomial. In order to achieve ideal value ($R^2 = 1$) coefficient of determination (R^2) and minimal value of NMSE, multiple regression was been applied. It turns out that the best results provide a second order polynomial (Figs. 6 and 7).

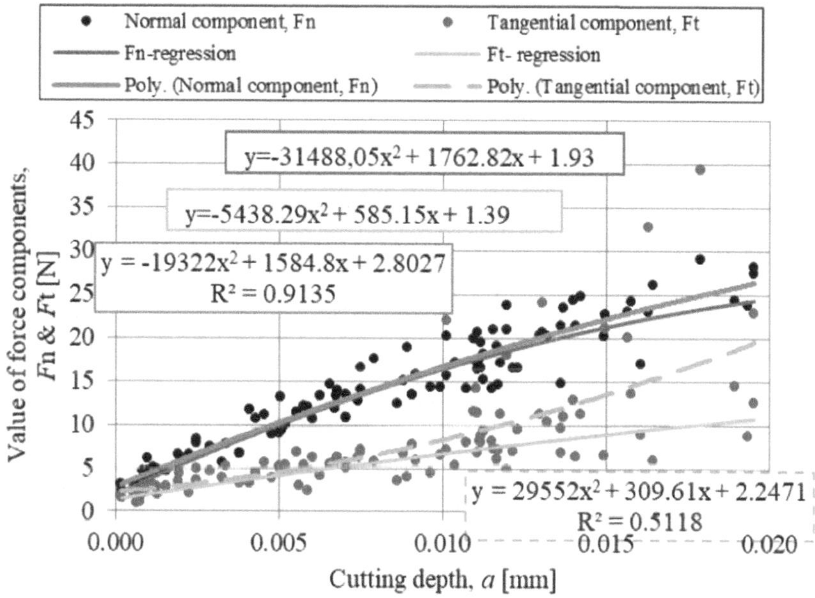

Fig. 6. Values of the normal F_n and tangential F_t cutting force components when micro cutting marble *Plavi tok* with tool radius R0.15 mm, fitted with polynomial regression.

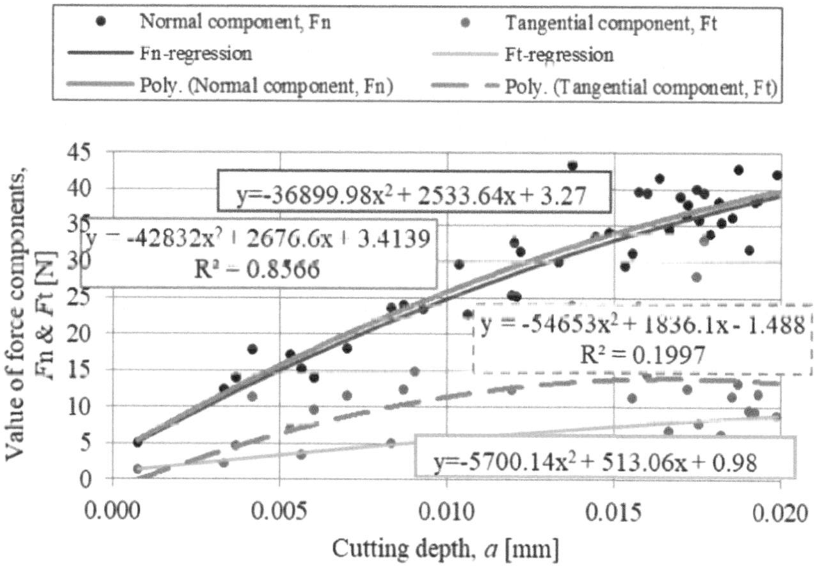

Fig. 7. Values of the normal F_n and tangential F_t cutting force components when micro cutting marble *Plavi tok* with tool radius R0.2 mm, fitted with polynomial regression.

When we compare results for tool R0.15 and R0.2 mm, better results were achieved on experimental data acquired by tool with tip radius R0.2 mm, especially when it comes to the regression of the tangential component.

4.2 Exponential Regression Model

The second used regression model was exponential. In order to achieve ideal value ($R^2 = 1$) coefficient of determination (R^2) and minimal value of NMSE, multiple regression was been applied. It turns out that the best results provide exponential equation presented on Figs. 8 and 9.

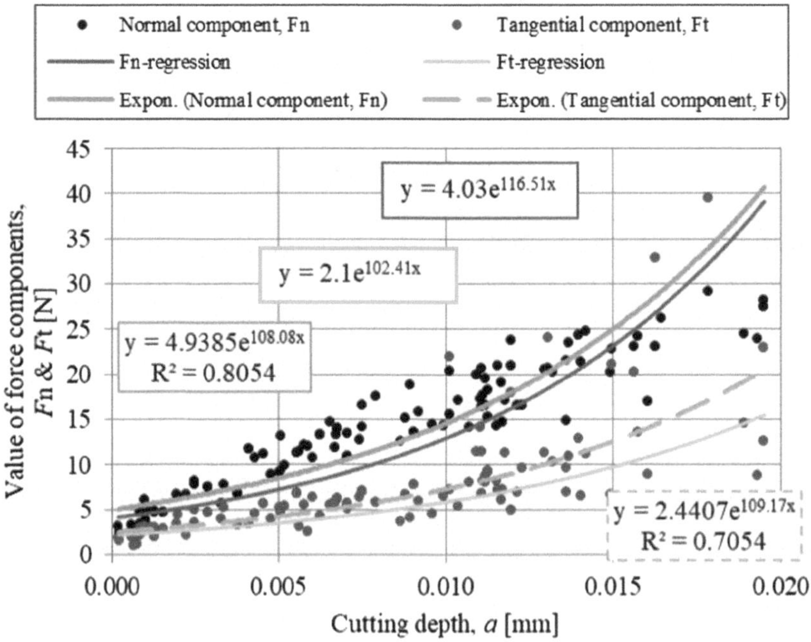

Fig. 8. Values of the normal F_n and tangential F_t cutting force components when micro cutting marble *Plavi tok* with tool radius R0.15 mm, fitted with exponential regression.

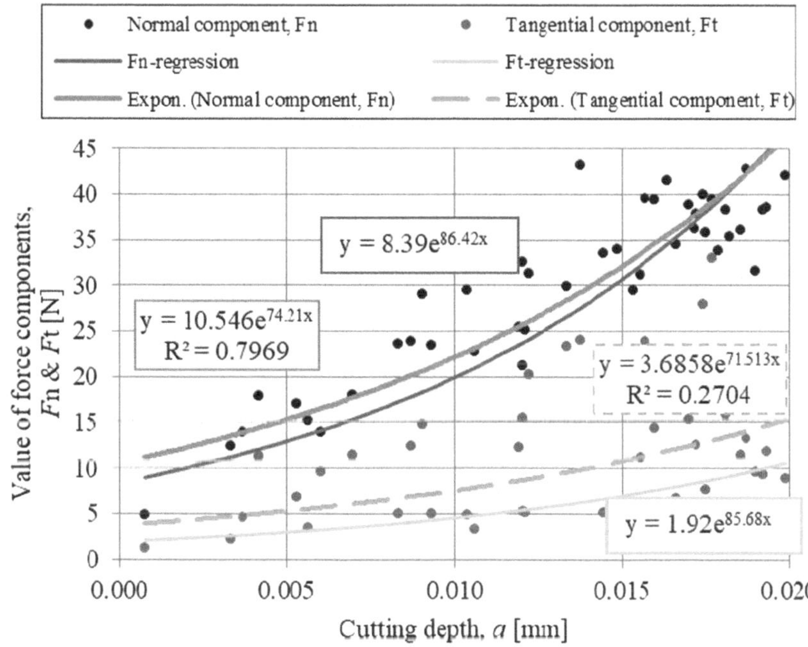

Fig. 9. Values of the normal F_n and tangential F_t cutting force components when micro cutting marble *Plavi tok* with tool radius R0.2 mm, fitted with exponential regression.

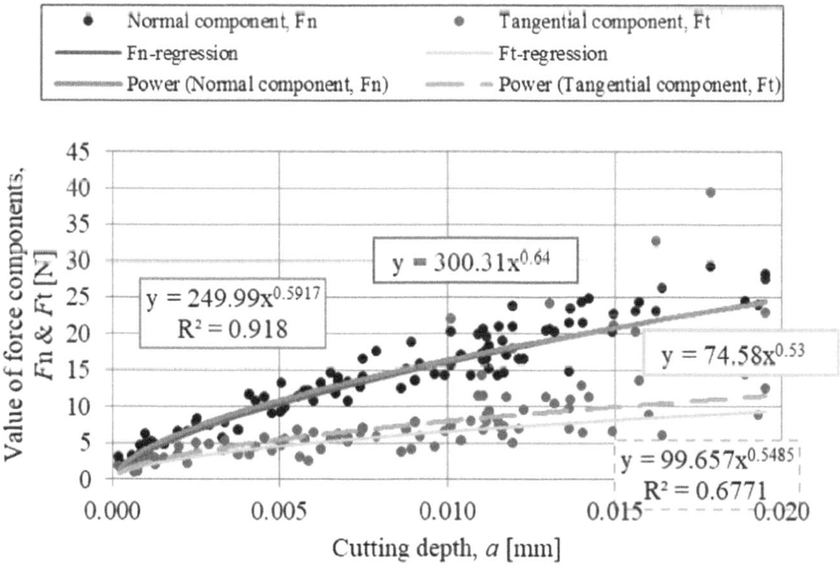

Fig. 10. Values of the normal F_n and tangential F_t cutting force components when micro cutting marble *Plavi tok* with tool radius R0.15 mm, fitted with power regression.

4.3 Power Regression Model

The third used regression model was power. In order to achieve ideal value ($R^2 = 1$) coefficient of determination (R^2) and minimal value of NMSE, multiple regression was been applied. It turns out that the best results provide a power equation presented on Figs. 10 and 11.

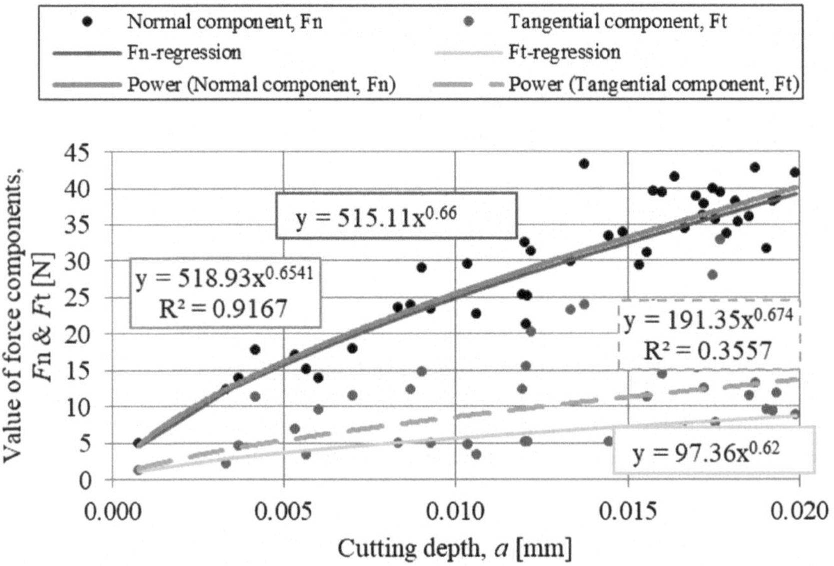

Fig. 11. Values of the normal F_n and tangential F_t cutting force components when micro cutting marble *Plavi tok* with tool radius R0.2 mm, fitted with power regression.

5 Optimal Regression Model

As it is said, the main goal of the conducted regression analysis was the determination of the optimal regression model for the MQC by two parameters R^2 and NMSE.

In our example, best R^2 was 0.91 (rounded to 2 digits). It means that 91% of our values fit the regression analysis model. In other words, 91% of the dependent variables (force components values) are explained by the independent variables (cutting depth values). The minimal value of NMSE is 0.6768.

The output of the regression analysis by both parameters, i.e. the optimal modeling of the two radii of the tool are shown in the Figs. 12 and 13

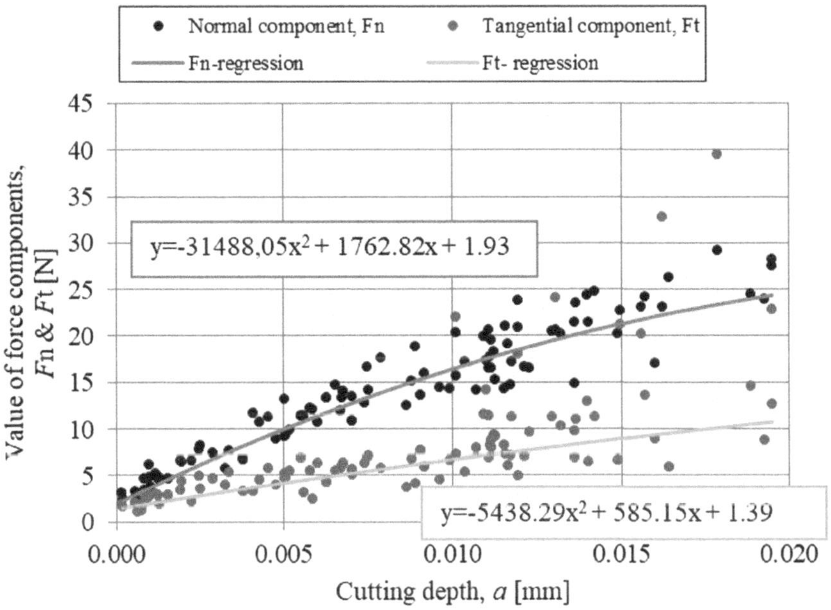

Fig. 12. Optimal regression model of the normal F_n and tangential F_t cutting force components when micro cutting marble *Plavi tok* with tool radius R0.15 mm, fitted with polynomial (second degree) regression.

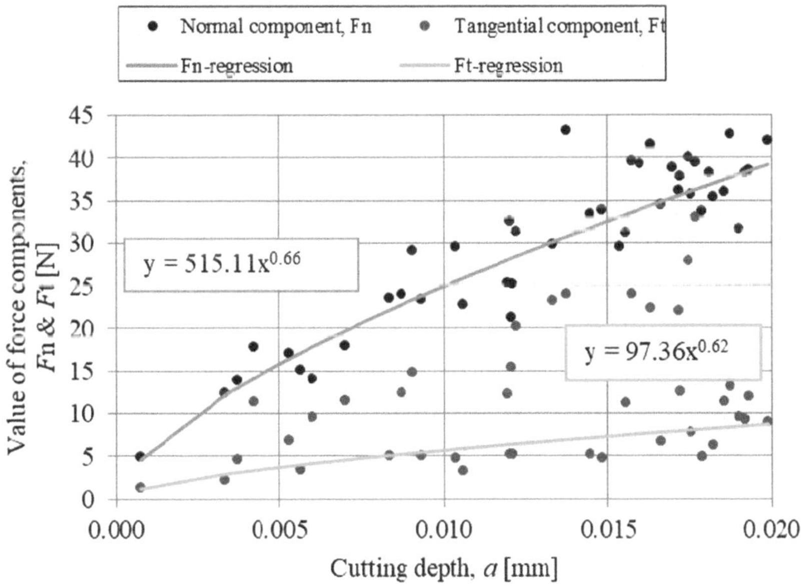

Fig. 13. Optimal regression model of the normal F_n and tangential F_t cutting force components when micro cutting marble *Plavi tok* with tool radius R0.2 mm, fitted with power regression.

6 Conclusion

Micro cutting of stone-based materials, such as marble, requires the understanding of the mechanisms that occur during chip formation. In order to define micro cutting mechanisms, knowledge of the nominal value of the cutting force is one of the key information that is at the same time parameter that represents quality characteristics of the micro cutting. Because marble belongs to materials that are highly heterogeneous, measurement of the components of the cutting force cannot accurately provide information of the nominal value. Depending on the hardness of the mineral that were being cut, the values of the components of the cutting force can highly vary.

In order to define nominal (regression) values of the cutting force in function of cutting depth, with minimal error, multiple regression models were conducted over the experimental data. It turned out that different regression models provide optimum results for different cutting conditions. With the use of tool with tip radius R0.15 mm, a second order polynomial regression model provide optimum results for the nominal values of the cutting force by evaluating two parameters R^2 and NMSE. Increase in the tool tip radius to R0.2 mm led to usage of the power regression model to achieve optimum results for the nominal values of the cutting force. Results of this optimal regression model acquire minimal value of $R^2 = 91\%$, as a first parameter and NMSE = 0.6768.

References

1. Huang, H., Li, Y., Shen, J.Y., Zhu, H.M., Xu, X.P.: Micro-structure detection of a glossy granite surface machined by the grinding process. J. Mater. Process. Technol. **129**(1–3), 403–407 (2002)
2. Pjević, M., Tanović, L.: Pregled stanja istraživanja u domenu mikro-rezanja krtih materijala. In: XIII Međunarodna konferencija Održavanje I proizvodni inženjering, KODIP - 2015, pp. 21–26, Inženjerska akademija Crne Gore, Budva (2015)
3. Tanović, L., Bojanic, P., Puzović, R., Milutinovic, M.: Experimental investigation of microcutting mechanisms in granite grinding. J. Manufact. Sci. Eng. **133**(2), 024501 (2011)
4. Lawn, B.R., Swain, M.V.: Microfracture beneath point indentations in brittle solids. J. Mater. Sci. **10**(1), 113–122 (1975)
5. Lawn, B., Wilshaw, R.: Indentation fracture: principles and applications. J. Mater. Sci. **10**(6), 1049–1081 (1975)
6. Anton, R.J., Subhash, G.: Dynamic Vickers indentation of brittle materials. Wear **239**(1), 27–35 (2000)
7. Chiaia, B.: Fracture mechanisms induced in a brittle material by a hard cutting indenter. Int. J. Solids Struct. **38**(44–45), 7747–7768 (2001)
8. Pjević, M., Tanović, L., Vučetić, F.: Experimental determination of brittle fracturing appearance during static indentation of materials based on stone. In: International Conference on Advanced Manufacturing Engineering and Technologies, pp. 177–184. Springer, Cham (2017)
9. Tanović, L., Bojanic, P., Puzović, R., Klimenko, S.: Experimental investigation of microcutting mechanisms in marble grinding. J. Manufact. Sci. Eng. **131**(6), 064507 (2009)

10. Venkatachalam, S., Li, X., Liang, S.Y.: Predictive modeling of transition undeformed chip thickness in ductile-regime micro-machining of single crystal brittle materials. J. Mater. Process. Technol. **209**(7), 3306–3319 (2009)
11. Son, S.M., Lim, H.S., Ahn, J.H.: Effects of the friction coefficient on the minimum cutting thickness in micro cutting. CIRP Ann.-Manufact. Technol. **45**(4–5), 529–535 (2005)
12. Mladenović, G., Bojanic, P., Tanović, L., Klimenko, S.: Experimental investigation of microcutting mechanisms in oxide ceramic CM332 grinding. J. Manufact. Sci. Eng. **137**(3), 034502 (2015)
13. Stojadinovic, S., Tanović, L., Savicevic, S.: Micro-cutting mechanisms in silicon nitride ceramics silinit R grinding. 中國機械工程學刊 **36**(4), 291–297 (2015)
14. Pjević, M., Tanović, L., Mladenović, G., Marković, B.: Experimental examination of the impact of tool radius on specific energy in microcutting of granite. J. Eng. Mater. Technol. **139**(4), 041004 (2017)
15. Pjević, M., Tanović, L., Ćosović, V.: Eksperimentalna identifikacija uticaja geometrije alata na bočno razaranje materijala kod mikro rezanja mermera. In: Proceedings of the 3rd International Scientific Conference, COMETa2016 – Conference on Mechanical Engineering Technologies and Applications, pp. 229–236. University of East Sarajevo, Faculty of Mechanical Engineering, East Sarajevo (2016)
16. Pjević, M., Tanović, L., Mladenović, G.: Uticaj putanje alata na kritičnu dubinu prodiranja kod mikro rezanja krtih materijala. In: XL JUPITER konferencija, 36. SimpozijumNU * ROBOTI * FTS, pp. 3.33–3.38, Mašinskifakultet Beograd, Beograd (2016)
17. Pjević, M., Tanović, L.: Experimental indentification of material removal mechanisms during micro cutting stone based materials. In: XIV International Conference Maintenance and Production Engineering, KODIP - 2017, pp. 11–16. Engineering Academy of Montenegro, Budva (2017)
18. Tanović, L., Bojanic, P., Popović, M., Belic, Z., Trifkovic, S.: Mechanisms in oxide-carbide ceramic BOK 60 grinding. Int. J. Adv. Manufact. Technol. **58**(9–12), 985–989 (2012)
19. Pjević, M., Popović, M., Tanović, L., Mladenović, G.: Experimental examinations of machinability of ceramic materials during micro processing. In: 22nd European Conference on Fracture - ECF22, p. 131, Society for Structural Integrity and Life – Prof. Dr. Stojan Sedmak (DIVK), Belgrade, Serbia (2018)
20. Subbiah, S., Melkote, S.N.: Evidence of ductile tearing ahead of the cutting tool and modeling the energy consumed in material separation in micro-cutting. J. Eng. Mater. Technol. **129**(2), 321–331 (2007)
21. Kim, C.J., Mayor, J.R., Ni, J.: A static model of chip formation in micro scale milling. J. Manufact. Sci. Eng. **126**(4), 710–718 (2004)
22. Vogler, M.P., DeVor, R.E., Kapoor, S.G.: On the modeling and analysis of machining performance in micro-end milling, part I: surface generation. J. Manufact. Sci. Eng. **126**(4), 685–694 (2004)
23. Altindag, R.: Correlation of specific energy with rock brittleness concepts on rock cutting. J. South Afr. Inst. Min. Metall. **103**(3), 163–171 (2003)
24. Cheng, J., et al.: Study on grinding force modeling and ductile regime propelling technology in micro drill-grinding of hard-brittle materials. J. Mater. Process. Technol. **223**, 150–163 (2015)
25. Lin, J.T., Bhattacharyya, D., Kecman, V.: Multiple regression and neural networks analyses in composites machining. Compos. Sci. Technol. **63**(3–4), 539–548 (2003)
26. Elkaseer, A.M., Popov, K.B., Dimov, S.S., Minev, R.: Material microstructure effect-based investigation of tool wear in micro-end milling of multi-phase materials. In: 7th International Conference on Multi-Material Micro Manufacture, pp. 188–191 (2011)

27. Bissacco, G., Hansen, H.N., Slunsky, J.: Modeling the cutting edge radius size effect for force prediction in micro milling. CIRP Ann.-Manufact. Technol. **57**(1), 113–116 (2008)
28. Malekian, M., Park, S.S., Jun, M.B.: Modeling of dynamic micro-milling cutting forces. Int. J. Mach. Tools Manufact. **49**(7–8), 586–598 (2009)
29. Afazov, S.M., Ratchev, S.M., Segal, J.: Modeling and simulation of micro milling cutting forces. J. Mater. Process. Technol. **210**(15), 2154–2162 (2010)
30. Merchant, M.: Mechanics of the metal cutting process I. Orthogonal cutting and a type 2 chip. J. Appl. Phys. **16**(5), 267–275 (1945)
31. Merchant, M.: Mechanics of the metal cutting process II. Plasticity conditions in orthogonal cutting. J. Appl. Phys. **16**(6), 318–324 (1945)
32. Weule, H., Huntrup, V., Tritschler, H.: Micro-cutting of steel to meet new requirements in miniaturization. CIRP Ann.-Manufact. Technol. **50**(1), 61–64 (2001)
33. Xie, J., Tamaki, J.: Parameterization of micro-hardness distribution in granite related to abrasive machining performance. J. Mater. Process. Technol. **186**(1), 253–258 (2007)

Assessment of Influence of Scanning Parameters on Uncertainty of Measurements Performed Using Laser Tracking System

Maciej Gruza(✉), Piotr Gąska, Wiktor Harmatys, and Adam Gąska

Laboratory of Coordinate Metrology, Cracow University of Technology, al. Jana Pawla II 37, 31-864 Cracow, Poland
maciej.gruza@mech.pk.edu.pl

Abstract. Laser tracking systems are the basic tool used in large-scale measurements in such fields like aviation or automotive industries. Except to the standard applications like the construction of features from individually measured points, they can be also used in the scanning mode measurement to obtain a trajectory of moving target. Such application can be met for example during the industrial robots accuracy verification. Evaluation of the accuracy of these systems, in particular when they operate in scanning mode, is difficult due to the multiplicity of parameters affecting the measurement uncertainty. Factors that have an impact on accuracy of the Laser Tracker system include user-defined parameters that determine the density of points acquisition during scanning. Authors try to assess the impact of these parameters using a methodology that utilizes coordinate machine with high repeatability. A Spherically Mounted Retroreflector system installed on the coordinate measuring machine quill allows the Laser Tracker system to track the movement of the machine along the programmed path. Coordinate measuring machine performing continuous movement enables scanning measurements of specific features and their evaluations. By repeating measurement procedure for changing scanning mode parameters it is possible to indicate differences in the accuracy of the system. Basing on the obtained differences authors propose the limits for tested parameters which allow to achieve the optimal ratio of point acquisition density and the measurement duration for scanning measurements performed on the laser tracking system.

Keywords: Laser tracking systems · Uncertainty estimation · Scanning mode

1 Laser Tracking Systems in Geometrical Metrology

Coordinate measuring systems carry multiple advantageous in the field of metrology, such as the universality of their applications, as well as speed, automatization capabilities, and increasingly, their ease of use. On the other hand, the most notable disadvantages associated with standard coordinate systems stem from their limited measuring space and lack of portability. In order to alleviate such drawbacks, several laser tracking systems have been developed, including Laser Tracker, and Laser Tracer. Systems of this type are relatively small in size and fully portable. In addition, their

V. D. Majstorovic and N. Durakbasa (Eds.): IMEKOTC14 2019, LNME, pp. 201–208, 2019.
https://doi.org/10.1007/978-3-030-18177-2_19

measuring space is several dozen times larger relative to standard coordinate measuring machines. Owing to their considerable benefits, laser tracking systems have been widely used in several industry sectors, such as aviation, shipbuilding, automotive industry, and in measurements of large-scale components. In terms of measurement accuracy, laser tracking systems are comparable to coordinate measuring machines, and even surpass them in certain tasks of large-scale measurements. Common challenges in the field of metrology of geometric quantities and the issues associated with accuracy determination in coordinate systems have been broadly discussed in literature [1, 2].

Main component of typical laser tracking system is a laser interferometer which measures distance from device reference point to target element. Depending on chosen solution laser beam direction can be changed by rotating the laser interferometer or using revolving mirror system. In both cases rotation is realized by two motors with mutually perpendicular rotation axes. A target tracked by device consist of retroreflector, most often placed in spherical housing, which reflects beam back to tracking system. Returning beam is split and directed to laser interferometer and to Position Sensitive Device (PSD) which is responsible for detection of target movements. Detector records each deflection of the laser beam from neutral position caused by the retroreflector movement and provides feedback for the motors that are responsible for changing the emitted beam direction. This enables continuous tracking of the retroreflector position without interrupting the laser beam. The rotations of motors are measured with angular encoders. Both measured angles, about vertical rotation axis (θ) and horizontal rotation axis (φ) together with distance (d) given by laser interferometer unambiguously determine the position of the measuring point in the spherical coordinate system which center is located in the intersection of the rotary axes of the tracking device (Fig. 1).

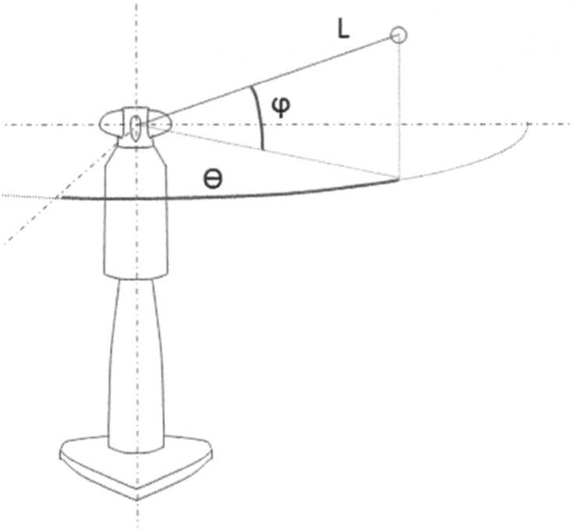

Fig. 1. The spherical coordinate system of Laser Tracker

These coordinates can be then converted into a Cartesian coordinate system. If the connection between retroreflector and tracking device breaks - the counting of interference fringes is stopped and base distance need to be set for interferometric measurements. It can be done using additional distance measuring system e.g. Absolute Distance Meter (ADM) or by placing retroreflector in special spot on machine called "bird-bath" which offsets from the origin of the machine coordinate system about known distance determined precisely during construction phase of device. More information about laser tracking system devices can be found in [3–5].

Assessment of the accuracy of Laser Tracker measurement is always a challenging task. Commonly known and generally accepted methods, such as the calibrated workpiece method and multiple measurements method, require usage of specific material standards, multiple repetition of measurements, and in case of multiple measurement method also changes in the orientation of measured object [6, 7]. Considering the basic area of laser trackers applications, i.e. long distance measurements, both methods are hard to use or in some cases, even impossible to apply. On the other hand it is often meet in measurement practice, to give information about the accuracy of the measurement in form of maximum permissible errors equation which depends on the measured distance. However, this approach is not the best solution, primarily because it treats all of assessed features as a length measurements.

Laser Trackers can work both in standard single point measurement mode and in scanning mode. Second mentioned mode is usually used for checking the trajectory of robots effectors. During such tests, the retroreflector is mounted in place of the manipulator actuator. Determining the accuracy of the measurement done in the scanning mode has not yet been standardized. The only part of the ISO 10360 standard that deals with the accuracy of measurement performed using Laser Tracker systems is [8], however, it refers to measurements carried out for individual measured points.

The authors' experience, shows that there is a need to develop a methodology for determining laser trackers accuracy that would be quick and simple to apply. The procedure proposed below could meet formulated requirements, however it should be noted that it won't give a full answer to the problem of determining the measurement accuracy of the considered systems, but rather it will indicate the measuring ability of Laser Tracker working in the scanning mode.

2 Experiment and Results

The device used in the research was the Laser Tracker Leica LTD 840 with a maximum permissible error for spatial point to point measurements given as (1):

$$E_{L,MPE} = 25 + 45 * L / 1000\,[\mu m] \quad (1)$$

where: L – measured length given in mm

Leica LTD 840 belongs to a group of Laser Trackers with a rotating mirror which reflects a beam emitted by the laser interferometer. In the research described in this article, a Coordinate Measuring Machine (CMM) with a positioning accuracy of an order of magnitude greater than the accuracy of the laser tracking system was used. The CMM utilized during experiments is Zeiss WMM 850S machine with mowing bridge, with measuring volume of X 800\ Y 1200\ Z 700 mm, which is located in air-conditioned room. During whole experiment the ambient conditions were monitored and the temperature in room varied between 19.7°C – 20.1°C. Machine maximum permissible errors equation can be given as (2):

$$E_{L,MPE} = 2 + 3 * L / 1000\,[\mu m] \tag{2}$$

where: L – measured length given in mm

During experiment the CMM is utilized as a reference system which reproduces the retroreflector path for the Laser Tracker system. As a target for tracking system an SMR retroreflector with a diameter of 0.5 in. was used which was attached to the probing system of CMM by means of a specially developed handle. In order to separate the Laser Tracker system from the influence of external conditions (ground vibrations) the measuring unit was placed on the CMM measuring table. The experiment setup is shown on (Fig. 2).

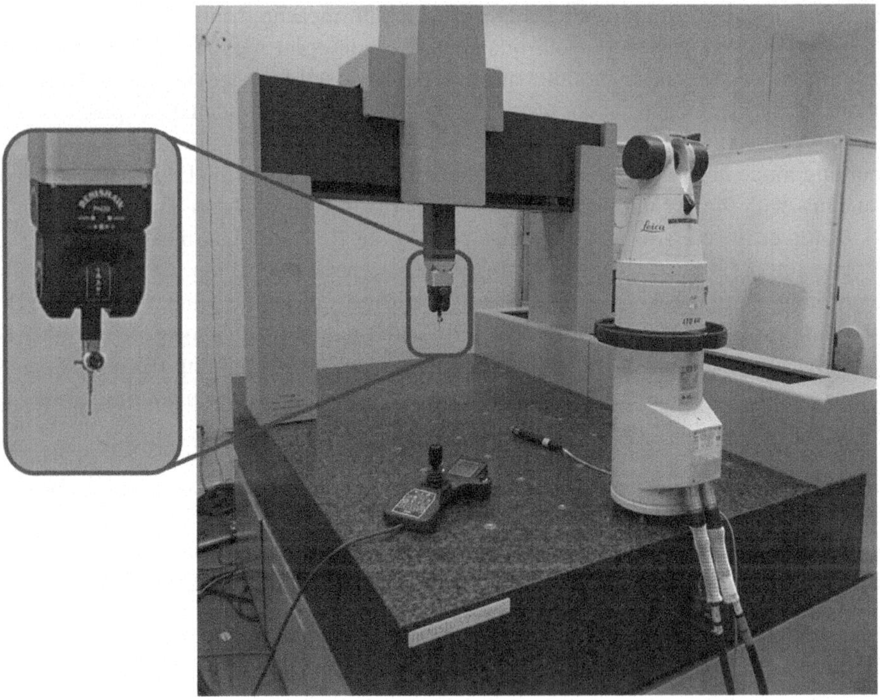

Fig. 2. The experiment setup: Leica LTD 840 Laser Tracker and Zeiss WMM 850S machine. On the left side retroreflector mounted attached to the probing system of machine

The first step before proceeding the main part of the experiment was to check the base repeatability of the system (R_{base}) by measuring 50 times the position of retroreflector resting in an unchanging position. Next on the basis of obtained coordinates the standard deviations of point reproduction were calculated. Obtained results are shown in Table 1. It was assumed then that R_{base} will equal to the highest obtained value.

Table 1. Standard deviation of point reproduction given in mm.

Coordinates	x	y	z
Standard deviation	0,00597	0,00077	0,00363

The trajectory of retroreflector was programmed in CMM software, as a cosine function with a known amplitude and period length. The cosine function was chosen due to easiness of its description using the Best Fit function which is utilized as the default evaluation method in the vast majority of metrological programs. The length of the cosine function period depends on the measuring volume of the CMM. To minimize the influence of dynamic effects connected with movement of CMM table during bridge displacement it was chosen to use mainly movements along X and Z axes of the machine.

In the experiment, the cosine function period was set at 200 mm and the amplitude A equaled to 0,6 mm. During measurements retroreflector travels along one period of cosine function.

Main parameters that can be defined by the operator during measurements done in scanning mode are connected with density of measured points gathered during scanning. The density of measured points can be specified by defining time interval between subsequent points (option called "fixed time" by software developer) or by giving the distance between points ("fixed distance"). For the distance parameter following intervals (Delta) were tested: 0.5; 2; 5; 10; 20 and 30 mm, while for the time parameter chosen values were set as: 0.1; 0.2; 0.5; 1; 2 and 5 s. A series of 10 measurements of the trajectory described by the cosine function was carried out for each value of both distance and time parameters. After each measurement, the points recorded during scanning were linearized using the BestFit function and the straightness error of the created line was determined. Then, the mean value of obtained straightness and associated standard deviation for each tested parameter value was determined. An example of the retroreflector path and its straightness error determined in the PC-DMIS program is shown in (Fig. 3).

The mean error values of straightness for each tested parameter value are shown in (Fig. 4) where Delta represents distance intervals and in (Fig. 5) where Delta represents time intervals. On both graphs, the lines defining the area created by subtracting and adding the doubled R_{base} value to the amplitude of the cosine function (which in the ideal case determines the straightness error of the trajectory of the retroreflector movement) were marked. The area of dispersion (± doubled standard deviation) associated with each mean value are marked as error bars.

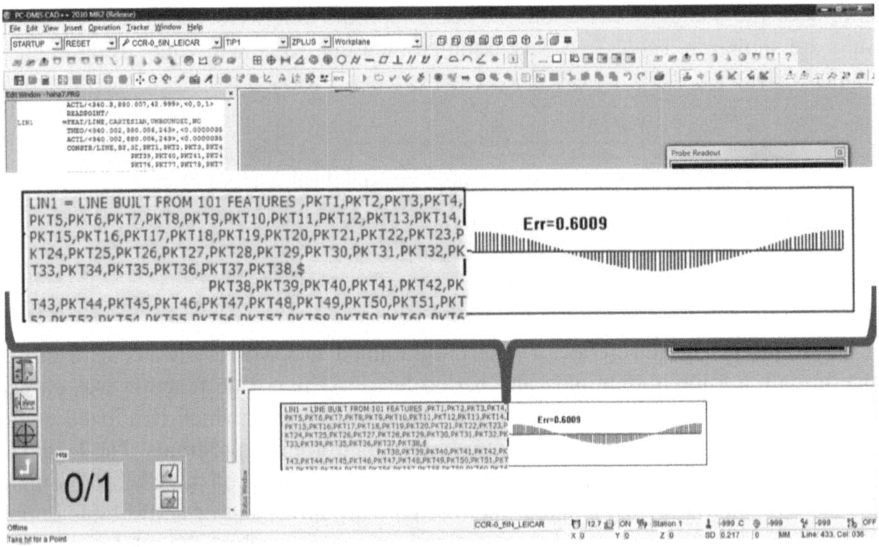

Fig. 3. Example of linearization result for measured retroreflector trajectory

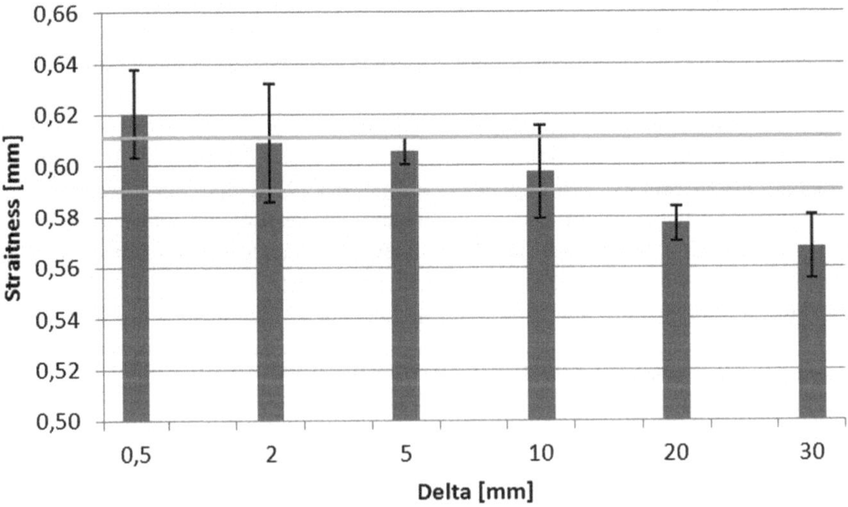

Fig. 4. Mean values of straightness with associated dispersion area obtained for different distance intervals (Delta). All values given in mm

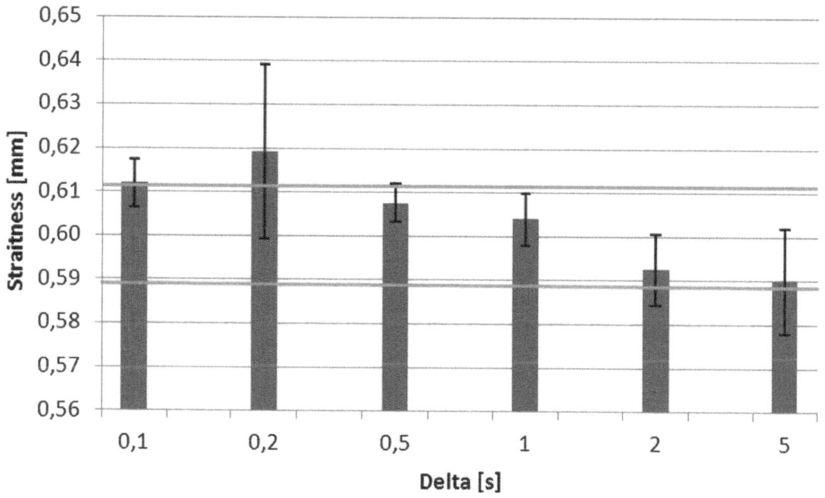

Fig. 5. Mean values of straightness with associated dispersion area obtained for different time intervals (Delta). All values given in mm.

3 Discussion

Analyzing the experiment results, it should be noted that the obtained straightness values are affected by both the repeatability of tested laser tracking system and the error caused by the use of the selected scanning parameter (Delta). The proposed methodology does not delimit the impact of these two sources of errors, however, it allows to formulate some interesting conclusions about the accuracy of the system working in scanning mode. According to the authors, each tested scanning parameter value for which the average straightness error exceeds the indicated area ($A \pm 2 * R_{base}$) should be considered as significantly influential on the accuracy of measurement. In the case of the presented experiment, such values are Delta distance parameters equal to 0,5; 20 and 30 mm and, for parameters expressing the time interval, Delta values that equal to 0,1 and 0,2 s. The results presented in chap. 2 show a clear decreasing tendency, which is caused by a decrease in the number of measuring points per cosine period, and thus a decrease in the straightness error. It does not mean that such values of scanning parameters will allow to obtain the optimal measurement accuracy, but in this case it means deterioration in the accuracy of trajectory reproduction of the retroreflector movement. In order to chose the optimal values of the scanning parameters in terms of the accuracy of the trajectory reproduction, the Delta values should be pointed for which mean straightness lies within range of $A \pm 2 * \sigma$. Moreover, when analyzing the dispersion areas associated for each mean value, those parameters should be indicated for which entire dispersion area is in mentioned above range. Regardless of the used type of Delta parameter, utilization of too small number of points during the measurement in the scanning mode leads to under evaluation of the form error with respect to the actual trajectory of the motion, which results in a reduced value of the mean straightness error. On the other hand the usage of too many points will lead to an increase of the impact of errors related to the repeatability of the machine.

Basing on the conducted experiment, it is possible to indicate the optimal values of the Delta parameter both for the distance and time intervals. The authors consider the proposed method of assessing the scanning parameters impact on system accuracy as promising, taking into account the limitations resulting from the method. However, in order to confirm the validity of the presented methodology, additional experiments are needed, in particular with the use of other laser tracking systems.

Acknowledgments. Reported research was realized as part of a project financed by the National Science Centre; grant no.: 2017/27/N/ST8/03063.

References

1. Jakubiec, W., Malinowski, J.: Metrologia wielkości geometrycznych. WNT, Warszawa (2004)
2. Sładek, J.: Modelowanie i ocena dokładności współrzędnościowych maszyn pomiarowych. Politechnika Krakowska, Kraków (2001)
3. Hocken, R., Pereira, P.: Coordinate Measuring Machines and Systems, 2nd edn. CRC Press, Boca Raton (2017)
4. Huo, D., Maropoulos, P.G., Cheng, C.H.: The framework of the virtual laser tracker – a systematic approach to the assessment of error sources and uncertainty in laser tracker measurement. In: Proceedings of DET2009, AISC, vol. 66, pp. 507–523 (2010)
5. Schmitt, R.H., Peterek, M., Morse, E., Knapp, W., Galetto, M., Härtig, F., Goch, G., Hughes, B., Forbes, A., Estler, W.T.: Advances in large-scale metrology - review and future trends. CIRP Ann. **65**(2), 643–665 (2016)
6. ISO/TS 15530-3:2004 GPS – Use of calibrated workpieces or standards
7. Trapet, E.: Uncertainty analysis of measurements with cmms using multiple measurement strategies. Report of EU Project EASYTRAC (2003)
8. ISO 10360-10 Part 10: Laser trackers for measuring point-to-point distances (2018)

On the Association of Datums and Measurements Using Conventional Measuring Devices - Topics Within the GPS Toolbox Project

Grigore Marian Pop(✉), Liviu Adrian Crisan, Mihai Tripa, Calin Neamtu, and Mihai Dragomir

Department of Design Engineering and Robotics, Faculty of Machine Building, Technical University of Cluj-Napoca, 103-105, Muncii Bulevard, Cluj-Napoca, Romania
grigore.pop@muri.utcluj.ro

Abstract. This paper presents the challenges encountered in elaborating two learning modules for the Erasmus+ project 2015-1-PL01-KA202-016875N "GEOMETRICAL PRODUCT SPECIFICATION (GPS) AND VERIFICATION AS TOOLBOX TO MEET UP-TO-DATE TECHNICAL REQUIREMENTS". The ISO GPS standards define the language that allows expressing all requirements for the geometry of a product considering the current possibilities of measurement. GPS specifications are continuously growing, and many additional tools appeared, especially new, so-called modifiers. The authors of this paper developed these courses allowing students to gain enough knowledge on subjects like metrology, measurements, design and technical drawings. The learning module 8 (first topic) includes subchapters that present the symbols, datum indications, figures and examples of 2D and 3D drawing indications of datums (datum feature indicators and identifiers, datum targets, specification of datums and datum systems) according to the latest versions of ISO 5459 and ISO 1101. Learning module 23 (second topic) provides information for reading the indications of three different types of conventional instruments (callipers, micrometers, dial gauges) as well as for choosing the right instrument for the required measurement.

Keywords: Datums · Datum systems · Association of datums · Geometrical tolerances · Callipers · Micrometers · Dial gauges

1 Introduction

The Technical Committee ISO/TC 213 has been working towards the development of international standards covering problems of dimensioning, geometrical tolerancing, surface texture and related metrology. The unified model for design, manufacturing and verification based on the improved GPS (Geometrical Product Specification) language and its new concepts like surface models, geometrical features, characteristics, specification uncertainty, correlation uncertainty and operations are aimed to cover all

V. D. Majstorovic and N. Durakbasa (Eds.): IMEKOTC14 2019, LNME, pp. 209–216, 2019.
https://doi.org/10.1007/978-3-030-18177-2_20

aspects of the product development up to launch on the market [1]. The ISO GPS standards define an internationally uniform symbol language, that allows expressing all requirements for the geometry of a product with the corresponding requirements for the inspection process in technical drawings, considering current possibilities of measurement and testing technology. This avoids ambiguities and inconsistencies during the planning of manufacturing and inspection processes [2].

The design and metrology engineers must have good knowledge on the GPS and verification methodology which should be gained during studies or refreshed and supplemented by specialized training courses on the new standards. The practice shows that the university curricula of the mechanical engineering faculties often include only limited classes on the GPS. This does not allow students to gain enough knowledge on the subject [2].

The curriculum content is divided among partners according to their competences field and necessary resources for preparation of the learning material. The following learning modules were developed:

1. Geometrical characteristics
2. Thirteen ISO fundamental principles
3. Linear sizes
4. ISO system of limits and fits
5. Angular sizes
6. Geometrical tolerances
7. Tolerances of form
8. Datums
9. Tolerances of orientation
10. Tolerances of location
11. Profile tolerancing
12. Run-out
13. Material modifiers
14. Other modifiers
15. General tolerances
16. Complex geometrical features
17. Roughness, waviness and primary profile
18. Tolerancing of assemblies
19. Dimensional chains
20. Measurement uncertainty
21. Decision rules
22. Guidelines for selection of measuring equipment
23. Measurements with use of conventional measuring devices
24. Verification by using measuring machines
25. Verification by using non-contact coordinate measuring systems
26. Measurements with use of form testers
27. Measurements in micro- and nanoscale
28. Verification by using surface measuring instruments
29. Verification by gauges
30. Calibration of measuring equipment

Three of these learning modules (8, 15 and 23) were developed by the authors of this paper and reviewed by the other partners of the project. This paper presents the modules 8 and 23.

2 Module 8, Datums

The most important GPS standards concerning Datums and Datum systems are ISO 1101 and ISO 5459.

The first subchapter of this learning module presents the learning aims, after completing these chapter, the user will be able to:

- understand and define general features and classification of datums;
- recognize and indicate datums, common datums and datum systems on technical drawings (2D and 3D);
- interpret and explain the association of single datums (the learner will be able to understand and establish a single datum from three datum targets on a planar surface, a single datum from a spherical surface, a single datum from a cylindrical surface);
- interpret and explain the association of datums and datum systems (the learner will be able to establish a common datum from two coplanar planes, from two coaxial cylinders as well as from a plane and a perpendicular cylinder).

The 2nd subchapter presents an overview about datums and datum systems.

The 3rd subchapter presents the terms and definitions applied in this learning module (situation feature, datum feature, associated feature, datum, primary datum, secondary datum, tertiary datum, single datum, common datum, datum system, association, invariance class, TED - theoretically exact dimension).

The 4th subchapter, datums general concepts, presents the table of invariance classes according to ISO 5459, the datum features, datum target features and modifier symbols.

The 5th subchapter deals with:

- datum feature indicators and identifiers;
- datum targets;
- specification of datum and datum systems;
- datum feature indicator layout according to ISO 5459;
- single datum, common datum and datum systems.

The 6th subchapter, *Association of datums ISO 5459*, presents association methods and examples of association of single datums, common datums and datum targets. Datums and datum systems are theoretically exact geometric features used together with implicit or explicit TED's (theoretically exact dimension) to locate or orientate tolerance zones for toleranced features. A datum consists of a set of situation features for an ideal feature (feature of perfect form). This ideal feature is an associated feature which is established from the identified datum features of a workpiece. Datum features may be complete features, or portions of a feature.

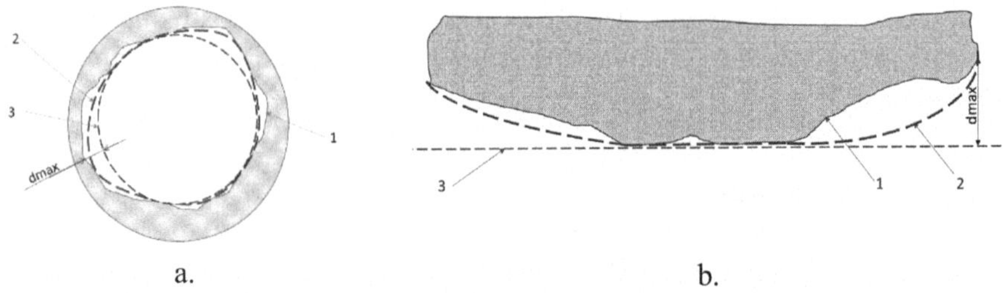

Fig. 1. a. Example of associated features for a cylindrical surface, where: 1 - real workpiece, 2 - filtered feature, 3 - associated feature the maximum inscribed cylinder minimizing the maximum distance (dmax) with the filtered feature; b. Example of associated features for a planar surface, where: 1 - real workpiece, 2 - filtered feature, 3 - associated feature, the outside material tangent plane minimizing the maximum distance (dmax) with the filtered feature Source: https://e-uczelnia.ath.bielsko.pl/ [2, 3]

According to ISO 5459, the associated features, used to establish the datums or datum systems, simulate contact with the real integral features in a way that ensures that the associated feature is outside the material of the non-ideal feature. Figures 1 and 2 present the procedures to establish an associated feature (Figs. 3, 4, 5, 6 and 7).

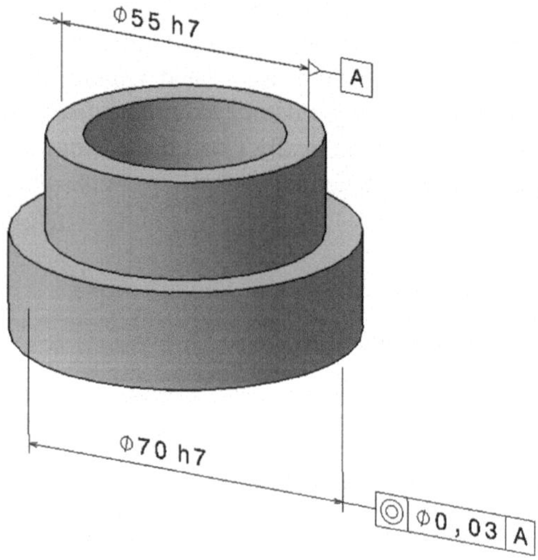

Fig. 2. 3D drawing indication of a coaxiality tolerance using a single datum Source: https://e-uczelnia.ath.bielsko.pl/ [2–4]

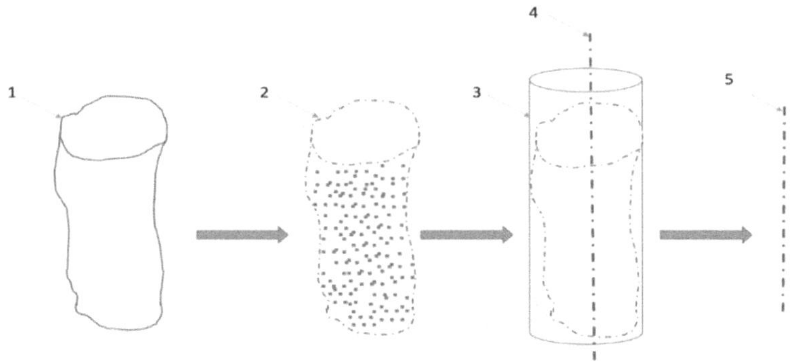

Fig. 3. Illustration of features used for establishing a single datum indicated in Fig. 2, where: 1 - datum feature: real integral feature, 2 - extracted integral feature, 3 - associated integral feature, 4 derived feature of associated integral feature, 5 - single datum: straight line (axis of the cylinder) Source: https://e-uczelnia.ath.bielsko.pl/ [2–4]

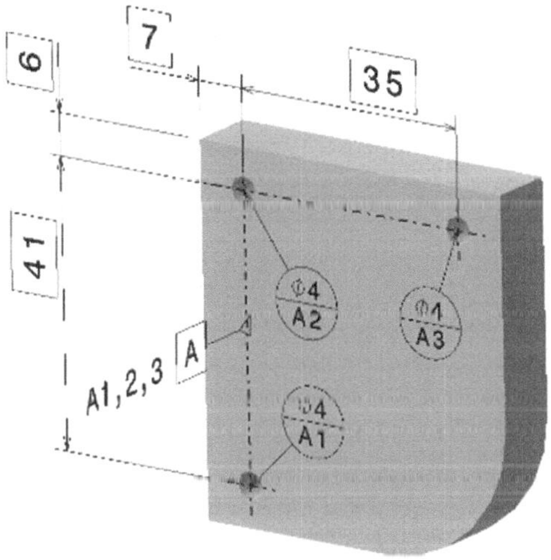

Fig. 4. 3D drawing indication for three datum targets on a plane Source: https://e-uczelnia.ath.bielsko.pl/ [2–4]

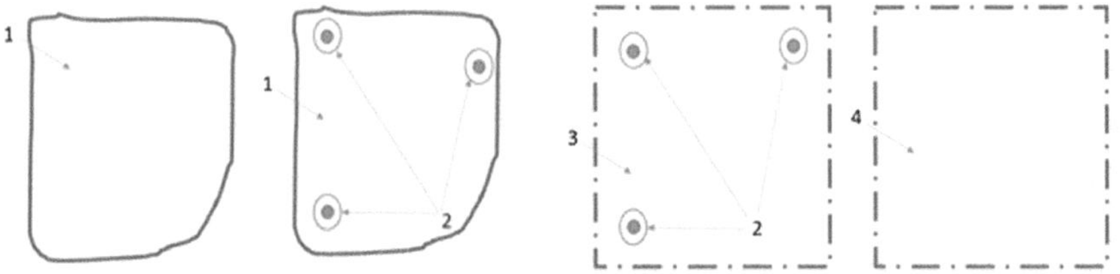

Fig. 5. Explanation for establishing a single datum from three datum targets on a planar surface (the situation feature is a plane), where: 1 - datum feature: real integral feature, 2 - datum targets taken on the datum feature, 3 - associated feature to the three datum targets, plane, 4 - single datum, plane Source: https://e-uczelnia.ath.bielsko.pl/ [2–4]

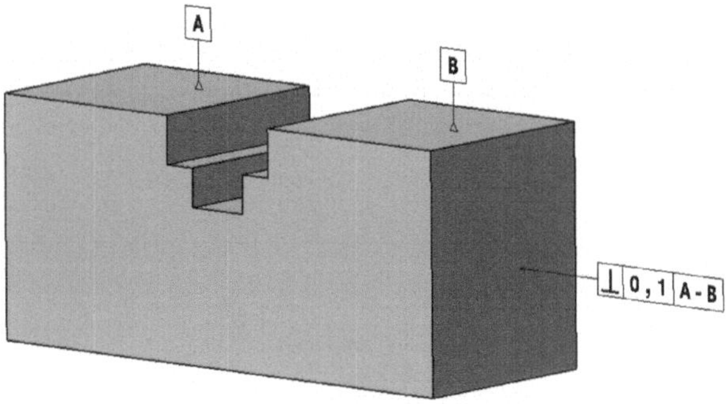

Fig. 6. 3D drawing indication of common datum, two coplanar planes Source: https://e-uczelnia.ath.bielsko.pl/ [2–4]

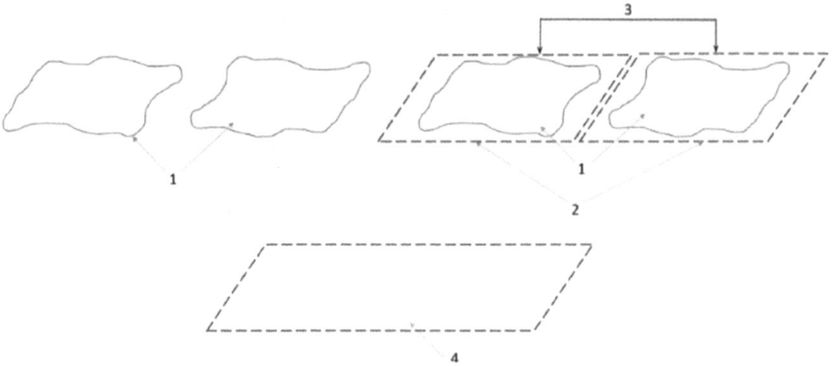

Fig. 7. Explanation for establishing a common datum from two coplanar planes (the situation feature is a plane), where: 1 - datum features: real integral features, 2 - associated features (two coplanar planes), 3 - orientation and location constraint (coplanarity between associated features), 4 - single datum: common plane. Source: https://e-uczelnia.ath.bielsko.pl/ [2–4]

The 7th subchapter presents a comprehension test with multi-choice questions.

3 Module 23, Measurements with Use of Conventional Measuring Devices

The first subchapter presents the learning aims, after completing these chapter, the user will be able to:

- identify the right conventional instrument for the required verification;
- read the indication of a vernier calliper;
- make verifications using digital, vernier and dial callipers;
- read the indications of an analogue indicating micrometer;
- make verifications using digital and analogue indicating micrometers;
- read the indications of dial indicators;
- make verifications using digital, and dial indicators.

The 2nd subchapter, measurements using callipers, presents reading examples and good practices using vernier callipers, dial callipers and digital callipers:

- incorrect and correct part positioning for outside measurements;
- "0" setting up to measure width parallel slots;
- "0" setting for external measurement;
- step measurement;
- external measurements using vernier calliper with measuring range: 0–1000 mm and 0–2000 mm.

The 3rd subchapter, measurements using micrometers, presents reading examples and good practices using micrometers:

- external measurements using a 0–25 mm range micrometer;
- external measurement using a 500–600 mm range micrometer;
- depth measurement;
- hole diameter measurement using a three-point contact, 70–100 mm range micrometer.

The 4th subchapter, measurements using dial gauges, presents reading examples and good practices for:

- parallelism measurement using a dial gauge;
- simplified method for radial circular run-out measurement using a dial gauge;
- differential measurement of a shaft diameter using a 0.01 mm resolution dial gauge;
- differential measurement of a hole diameter using a 0,001 mm resolution dial gauge;
- measuring scheme of an angle using the sine bar and a 0.01 mm resolution dial gauge.

The 5th subchapter presents a comprehension test with multi-choice questions.

4 Conclusions

The authors developed a high-quality course, covering the up-to-date state of knowledge in the area of GPS, which can be used as a supplement for the existing engineering studies curricula.

Students were not the target group of this project, but they will also benefit from these courses as the project partners are planning to use the developed material in the lectures where the GPS subject is included.

The e-learning modules developed to provide high-quality teaching materials aims to bridge the gap in understanding that exists in the area of GPS, a vital component of manufacturing and quality management.

By transferring and applying this knowledge in the industry will lead to an increase of quality, uprising the competitiveness of the enterprises.

The students and engineers willing to develop their knowledge on GPS can easily access these didactic resources using their smartphones, pc or tablets, making the teaching process more dynamic and more attractive than traditional methodologies.

References

1. Humienny, Z.: State of art in standardization in GPS area. CIRP J. Manufact. Sci. Technol. **2** (1), 1–7 (2009)
2. Płowucha, W., et al.: Geometrical product specification and verification as toolbox to meet up-to-date technical requirements, 2015-1-PL01-KA202-016875N. https://e-uczelnia.ath.bielsko.pl
3. EN ISO 5459:2012: Geometrical product specifications (GPS) - geometrical tolerancing, datums and datum systems
4. EN ISO 1101:2017: Geometrical product specifications (GPS) - geometrical tolerancing, tolerances of form, orientation, location and run-out
5. Griffiths, B.: 5 - limits, fits and geometrical tolerancing, engineering drawing for manufacture, pp. 88–110 (2003). https://doi.org/10.1016/B978-185718033-6/50019-7
6. Morse, E., et al.: Tolerancing: managing uncertainty from conceptual design to final product. CIRP Ann. **67**(2), 695–717 (2018). https://doi.org/10.1016/j.cirp.2018.05.009
7. Henzold, G.: Geometrical Dimensioning and Tolerancing for Design, Manufacturing and Inspection: A Handbook for Geometrical Product Specification Using ISO and ASME Standards, 2nd edn. Elsevier, Amsterdam (2010). ISBN 978-0-7506-6738-8
8. ISO 3611:2010: Geometrical product specifications (GPS) - dimensional measuring equipment: Micrometers for external measurements - design and metrological characteristics
9. EN ISO 13385-1:2011: Geometrical product specifications (GPS) - dimensional measuring equipment - Part 1: callipers; design and metrological characteristics
10. EN ISO 13385-2:2011: Geometrical product specifications (GPS) - dimensional measuring equipment - Part 2: calliper depth gauges; Design and metrological characteristics

Practical Aspects in the Application of Geometrical Product Specifications and Verification (GPS) in the Micro and Nano-Scale Manufacturing

N. M. Durakbasa[1] and G. Poszvek[1,2(✉)]

[1] Department of Industrial Metrology and Adaptronic Systems Institute of Production Engineering and Photonic Technologies, TU Wien, BA-09, 1060 Vienna, Austria
guenther@poszvek.net
[2] Norwegian University of Science and Technology - NTNU, Trondheim, Norway
https://aum.ift.tuwien.ac.at/

Abstract. The exciting task of penetrating into the micro and nano range represents a great challenge for the engineer. In this work we try to get closer to the area by means of the geometrical product specifications and verification (GPS) system. In order to familiarize future engineers with the application in this area, it is necessary to deal with these topics. Generally technologies in that size seeks to explore and exploit the novel and unique properties of materials when their size is reduced to the nanoscale (approximately 1–100 nm). Also the production in reduced scale shows the limits of the manufacturing methods. Materials developed for a particular purpose or function are termed engineered nanomaterials (ENMs) to distinguish them from other nanoscale materials, which are produced naturally or an anthropogenic process [1]. ENMs can take various forms and are generally categorized as nanoparticles. The objects produced with the help of electron beam lithography serve on the one hand to get to know the technology and on the other hand to transfer the related experience into the engineering education.

Keywords: Geometrical Product Specifications and Verification - GPS · Cleanroom · Electron beam lithography

1 Introduction

How to get in touch as a mechanical engineer in the range of micro and nano dimensions? Starting with the geometric product specification (GPS), it was the declared aim to use the same standards in the dimensions of micro and nano.

"GPS should work on any scale. For example, ISO 1101:2017 has the tools to specify the wavelengths that should be included in geometrical specifications. I also believe that the definitions for size in the ISO 14405 series work regard-less of the scale of the objects." This statement was given by Henrik S. Nielsen, an expert in

The original version of this chapter was revised: The chapter has been updated with proper acknowledgement of the missing citations. The correction to this chapter is available at https://doi.org/10.1007/978-3-030-18177-2_29

V. D. Majstorovic and N. Durakbasa (Eds.): IMEKOTC14 2019, LNME, pp. 217–239, 2019.
https://doi.org/10.1007/978-3-030-18177-2_21

standardization and Chairman of the Technical Committee 213 "Geometrical Product Specifications and Verification" ISO International Organization for Standardization.

So there really is not any scale limit for the applicability of GPS. It may be more crucial to define wavelengths, etc., than in normal manufacturing, but the tools are there to do it. So the first step was to find out if there is a general model used in the standards of GPS. How this was done is written in the section methods. It is very well to manufacture things with atomic precision, but how do we know we have done it? So there are different ways to come in that tiny dimensions. So let me call it nanofacture like it is call in [2]. In Fig. 1 you can see these different approaches.

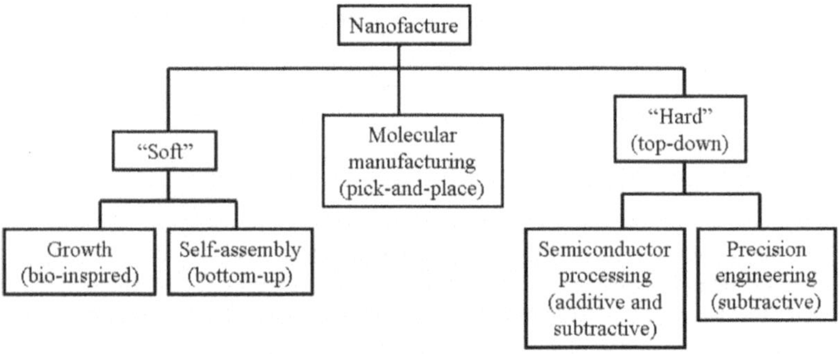

Fig. 1. Different modes of nanofacture. Source: [2]

So generally we talk about two approaches for the manufacture of products: top-down and bottom-up. The top-down approach often uses the traditional workshop or micro-fabrication methods where externally controlled tools are used to cut, mill, and shape materials into the desired shape and order. Different patterning techniques, such as lithography belong to this category. Getting more and more in touch with this small dimension manufacturing processes lots of new technologies have to be considered. So using the Electron Beam Lithography (EBL) as an first possibility some information will be given about this technique. Before starting the work in the laboratory there are some information about clean room conditions and chemical behavior. Normally you can see on the first sight in what kind of dimension a work is done in the laboratory when you perceive the surroundings. Working in the micro and nano field is done under clean room conditions and usage of cleaning chemicals. This will be also seen in the methods part. Getting in touch with scanning electron microscope (SEM) you are making your sample visible. How you do this is also written in the methods part.

In the last section you will find the results and some proposals will be discussed. How the work can be considered of going on and the ways how progress can be done in that field of research.

2 Methods

Now going a little deeper in the touch with the field of micro and nano scale processes are now described by the way of producing a sample.

- conditions around producing
- creating model
- technologies used
- make it visible

Conditions Around Producing

The environmental conditions determine to a large extent with what accuracy can be made. Thus, for the time being, the environment must be determined or prepared so that a production in the micro and nano range is possible. These conditions are also generally defined by the cleanroom. A cleanroom is designed to reduce the amount of particles to an acceptable and controlled level. A required low number of particles in the surrounding air. On the one hand to meet the requirements of a clean room and on the other hand to the high-precision metrology, a high accuracy in the sub-micron and nanometer range for international recognition of the results in compliance with international standards is required. Only the conveniently designed laboratory with consistent and reproducible environmental conditions may ensure measurements with high accuracy fulfilling the tasks with the smallest measurement uncertainties. The most important environmental measurement influences are:

- temperature (thermal conduction, convection and radiation)
  vibrations
- humidity
- pollution

The structural organization of a cleanroom ensures that these disturbing influences will be reduced and kept constant. The NTNU NanoLab demonstrates the demands for such a building and features of technical realization. The standards and guidelines regulate the particle concentration (EN ISO 14644 and VDI 2083). The amount of particles in the air determines the classification of a cleanroom. There are two common classification systems, ISO and the older USFED, as seen in Fig. 2. An US FED class of 10.000 means that the areas have a controlled environment with a particle count of less than 10.000 particles larger than 0.5 μm in any given cubic foot of air.

The smallest particle size visible to the naked eye is 10 μm, 20 times larger than the particles counted! In Fig. 3 are more informations about the classification in the ISO standards.

Personnel represent a critical source of contamination in cleanrooms. They release:

- ionic and molecular contamination
- microrganisms
- particles of various sizes

maximum numbers of particles per ft³ >0,5 µm	cleanroom class US FED STD 209	ISO class	NanoLab areas
1	1	3	-
10	10	4	-
100	100	5	thin film / dry etch
1.000	1.000	6	lithography / EBL
10.000	10.000	7	all other cleanroom areas
100.000	100.000	8	gowning room
1.000.000	1.000.000	-	office space

Fig. 2. Cleanroom classification. Source modified: [14, 15]

	particle je m³					
class	0,1 µm	0,2 µm	0,3 µm	0,5 µm	1,0 µm	5,0 µm
ISO 1	10					
ISO 2	100	24	10			
ISO 3	1.000	237	102	35		
ISO 4	10.000	2.370	1.020	352	83	
ISO 5	100.000	23.700	10.200	3.520	832	
ISO 6	1.000.000	237.000	102.00	35.200	8.320	293
ISO 7				352.000	83.200	2.930
ISO 8				3.520.000.	832.000	29.300
ISO 9				35.200.000	8.320.000	293.000

Fig. 3. Cleanroom classification ISO. Source modified: [14]

Particulate contamination arises from

- the respiratory tract: it is released by speaking, ex-haling and sneezing through the mouth and nose in droplet form
- the skin surface
- the clothing

Particle emission arising from the skin surface is attributable to abrasion of scales. The entire skin surface (approximately 1.75 m^2) of humans is renewed within 5 days by desquamation. Skin particles, and thus also microorganisms, are restricted to a certain extent by the corresponding choice and care of cleanroom garments. High concentrations of microorganisms are determined above all on hands (100/cm^2 to 1000/cm^2) as well as on the forehead (104/cm^2 to 105/cm^2), the scalp (106/cm^2) and under the armpits (106/cm^2 to 107/cm^2).

Cleanroom behavior: The main particle source in a cleanroom is the people working there. Dressing in cleanroom garments (Fig. 4) reduces the amount of particles reaching the air, but your behavior is still of utmost importance to reduce the amount of particles. Figure 5 shows how many particles that are generated just by moving.

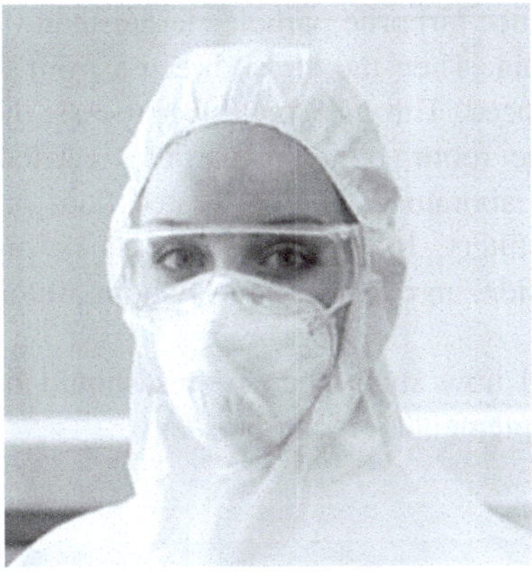

Fig. 4. Cleanroom garments. Source modified: [16]

activity	particles generated (>0,5 µm/minute)
sitting or standing	100.000
small motions by head or arm	1.000.000
rising up	2.500.000
slow walk	5.000.000
fast walk	10.000.000
gymnastis	25.000.000

Fig. 5. Number of particles generated when moving. Source modified: [17, 18]

You should move smoothly inside the laboratory to avoid unnecessary turbulence and particle generation. Correct cleanroom behavior together with wearing correct cleanroom garments is your main contribution to keeping the particle amount down.

The principle of laminar flow: The foundation for the development of cleanroom technology was laid in the mid-sixties in the USA when the principle of low-turbulence displacement flow (laminar flow) was introduced. The ambient air is drawn in by means of a radial fan and forced through the filter and laminarizer.

This creates a laminar flow, the down flowing air flows in parallel streamlines. Particles are detected by the parallel airflow and transported out of the box. The air can escape through the perforated bottom of the lab. In the Figs. 6 and 7 the air flow in a cleanroom is visible and the flow principle is demonstrated.

The structural organization of a clean room ensures that these disturbances are reduced and kept constant. The NTNU NanoLab meets the requirements for such a building and the characteristics of the technical realization. For this purpose, HEPA[1] (high-efficiency particulate absorber) filters have been installed in the ceiling of the

[1] HEPA = High Efficiency Particulate Air filter originally called high-efficiency particulate absorber.

building. These filters reduce airborne particles larger than 0.3 microns and thus filter approximately 99.997% out. Then the air is directed from the ceiling to the floor in vertical flow at reduced speed. Through this flow process, the smallest particles of air are transported out of the room, both the particles carried into the room and the particles produced in the laboratory. The perforated floor provide for the removal of the flow directly to the filters. In order to suppress or prevent the penetration of the particles from the outside, an overpressure of about 30 Pa is generated in the room [17, 18].

[18] describe in detail how such clean rooms should be constructed and in the NanoLab the practical implementation can be seen.

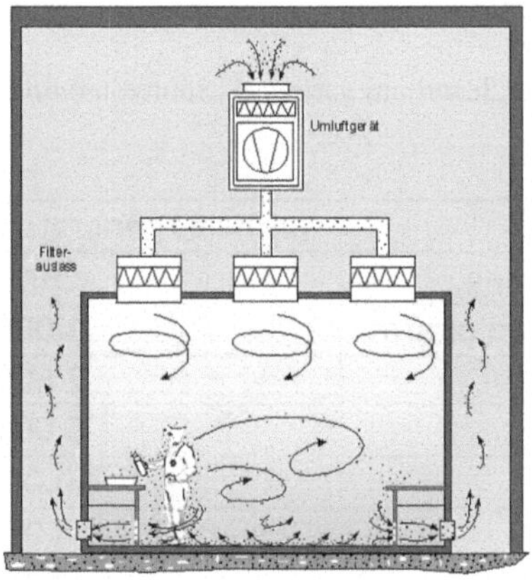

Fig. 6. Turbulent cleanroom source: [8]

Fig. 7. Laminar flow cleanroom source: [8]

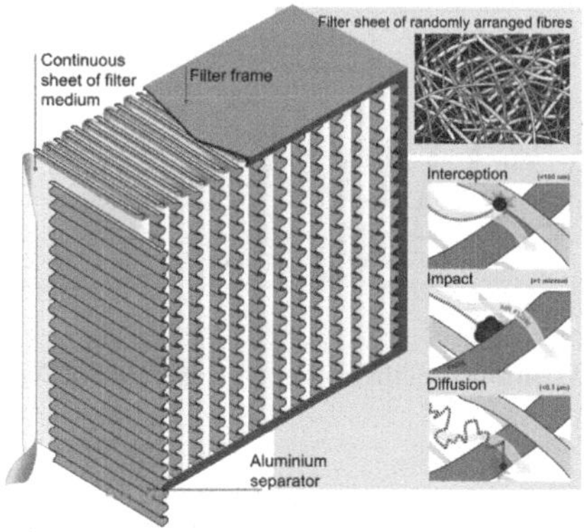

Fig. 8. HEPA filter with functional description source: [9]

Summing up there is the outstanding characteristic of the NanoLab: HEPA filters installed in the ceiling filter away 99.997% of all particles larger than 0.3 μm. Through holes in the floor the air is returned to the HEPA filters. A constant overpressure at 30 Pa.

A very low level of vibration (VC-E)[2].

A stable temperature at 19° ± 1 °C

A relative humidity of 43% ± 5%.

Creating a Model

There are a lot of examples given in the standard ISO 1101 but no real general model [3]. The decision to creature a sample with some functions of the GPS system was done. Using position specification of a centerline as it can be seen in Figs. 9 and 10. In the Fig. 11 you can see the tolerancing.

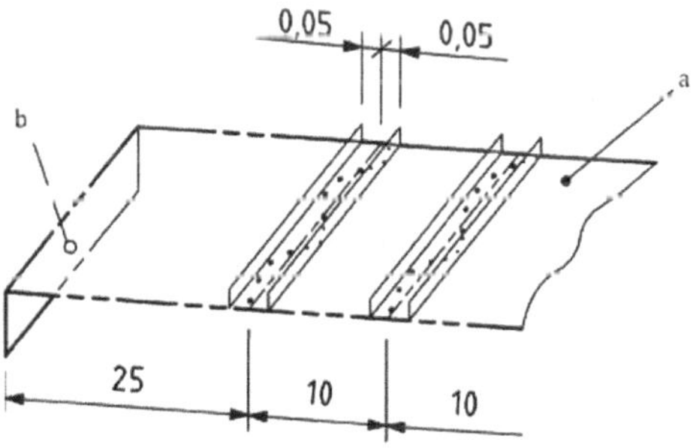

Fig. 9. Position specification of a centerline 3D source: [3]

[2] Vibration Criterion - E 3.12 μm/s Amplitude.

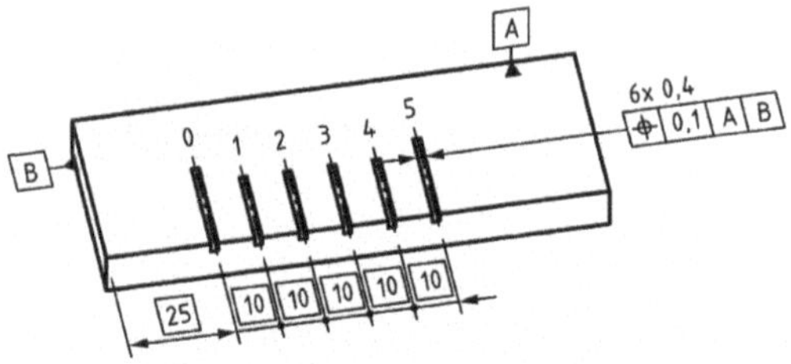

Fig. 10. Position specification of a centerline 2D source: [3]

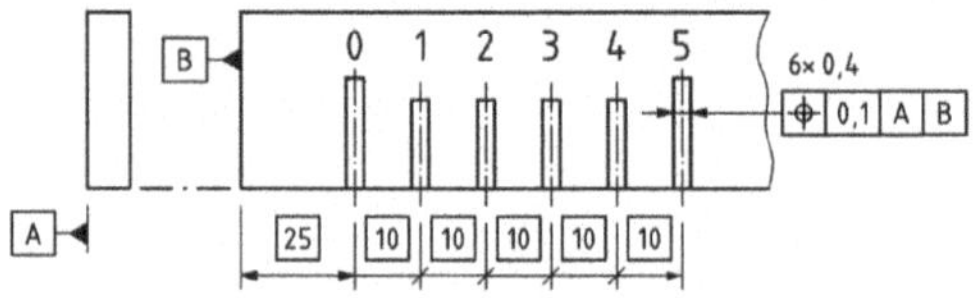

Fig. 11. Tolerance zones source: [3]

A united geometry element as shown in Fig. 12 and his tolerancing in Fig. 13 symmetrical to the theoretically exact location of the median plane to the reference axis a.

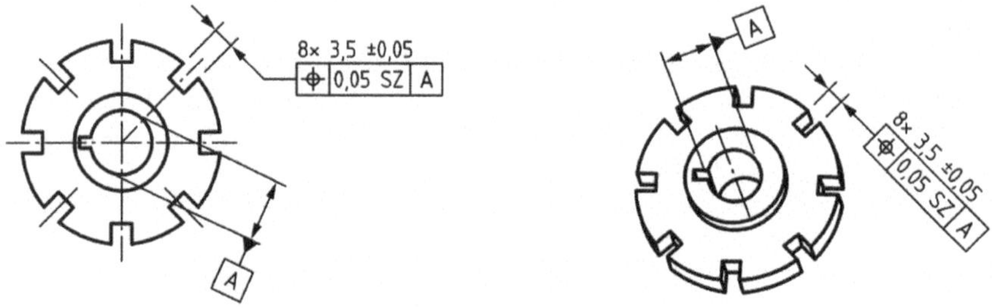

Fig. 12. United geometry element 2D and 3D source: [3]

A position specification (p. s.) of a midplane as it can be seen in Figs. 14 and 15 with tolerancing in Fig. 16. The extracted centerline of each hole must be within a cylindrical zone of diameter 0.1 whose axis coincides with the theoretically exact location of the considered hole relative to the reference planes C, A and B. This three samples have been taken out of the standard.

Here again it is pointed out that these selected examples are not representative of all possibilities of the GPS but only for a first attempt to understand the technology and to be able to apply it. Doing this, there was the first point to learn that there are limits in the CAD programs. Drawing as it is used in the construction field is not the same when you decide to draw in the dimension of micro and nano. There are limitations given by the program itself. So using SolidWorksTM it is limited to draw under 100 nm.

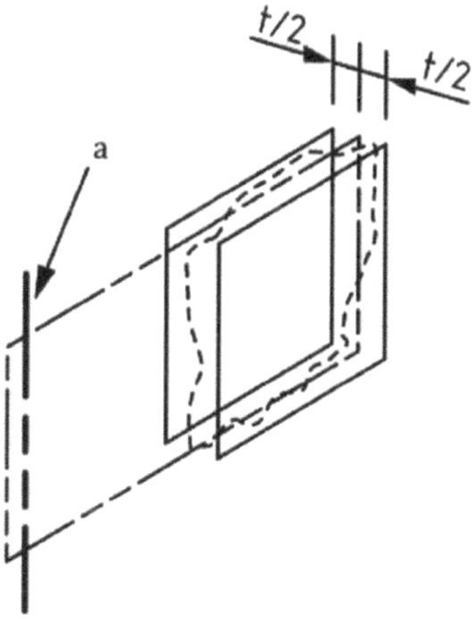

Fig. 13. Tolerance zones - united geometry element source: [3]

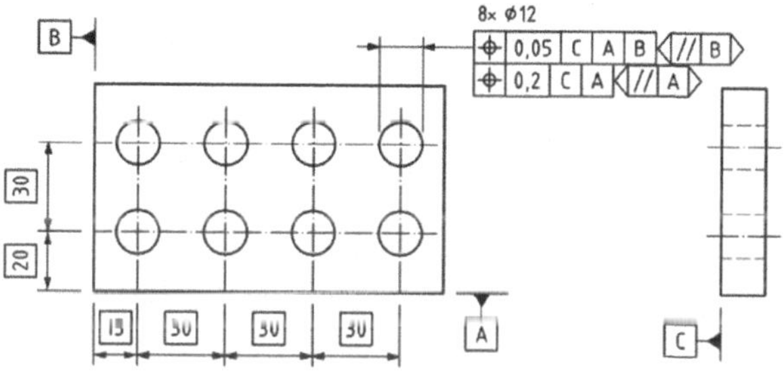

Fig. 14. P. s. of a midplane 2D source: [3]

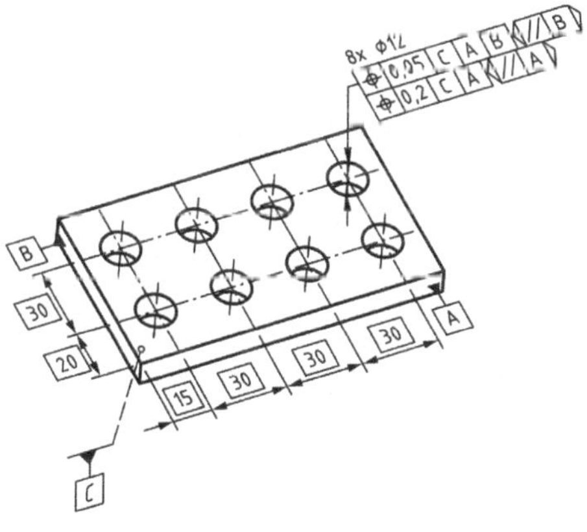

Fig. 15. P. s. of a midplane 3D source: [3]

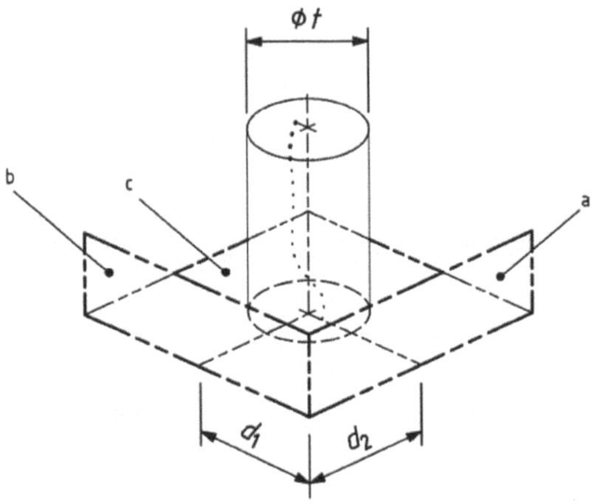

Fig. 16. Tolerance zones source: [3]

The first compromise is to draw on one hand with other specialized software tools or to use automatic scale function. Using this means the model is not really drawn in one by one dimension as design engineers normally used to do their work. If you need than a given dimension it is no problem when the calculation can be done powered by ten. All other calculations will get a slightly rounding error and comparing to the standards is that not what we really expect. As practical engineers we should remember to use the series of preferred numbers. The series of preferred numbers according to DIN 323 (German Standards) are a universal number system agreed by international standards (ISO 3, ISO 17, ISO 497), which serves a comprehensive order and simplification in the technical and economic creation. These numbers are for the selection or grading of sizes of any kind (e.g. lengths, areas, volumes, forces, pressures, moments, voltages, speeds, powers) with the aim of keeping a practically required number of numbers to a necessary minimum. So by doing some miniaturization we should keep this in mind. The program I use is CleWin 4 from the company PhoeniX Software. CleWin 4 is a physical and hierarchical layout editor. The basic drawings are boxes, circles, wires and polygons. Rings and spirals are automatically generated as wires. Text can be included and is realized as polygons. Complex polygons can be created by using Boolean operations. There is a resolution of 1 μm. For designing a gear shaft you have to recognize some geometric relationships. This is not doing some construction like a 3D model, it's more a creating of a pattern. This is what you have to learn, when you decide to make your own model in the micro or nano scale. In Fig. 17 you can see the result of the drawing. It doesn't look like a technical drawing should look like but it is exact what was expected. The dimensions are not visible like in a drawing and you have to look carefully in which dimension you are drawing. There is no scale to see on the surface, you have to switch the grid size, so you can find out the information about the dimension. Normally it is set to μm (micro meter) but could be changed in nm (nano meter). There is than the possibility to generate a pattern as you can see in the Fig. 18. Here it is useful to refer to the information of a later chapter, that one of the resist, because it is essential to make a pattern for a positive or negative resist.

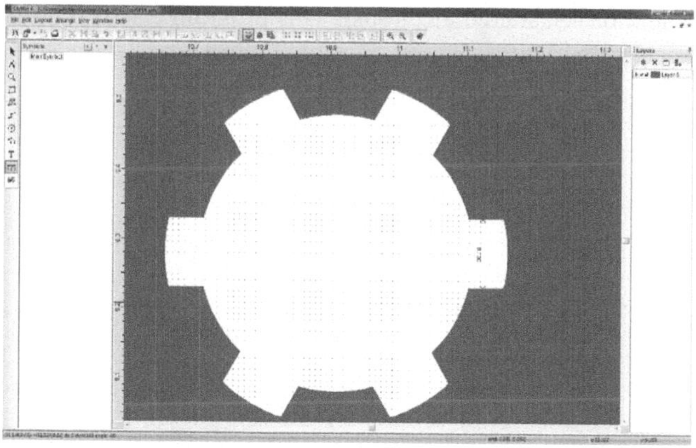

Fig. 17. Design splined shaft

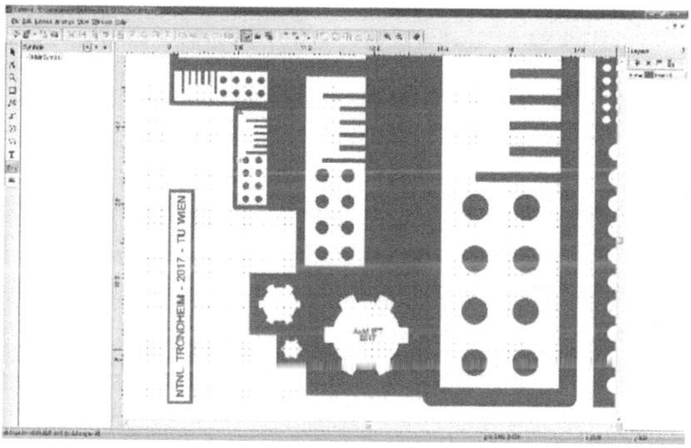

Fig. 18. Design a part of the pattern

Technologies Used

Electron beam lithography (EBL) is the most important and also the most versatile nanofabrication technique. Electrons work very much like photons for nanofabrication. They carry energy and transfer the energy into the energy-sensitive polymer materials called electron resists. The resists exposed by electron beam are developed into surface relief patterns which act as masks for subsequent transfer into the substrate materials [1].

So we first have to get a little understanding about electron beam. To explain how an electron beam works and to understand the challenges by using such a system a picture from a simulation will help us. As you can see in Fig. 19 how the electron beam is spread out. In the Fig. 19 you see the scattering simulation of 30 kV electron beam interacting with 500 nm SU-8 on a silicon wafer, displaying a certain degree of forward scattering. The CASINO acronym has been derived from the words "monte CArlo SImulation of electroN trajectory in sOlids". This program is a Monte Carlo simulation of electron trajectory in solid specially designed for low beam interaction in a bulk and thin foil. This complex single scattering Monte Carlo program is specifically designed for low energy beam interaction and can be used to generate many of the recorded signals (X-rays and backscattered electrons). This program can also be efficiently used for all of the accelerated voltage found on a field emission.

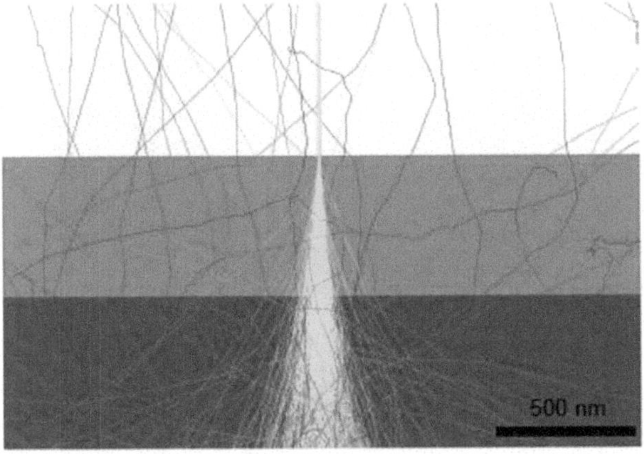

Fig. 19. CASINO scattering simulation of an electron beam. Source: [4]

Practical that means an electron beam loses his bundling of the rays. Later on we will see how these electrons are named and that the system of electrons works in the same way in a scanning electron microscope.

There are a lot of technologies used and there is given a short overview about it. It will be noted that there is now no complete list of all technologies, but a number of representative solutions are shown. To get a better overview how the following structure is coming up there will be a chronological listing what to do to produce a wafer with own structure.

- the wafer – silicon
- cleaning of the wafer
- resist - the material
- get the pattern on the wafer
- developing of the wafer
- measuring and validation

The Wafer – Silicon

A wafer is a thin slice of material, such as a crystalline silicon. Starting with some information about silicon given in the lectures of [5]. There was given much information about silicon such as crystal structure, Miller indices and properties:

Density: 5×1022 atoms/cm^3
Band gap: 1,11 eV = 1,776 10^{-19} J
Size: 2 in. (51 mm)
Thickness: 275 μm

Silicon has diamond structure as you can see in Fig. 20. The wafers used later on are in the view in {1, 1, 0} direction. So in Fig. 21 the Miller indices are shown and Fig. 22 you see the {1, 1, 0} direction. Wafers under 200 mm (8 in. or 8″) diameter have flats cut into one or more sides indicating the crystallographic planes of the wafer. 2″ (2 in.) wafers with a {110} face are used in this work. This is a help to cut the wafer in smaller pieces starting at the edge of the cut, so that creating more test samples and not always using a full wafer. How to produce a wafer with the Czochralski process see Fig. 23.

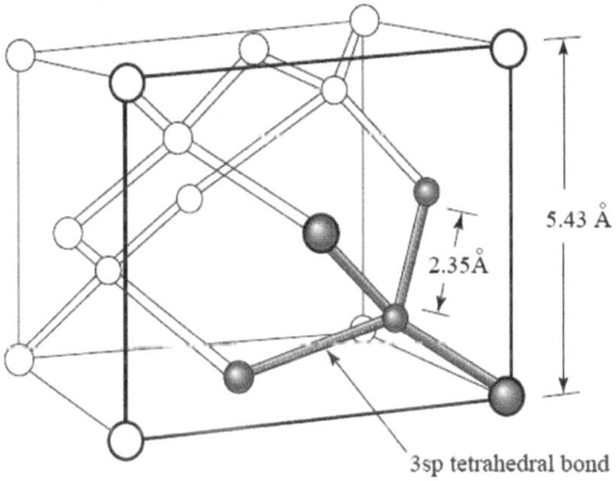

Fig. 20. Crystal. Source: [5]

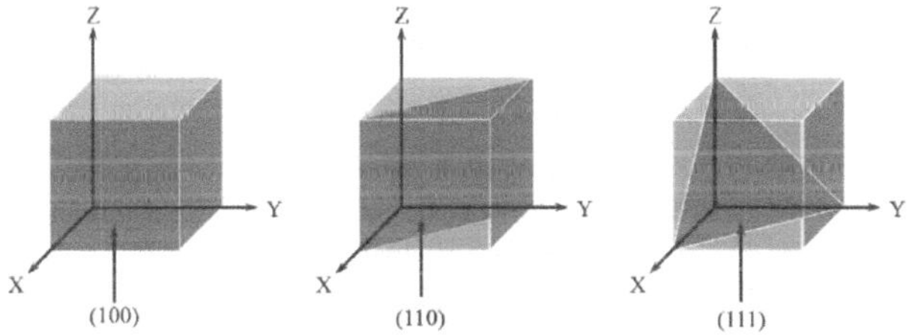

Fig. 21. Miller indices. Source: [5]

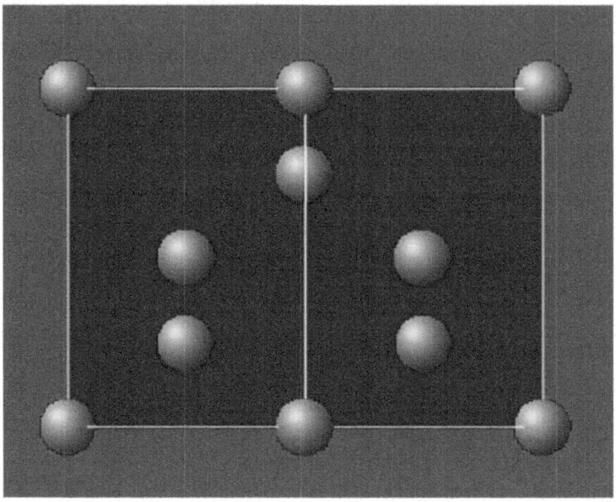

Fig. 22. {1, 1, 0} direction

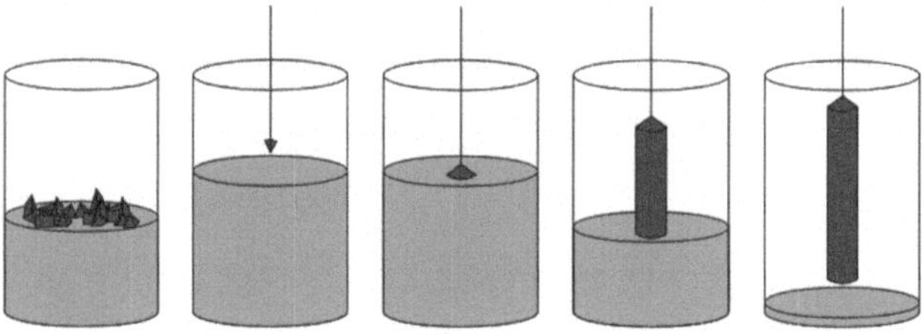

Fig. 23. Czochralski process. Source: [11]

Cleaning of the Wafer. The cleaning of the wafer starts in an Acetone bath for 2 min followed by rinse in IPA (Isopropyl alcohol) for 1 min and blow dry with Nitrogen (N2). Dehydration bake at 150 °C for 5 min on a hot plate. At the end an UV Ozone cleaner will take away the contamination from the surface. Instead of the UV ozone cleaner a plasma cleaner with 50% flow and 50% power for 2 min can be used.

Resist - The Material Photoresist. There are two types of photoresist, positive and negative resist, which are used in different applications. In positive resist, the exposed areas are insolubly and in negative resist the exposed areas are insolubly for wet chemical development. In Fig. 24 you can see it.

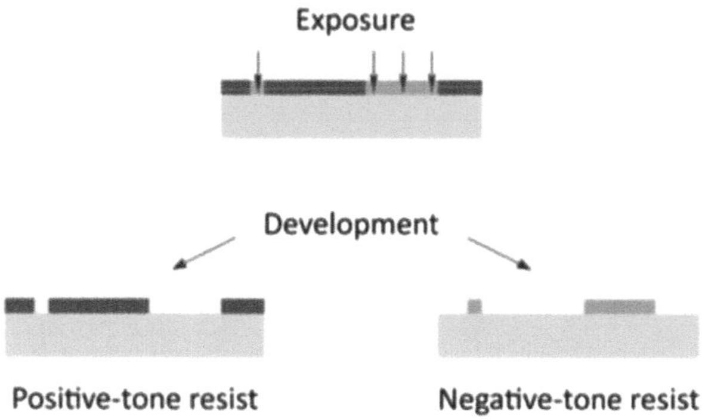

Fig. 24. Positive and negative resist. Source: [19]

Which kind of resist is taken depends on what will be done with this pattern respectively what is the aim of the structure. Working with the resist CSAR 62 there are some general informations and technical informations too. The positive E-Beam resists AR-P6200 is called CSAR 62. The CSAR 62 e-beam resists has a highest resolution and a high-contrast for the production of integrated circuits and masks. Here is some characterization of the resist:

- e-beam; layer thickness 0,05–1,6 μm (6000–1000 rpm)
- high sensitivity which can be adjusted via the developer
- highest resolution (<10 nm) and very high contrast
- highly process-stable, high plasma etching resistance
- developing of the wafer
- easy fabrication of lift-off structures
- poly (α-methyl styrene-co-α-chloroacrylate methylester)
- safer solvent anisole

get the pattern on the wafer Before patterning the wafer must be coated with the resist. So the wafer is placed in a so call spin coater and some drops are placed in the middle of the wafer. Speed and acceleration will bring the needed thickness of the resist. On Fig. 25 it is possible to see how it works.

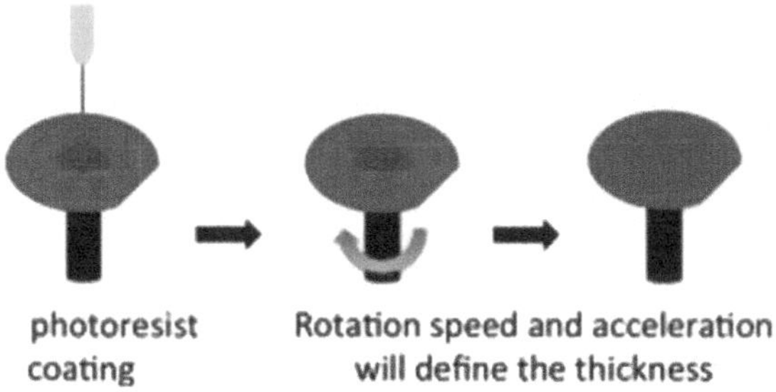

Fig. 25. Spin coater principle. Source: [12]

Figure 26 shows the spin curve of our resist as you can find it in the technical data. Here you can nicely see the relationship between speed and thickness of the resist.

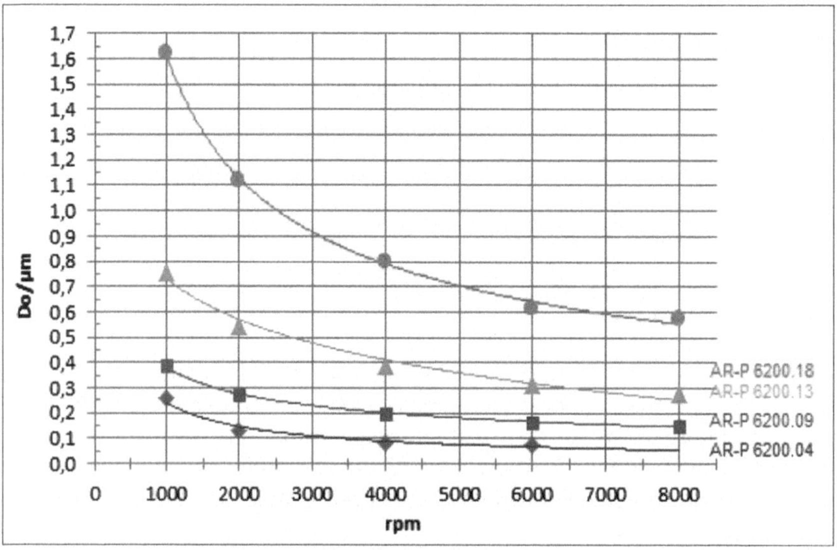

Fig. 26. Spin curve. Source: [12]

After this procedure the coated wafer has to be soft baken on a hot plate and afterwards a cooling to room temperature by means of metal block. In the next step we have to measure by a Reflectometer the thickness of the resist. The principle how a Reflectometer works is shown in Fig. 27.

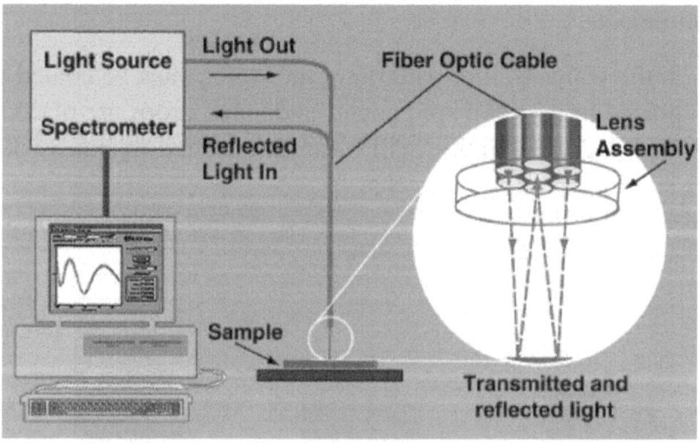

Fig. 27. Principle of a Reflectometer. Source: [23]

Now the wafer is ready for use in the exposure process.

The required e-beam exposure dose for structural imaging mainly depends on the desired minimum structure size, the developer, the acceleration voltage (100 kV), and the film thickness.

EBL (electron beam lithography) is now used to produce the pattern. It is the principal production method for fabricating the masks.

Electron beam lithography is a maskless form of lithography which is very popular in the research community, where the demand for high resolution and prototyping is ever increasing. In EBL, a beam of electrons is scanned in a patterned fashion across a sample covered with electron sensitive resist. Sub micrometer structures are created in the resist and subsequently transferred to the substrate material. The resolution of electron beam lithography is not limited by the wavelength of light as photolithography, but by other process parameter such as resist material and energy of the incoming electrons. The EBL setup at NTNU NanoLab is a Hitachi Field emitter 4300 SEM with state of the art EBL electronics from Raith. The laser interferometer stage enables stitching between write fields with less than a 20 nm shift. Masks are written in CleWin 4. With the resist and processes at NTNU NanoLab, 30–40 nm structures should be feasible.

How to find the applicable base dose?

In the BEAMER v5.2.0 manual a guide is given and here it is now rendered in a meaningful way:

"*BEAMER dose values are relative doses and thereby independent of the resist sensitivity. The dose reference point is called the base dose and is defined as the dose that results in correctly sized large features. This base dose needs to be determined for the different stacks and machine settings that are used in your process. The base dose is independent of layout design and is applicable for arbitrary layouts once determined. To determine the base dose one can use a 1:1 line/space pattern with a line width of 100 nm on a 200 nm pitch. The size of the layout needs to be such, that proximity effects from the border are not seen any more in the centre of the test structure. In other words, the centre of the pattern should be homogeneously sitting atop the backscattered energy. This means the size of the line-space pattern should be at least 4β by 4β in size. Typically a 120μm by 120μm area for bulk Si at 100kV will be sufficient. For 50kV, a 50μm by 50μm square of lines and spaces can be used instead.*

This pattern needs to be exposed at different doses to obtain a line-width (at the centre) vs dose curve. Inspection of the exposed structures at the centre will reveal the base dose, which is the dose that results in equal lines and spaces. Dose B is identified as the basic dose, as shown in Fig. 28*!*" [13]

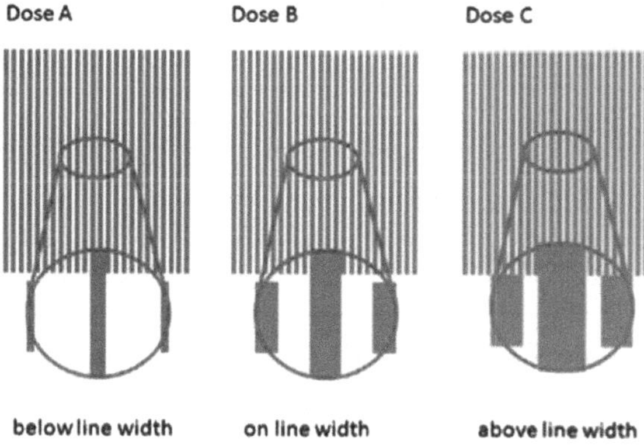

Fig. 28. Base dose pattern. Source: [13]

Trusting in the experience of the laboratory employees at NTNU there has been given two variants.

variant I:
Dot Number 500000 [dots]
Field Size 100 [micron]
Feed pitch 8 dots 1.6 [nm]
Scan pitch 8 dots 1.6 [nm]
Dose 400 [microC/cm^2]
Beam Current 1.00e−009 [A]
Dose Time 0.01024 [micro sec/dot]
variant II: ⇐ selected variant
Dot Number 500000 [dots]
Field Size 100 [micron]
Feed pitch 8 dots 1.6 [nm]
Scan pitch 8 dots 1.6 [nm]
Dose 500 [microC/cm^2]
Beam Current 1.00e−009 [A]
Dose Time 0.01280 [micro sec/dot]

If one is to set up an electron beam lithography process for the first time, the number of interdependent adjustment features seems to be very numerous and all must be considered. It is therefore essential to choose the right beam current and pitch. The design of the system determines the diameter of the beam and this is determined by the normal distribution of the electrons. Now, if the beam is held in position for a longer period of time, the entry or area receiving an increased dose. This is called an increase in the dose and this is greater than the clearing dose (Fig. 29). Another effect enlarges the entry in the border area, the backscattered electrons are the reason for this. Figure 29 is a mix of memory production and a retouched image, meaning that it has been digitally altered from its original version. The original is the intellectual property of NTNU and is not publicly available and can only be viewed by users of NTNU NanoLabs.

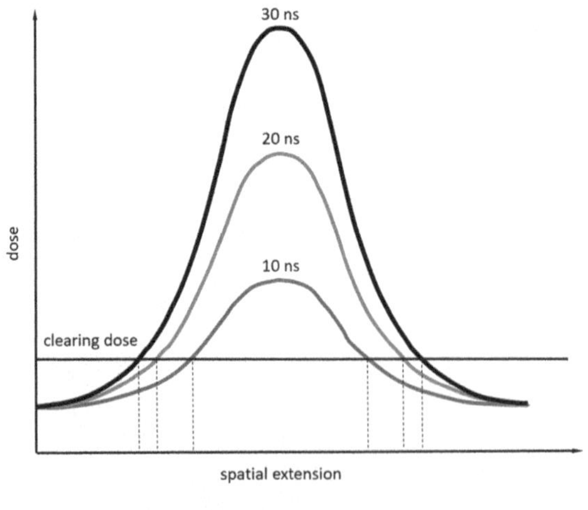

Fig. 29. Spin curve

In the following paragraph is inspired by the tutorial - beam current and pitch written by Einar Digernes, original text is only available at the NanoLab, and the Beamer manual.

The distance between each spot is called pitch, sometimes it is named dot size.

If you want to produce periodical square dot patterns its complete clear that you should use a large pitch or dot size. The next parameter you are able to change is the so called dose time. That time is responsible for the spread out of electron so that the diameter of the spots could be controlled. Most often complicated patterns with different geometries are desired. From this one can conclude that it will be very often necessary to change both the pitch size and the diameter. Since it is a stated goal to produce patterns with small dimensions, one will be tempted to reduce the distances accordingly. As an unwritten rule of thumb, the dot size should have 80% of the beam diameter. In other words, when choosing a diameter of 5 nm for the beam, the result of the pitch should be 4 nm.

This results in two ways to obtain an overlap in the irradiated areas, so that there is no unexposed area between the spots. As illustrated in Fig. 30 you can see, when you choose smaller pitch the number of the spots increase and this is the first way to do. The second way is to increase the dose time or beam current, so that the diameter of the spot increases. This could be also seen in the Fig. 30. When, however, the writing speed for a pattern should be correspondingly high, this can only be done by increasing the diameter.

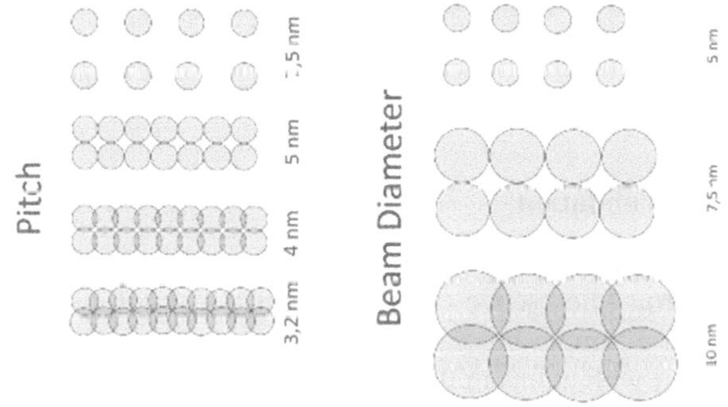

Fig. 30. Pitch

To these results, which are summarized here in a few words and are available in the documents of NanoLab, one comes by consistently following the instructions explained in the BEAMER manual [13].

Figure 30 is a mix of memory production and a retouched image, meaning that it has been digitally altered from its original version. The original is the intellectual property of NTNU and is not publicly available and can only be viewed by users of NTNU NanoLabs.

Developing of the Wafer. To develop resist films, various developers are recommended. As it can be seen from the data sheet of the company Allresist, the developers

AR 600-546, 600-548 and 600-549 are offered here. The developers differ by increasing the sensitivity and the data sheet also details the technical specifications. There are recommendations for the development time and it is also indicated in the data sheet on the appropriate temperatures. It is also important to note that small structures can be destroyed by heavy rinsing (see Fig. 31 below). The post-baking of the wafer is to be seen as the completion of the process. All information about development can be found at [6].

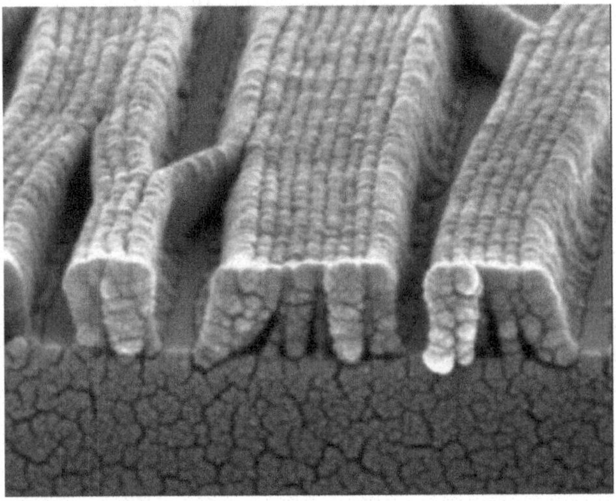

Fig. 31. Danger of collapsed lines after too rigid rinsing. Source: [10]

Measuring and Validation

Make it Visible

The Scanning Electron Microscope (SEM) is another type of electron microscope, it can create images of a sample by scanning the surface with a focused electron beam. The electrons interact with atoms in the sample, producing various signals that contain information about the sample's surface topography and composition. The electron beam is scanned in a raster scan pattern, and the beam's position is combined with the detected signal to produce an image. SEM can achieve resolution near by 1 nm. Specimens can be observed in vacuum. In the NTNU lab a SEM called Apreo is used to make tiny structures visible.

The most commonly used SEM mode is the detection of secondary electrons, which are the electrons emitted by atoms excited by the electron beam.

Among other things, the number of detectable secondary electrons depends on the topography of the sample.

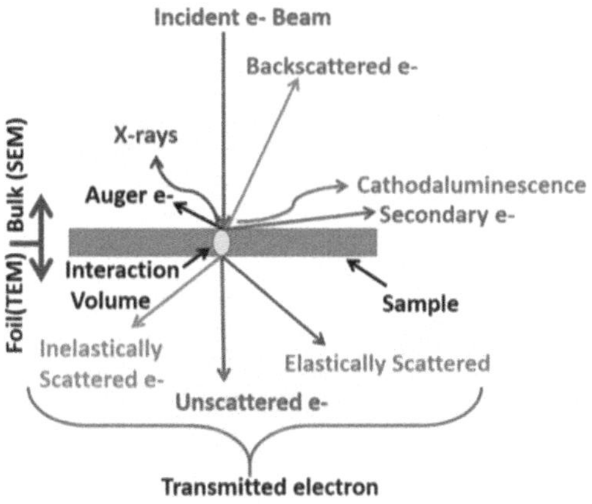

Fig. 32. Physical processes. Source: [19]

By scanning the sample and collecting the secondary electrons emitted by a particular detector, an image is created that indicates the topography of the surface. By scanning the sample and collecting the secondary electrons that are emitted using a special detector, an image displaying the topography of the surface is created. In the following Fig. 32 the physical processes and the names of the electrons are given.

In Fig. 33 the schematic of an SEM is demonstrated and in Fig. 34 the SEM and the needed hardware is shown.

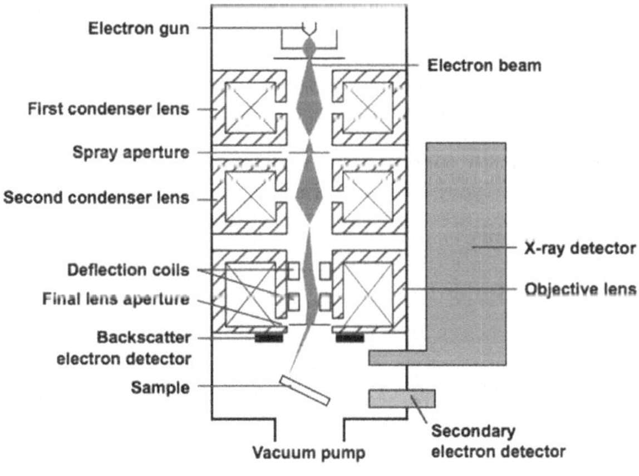

Fig. 33. Schematic of a SEM. Source: [21]

Fig. 34. SEM Apreo. Source: [22]

3 Realization

The next step is to do the practical implementation. As described in the preceding parts of the work, a corresponding pattern is now produced. The SEM then generates the corresponding images. So that the sample can be displayed in the SEM, it must first be provided with a metal coating. Gold is in this case in use and the pattern is provided with a 4 nm thick touch.

4 Conclusion and Outlook

An overview of many relevant parameters to achieve optimal electron beam writing speed, precision and feature quality has been given. Although these parameters have been tried to optimize for CSAR62 GPS samples writing in this work, the general trends in deciding appropriate instrument settings are relevant for many other EBL processes as well.

As an application example, well-defined, defect free arrays of CSAR62 GPS samples were produced on silicon by optimizing all of the above parameters. The arrays were produced over large areas of several cm, and written several times.

A look into the future shows that there is still a lot to do. It is the whole image processing with a software still open. ImageJ can display, edit, analyze, process, save, and print 8-bit color and grayscale, 16-bit integer, and 32-bit floating point images. ImageJ can calculate area and pixel value statistics of user defined selections and intensity-thresholded objects. It can measure distances and angles. This part will be done as the next work. Then an explanation about the big connections can be made. This is now the task of the following months and has already begun.

5 Support from NorFab

"The Research Council of Norway is acknowledged for the support to the Norwegian Micro- and Nano-Fabrication Facility, NorFab, project number 245963/F50".

I would like to express my greatest gratitude to the whole team of the NanoLab, I would not have been able to do the work without their support. Thank you, I would also like to say to Professor Jan Torgersen, who has actively supported me in my endeavours.

References

1. Cui, Z.: Nanofabrication by electron beam. In: Nanofabrication. Springer, Cham (2017)
2. Ramsden, J.: Nanotechnology. Jeremy Ramsden & Ventus Publishing ApS (2009)
3. ISO 1101: Geometrische Produktspezifikation (GPS) Geometrische Tolerierung Tolerierung von Form, Richtung, Ort und Lauf. Deutsche Fassung DIN EN ISO 1101:2017
4. Muller Beckwith, K.: EBL tricks: improving EBL performance for writing nanostructure arrays over large areas. Department of Physics, NTNU (2013)
5. Berstein, R.: TFE4162 & FE8130 MEMS DESIGN, Chapter 3 lesson ntnu autum 2017
6. Allresist GmbH: http://www.allresist.de. Accessed 15 Nov 2017
7. VDI 2083: VEREIN DEUTSCHER INGENIEURE. Cleanroom technology Personnel at the clean work place (2007)
8. Reinraum https://de.wikipedia.org/wiki/Reinraum. Accessed 15 Nov 2017
9. HEPA: https://en.wikipedia.org/wiki/HEPA. Accessed 15 Nov 2017
10. Filmetrics Inc. Rev. 02.06 Printed in USA 2006: https://www.lmetrics.com. Accessed 15 Nov 2017
11. Czochralski process: http://www.top-alternative-energy-sources.com/Czochralski-process.html. Accessed 15 Nov 2017
12. Spin-coated: https://www.elveflow.com/microfluidic-tutorials/soft-lithography-reviews-and-tutorials/introduction-in-soft-lithography/pdms-membrane-thickness-of-a-spin-coated-pdms-layer/. Accessed 15 Nov 2017
13. BEAMER v5.2.0 Manual: http://pegasus.kavli.tudelft.nl/caddoc/BEAMER-5.2.2/manual/manual.htm?new features.htm. Accessed 15 Nov 2017
14. ÖNORM EN ISO 14644-1: Cleanrooms and associated controlled environments (2016)
15. U.S. Federal Standard: 209E "Airborne Particulate Cleanliness Classes in Cleanrooms and Clean Zones" (1992)
16. Cleanroom garment: https://labproinc.com/blog/cleanroom-and-apparel-8/post/how-to-choose-sterile-clothing-for-cleanroom-use-19. Accessed 15 Nov 2017
17. VDI 2083-15 Cleanroom technology - Personnel at the clean work place.
18. Gail, L., Gommel, U (Hrsg): Reinraumtechnik - Textile Reinraumbekleidung. Springer-Verlag GmbH (2018)
19. Hensel, R., Braun, H-G.: Resist: https://www.ipfdd.de/fileadmin/_migrated/content_uploads/Handout_EBL.pdf.Accessed 15 Nov 2017
20. Mahapatro, A. K.: Physics; Characterization Techniques for Materials II; Surface Morphology; Scanning Electron Microscopy
21. SEM: https://commons.wikimedia.org/wiki/File:Schema_MEB_(en).svg. Accessed 15 Nov 2017
22. SEM Apreo: https://cemas.osu.edu/capabilities/apreo-lovac-sems.Accessed 15 Nov 2017
23. THIN-FILM MEASUREMENT; Filmetrics, Inc.: https://files.filmetrics.com/pdf/Filmetrics%20Tutorial%20-%20Thickness%20Metrology%20Guide%20v3N.pdf.Accessed 15 Nov 2019

Profile and Areal Surface Characterization of Additive Manufacturing Polymer and Metal Parts

Binnur Sagbas[1(✉)] and Numan M. Durakbasa[2]

[1] Yildiz Technical University, Istanbul, Turkey
bsagbas@gmail.com
[2] TU Wien, Vienna, Austria
numan.durakbasa@tuwien.ac.at

Abstract. Additive manufacturing (AM) has great potential on manufacturing both polymer and metal parts as final product. However, optimization of surface texture quality has not been fully achieved yet. AM surfaces present variable textures that differs according to the AM methods and process parameters. Because of their unusual texture, it is important to evaluate surface properties of metal and polymer parts produced by different AM processes. In this study, it is aimed to define 2D and 3D surface texture of poly-lactic acid (PLA) polymer and AlSi10Mg metal parts manufactured by Fused Deposition Modelling (FDM) and Direct Metal Laser Sintering (DMLS) respectively. ISO 4287 Ra, Rq and Rz profile roughness parameters were defined by tactile method while ISO 25178 Sa, Sq and Sz areal parameters were measured by optical method. Measurements were taken from up-skin and down-skin of the samples which were in 30 mm diameter and 4 mm thickness. Results were evaluated within themselves and literature studies. Differences between profile and areal surface characterization and need of new specifications were also discussed.

Keywords: Surface metrology · Additive manufacturing · FDM · DMLS

1 Introduction

Additive manufacturing (AM) techniques provide opportunity to creation complex geometries with hallow surfaces and it reduces need for assembly and waste of materials [1]. Although, these technologies have been started to use for manufacturing final products, surface quality is still concern. At that point surface metrology plays important role for characterizing surface texture properties and developing additive manufacturing process conditions [2].

Because of the different nature of the AM processes, surface texture properties of AM parts differ from conventionally manufactured products [3]. Due to these differences, metrology of AM surfaces needs to define proper measurement techniques and parameters for both metal and polymer based AM parts [4]. Also tables and standards need to be developed for selecting suitable specification tolerance at design step and inspection after manufacturing.

V. D. Majstorovic and N. Durakbasa (Eds.): IMEKOTC14 2019, LNME, pp. 240–246, 2019.
https://doi.org/10.1007/978-3-030-18177-2_22

Roughness of the surfaces can be measured by 2D (across a line) or 3D (across an area) techniques such as tactile mechanical and non-tactile optical profilometer respectively [5]. Because of the three dimensional nature of the functional surface, 3D areal surface characterization is more representative than 2D profile parameters. For this reason recent studies mostly focused on about 3D surface measurement for additive manufacturing parts. For optical 3D methods, confocal microscopy, focus variation microscopy and coherence scanning interferometry are the most widely used technologies [6]. Moreover, some studies reported that focus variation microscopy was an effective technology for surface characterization of additive manufacturing parts because of their highly variable surface texture [7–9]. Although, in research studies usage of 3D areal parameters is becoming more and more common every day, their measurement has not been widely propagated in manufacturing area. This is because of the lack of enough experience and standards on areal surface characterization [10].

The present study focused on to define 2D and 3D surface texture of poly-lactic acid (PLA) polymer and AlSi10Mg metal parts manufactured by Fused Deposition Modelling (FDM) and Direct Metal Laser Sintering (DMLS) respectively. ISO 4287 [11] Ra, Rq and Rz profile roughness parameters were defined by mechanical stylus profilometer while ISO 25178 [12] Sa, Sq and Sz areal parameters were measured by optical scanning method. Differences between profile and areal surface characterization were discussed by statistical evaluation.

2 Materials and Methods

In this study two different sample groups such as Polylactic Acid and AlSi10Mg were used for evaluating 2D and 3D surface roughness properties. PLA samples were manufactured by FDM method, in 30 mm diameter and 4 mm thickness, while AlSi10Mg samples were manufactured by DMLS in same dimensions. In previous study optimal process parameters were investigated for PLA by FDM method [13]. The samples were manufactured by using these parameters such as 3 shell number, 50% infill percentage, rectangular infill geometry and 0,25-layer thickness.

AlSi10Mg alloy samples were manufactured by DMLS with EOS M290 which uses 400 W Yb-fibre laser and allows a fast and flexible production of parts directly from CAD data. AlSi10Mg powders, provided by EOS and produced by gas atomization method, were melted under argon atmosphere. The medium size of the powder is about 30 μm. SEM image of the powder can be seen in Fig. 1. Also, Chemical composition was reported in Table 1 as stated by the producer in its data sheets.

After manufacturing, surface of AlSi10Mg samples were post processed by sand blasting and all the samples were cleaned in ultrasonic bath with ethyl alcohol and deionized water.

Surface roughness of the samples was measured by Taylor Hobson Form Talysurf Intra mechanical profilometer and AEP Nanomap 1000WLI optical profilometer. Probe diameter of contact stylus was 5 μm while traverse length was 12.5 mm and cut off wavelength was 2.5 mm. Magnification of optical profilometer objective was 10x where measurement area was 1064 × 1064 μm and optical resolution was 0.92 μm.

Table 1. Chemical composition of AlSi10Mg

Al (%)	Si (%)	Fe (%)	Cu (%)	Mn (%)	Mg (%)	Ni (%)	Zn (%)	Pb (%)	Sn (%)	Ti (%)
Balance	9.0–11.0	0.55	0.05	0.45	0.2–0.45	0.05	0.10	0.05	0.05	0.15

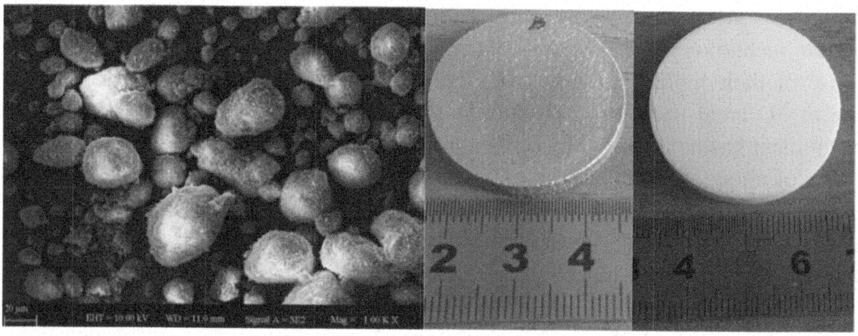

Fig. 1. SEM image of the AlSi10Mg powder and photograph of coupons.

Measurements were taken from five different region and the mean values were reported. Statistical analysis was applied to define if there is significant differences between up-skin and down-skin roughness values of the surfaces. Also regression analysis was applied to define relation between 2D and 3D roughness values.

3 Results and Discussion

Five different measurements were taken from the different region of the sample surfaces. Measurements were taken from random directions on metal samples where they were taken perpendicularly to the direction of the layers. However, lots of different surface parameters have been recorded in each measurement by the software of the profilometer, the most common 2D profile parameters such as Ra, Rq, Rz and 3D areal parameters; Sa, Sq, Sz were recorded. Mean values of ISO 4287 [11] profile parameters and ISO 25178-2 [12] areal parameters measurement results were reported in Table 2.

Table 2. 2D and 3D roughness measurement results

Surface finishing	Ra (μm)	Rq (μm)	Rz (μm)	Sa (μm)	Sq (μm)	Sz (μm)
AlSi10Mg-Up	7.26	8.63	30.28	22.35	28.57	91.42
AlSi10Mg-Down	7.86	8.97	31.12	23.48	29.56	93.59
PLA-Up	4.46	5.87	23.48	9.18	13.76	55.33
PLA-Down	8.61	10.35	45.85	17.39	24.85	95.16

The Sa parameter is the arithmetic mean of absolute height values, z(x, y), within a sampling area. It represents the surface roughness of the sample surface. It is defined by the following equation.

$$S_a = \frac{1}{A}\int_A |z(x,y)|dxdy \tag{1}$$

Although, Sa parameter is common and widely evaluated parameter for characterizing surface roughness of a sample, it doesn't provide enough information about the shape of the surface deviations. For instance, the surfaces have similar Sa values may have different form of deviations [14]. The Sq parameter is the root mean square height of the surface deviations. Being the standard deviation, the Sq parameter has more statistical significance than Sa (Leach 2009) [15]. It is represented with the following equation.

$$S_q = \sqrt{\frac{1}{A}\int\int_A z^2(x,y)dxdy} \tag{2}$$

Sz is the mean value of the five maximum height and five minimum height values within the evaluation area. It is used for evaluation of frictional force, contact strength surface strength and treatability.

3D images of the samples taken by optical profilometer can be seen in Fig. 2.

Up-skin and down-skin roughness values of the AlSi10Mg samples were not so different because same post processing procedure were applied to these two surfaces but as can be seen from figure b and c, roughness values of both side of PLA sample were highly different from each other. Up-skin of PLA sample was rougher than down-skin. This is because of being first layer of the product, sticking to the plate and temperature differences. For statistical evaluation of the roughness difference between surfaces of the same sample, t-test was applied by Minitab 16. Box plots of the up and down skin of the samples can be seen in Fig. 3 Roughness differences between two surfaces of AlSi10Mg sample was not significant (p value = $0.650 > 0.05$ and p value = $0.212 > 0.05$) while it was significant for PLA surfaces (p value = $0.00 < 0.05$.)

Although 2D and 3D surface parameters defines the texture of the same surface, direct comparison of these parameters is not sensible because of different analysis bandwidth and lack of distinct specifications for 3D surface measurement. For instance, Ra parameter is calculated from roughness profile by filtering spatial frequencies with a profile filter such as Gaussian filter, while areal roughness parameter Sa is calculated by a 3D filter on S-L or S-F surfaces. Profile filter considers a line along the X axis while areal filter considers all directions. For determination of a relation between Ra and Sa values, Regression analysis were applied to the Ra values of the samples. Regression graphs can be seen in Fig. 4. For Al samples Ra and Sa relationship was significant (p value = $0.013 < 0.05$) while regression coefficient, $R^2 = 0.75$. For PLA surfaces, relationship between Ra and Sa was significant (p value = $0.001 < 0.05$) and regression coefficient, $R^2 = 0.92$. Because of the high deviation of the roughness values,

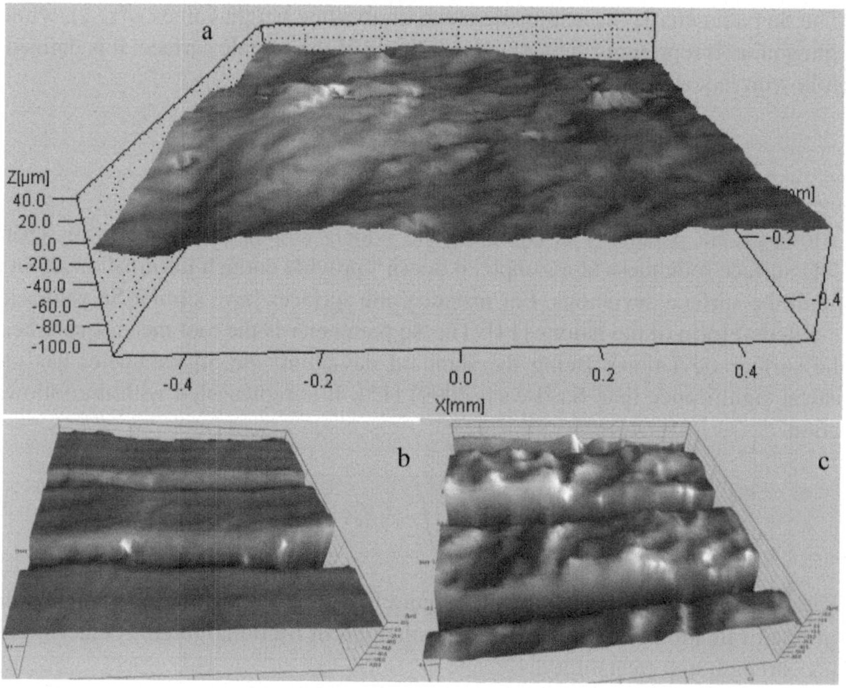

Fig. 2. Optical profilometer images of the sample surfaces. (a) AlSi10Mg, (b) PLA-up skin, (c) PLA-down skin

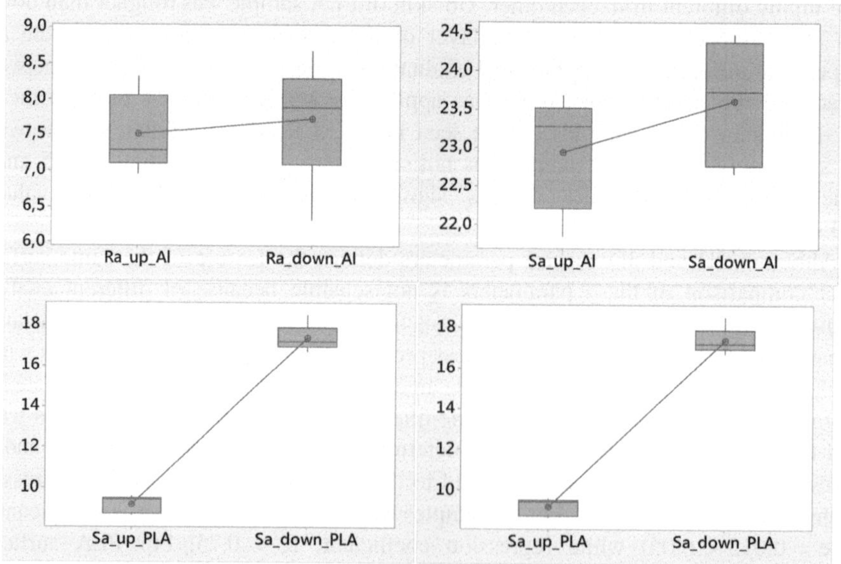

Fig. 3. Box plots of the Ra and Sa values of up and down skin of the samples

regression coefficient for Al samples was lower than PLA samples. Harcarik and Jankovych [10] studied about relation between profile and areal surface texture parameters and they reported high regression coefficient between those parameters for differently manufactured surfaces. As a result, it is worth to say that, with comprehensive and systematical roughness measurements, it is promising issue to obtain more significant and reliable models for predicting roughness values.

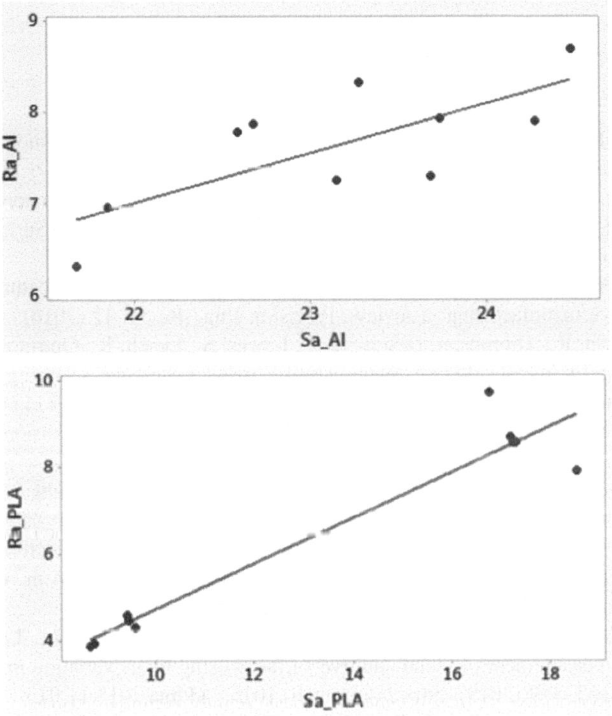

Fig. 4. Linear Regression Plots of Al and PLA samples Ra vs. Sa values

4 Conclusion

Roughness has important effects on functionality of surfaces. Evaluation of new technology surfaces has great importance for improvement of product quality and process efficiency. AM surfaces may have directional features, high slope angles and high aspect ratio textures because of different process parameters and surface post processing. Proper tables and procedures need to be developed for surface texture evaluation of AM parts.

This study evaluated 2D and 3D roughness parameters of polymer and metal AM samples and defined statistical relation between these parameters. For further studies repeated 2D and 3D roughness measurements may be taken from different types of AM surfaces systematically to develop reliable model between 2D and 3D roughness parameters.

Acknowledgments. Direct Metal Laser Sintered AlSi10Mg parts were manufactured at Aluminum Test Training and Research Center, Fatih Sultan Mehmet Vakif University, Turkey. Mechanical Profilometer measurements were taken at Precision Metrology Laboratory, TU Wien, Vienna, Austria. PLA samples production and optical surface measurements were applied at Yildiz Technical University laboratories.

References

1. Gibson, I., Rosen, D.V., Stucker, B.: Additive Manufacturing Technologies, 2nd edn. Springer, New York (2015)
2. Taylor, J.S.: Surface characteristics of additive-manufactured components. In: 15th International Conference on Metrology and Properties of Engineering Surfaces, University of North Carolina, Charlotte (2015)
3. Townsend, A., Senin, N., Blunt, L., Leach, R.K., Taylor, J.S.: Surface texture metrology for metal additive manufacturing: a review. Precision Eng. **46**, 34–47 (2016)
4. Gomez, C., Su, R., Thompson, DiSciacca, J., Lawes, S., Leach, R.: Optimization of surface measurement for metal additive manufacturing using coherence scanning interferometry. Opt. Eng. **56**(11), 111714 (2017). https://doi.org/10.1117/1.OE.56.11.111714
5. Alsoufi, M.S., Elsayed, A.E.: Surface roughness quality and dimensional accuracy—a comprehensive analysis of 100% infill printed parts fabricated by a personal/desktop CostEffective FDM 3D printer. Mater. Sci. Appl. **9**, 11–40 (2018). https://doi.org/10.4236/msa.2018.91002
6. Thompson, A., Senin, N., Giusca, C., Leach, R.: Topography of selectively laser melted surfaces: a comparison of different measurement methods. CIRP Ann. **66**(1), 543–546 (2017)
7. Newton, L., Senin, N., Gomez, C., Danzl, R., Helmli, F., Blunt, L., Leach, R.: Areal topography measurement of metal additive surfaces using focus variation microscopy. Add. Manuf. **25**, 365–389 (2019). https://doi.org/10.1016/j.addma.2018.11.013
8. Senin, N., Thompson, A., Leach, R.K.: Characterisation of the topography of metal additive surface features with different measurement technologies. Meas. Sci. Technol. **28** (2017). https://doi.org/10.1088/1361-6501/aa7ce2
9. Cabanettes, F., Joubert, A., Chardon, G., Dumas, V., Rech, J., Grosjean, C., Dimkovski, Z.: Topography of as built surfaces generated in metal additive manufacturing: a multi scale analysis from form to roughness. Precision Eng. **52**, 249–265 (2018)
10. Harcarik, M., Jankovych, R.: Relationship between values of profile and areal surface texture parameters. Mod. Mach. Sci. J. **6**, 1659–1662 (2016)
11. ISO 4287:2010-07 Geometrical Product Specifications (GPS) - Surface texture: Profile method - Terms, definitions and surface texture parameters
12. ISO 25178-2:2012 Geometrical product specifications (GPS) — Surface texture: Areal — Part 2: Terms, definitions and surface texture parameters
13. Sagbas, B.: Surface texture characterization and parameter optimization of fused deposition modelling process. Düzce Univ. J. Sci. Technol. **6**, 1028–1037 (2018)
14. Leach, R.K.: Fundamental Principles of Engineering Nanometrology, pp. 211–229. Elsevier Inc. (2010)
15. Leach, R.: Characterisation of Areal Surface Texture. Springer, Heidelberg (2013)

Towards Traceable Dimensional Measurements by Micro Computed Tomography

Ágota Drégelyi-Kiss(✉)

Bánki Donát Faculty of Mechanical and Safety Engineering, Óbuda University, Budapest, Hungary
dregelyi.agota@bgk.uni-obuda.hu

Abstract. Industrial CT equipment are increasingly used to determine the dimensions of geometric features (e.g. diameter, position, cylinder) in the industry. In the course of using this new measurement technology in the field of quality control it has to be verified the measured data. In this article the measurement error and uncertainty of the micro computed tomography are investigated by the performing designed experiments. The setting parameters of the CT are changed systematically (RSM method, CCD design) to determine the distance of two ruby spheres. These spheres are the parts of a calibrated ball bar which is connected to an Al test piece during the experiments. The purpose of this work is to make emphasis of the adjustment of the voxel data in case of dimensional measurements performed by industrial computed tomography.

Keywords: Dimensional measurements · RSM method · Computed tomography

1 Introduction

The traceability of measurements is one of the most important issue in case of accurate dimensional measurements. The traceability can be reached by the calibration of the measurement method. In case of dimensional measurements performed by computed tomography the calibration process has been not standardised yet. The different manufacturers offer different solutions to solve this problem [1].

The dimensional measurement of a certain part by industrial CT has several difficulties. The measurement process for dimensional size determination by CT consists of the following step: adjustment of the CT device by a standard, parameter setting of CT equipment, parameter setting for voxel model, surface determination and performing 3D measurements by software [2]. In order to reach accurate dimensional measurement for a certain part it is proposed [3] that let the standard scan together with the part simultaneously. This is important because dimensional measurement values have become more accurate, the values are adjusted with this calibrated ball bar value during the dimensional measurements phase during the 3D measurements performed by the used software.

The accuracy and precision of dimensional measurements in case of industrial CT are affected by several parameters such as mechanical magnification, number of 2D

V. D. Majstorovic and N. Durakbasa (Eds.): IMEKOTC14 2019, LNME, pp. 247–254, 2019.
https://doi.org/10.1007/978-3-030-18177-2_23

X-ray images during the scanning and the scanning parameters (power, filters, timing) [4–8]. Previous studies show the effects of the reconstruction process [1], the effect of different materials [9]. In this article the setting parameters of the CT equipment are investigated, the effect of the magnification, the number of views and the setting base parameters. The experiments were performed in the base of Design of Experiments (DOE), with RSM method. The purpose of this research is to show the effects of the chosen parameters on the size of the calibrated small ball bar within real measurement environment.

2 Materials and Methods

2.1 Standards

Two ball bars were used during the investigations. One ball bar with 99.9276 mm calibrated ruby sphere distance was used for the adjustment of the CT equipment. This process was repeated once the magnification had changed. The other, small ball bar with 15.9329 mm calibrated ruby sphere distance was scanned together with the test part. The distance between the two ruby spheres was determined as the output parameter of the experiments.

2.2 Test Work Piece

A test work piece was designed and produced from aluminium. The overall size of this part is 90 mm × 90 mm × 90 mm. The material is AlMgSi1. The manufacturing was prepared in MSN-500 type milling machine. The ball bar which is used for the correction in the dimensional measurement phase was connected to the test piece in the middle of the upper part of the cube (Fig. 1).

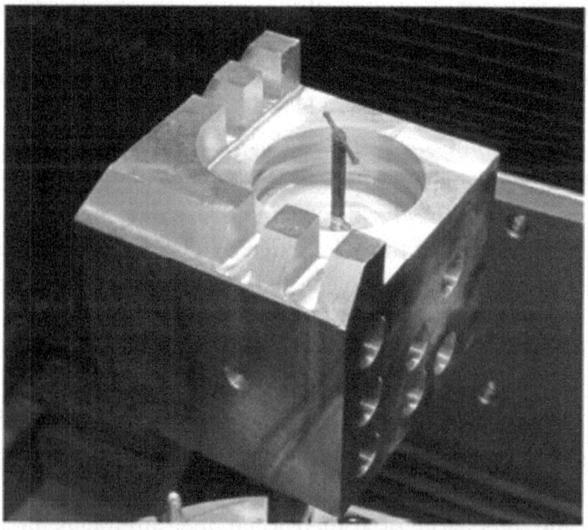

Fig. 1. The aluminium test piece with small ball bar (size: 90 × 90 × 90 mm)

Table 1. The specification of the CT system used according to the manufacturer

System component	Value/attribute/type
Max. voltage/power	300 kV/500 W
Min. voxel size	Down to 1 micron (microfocus tube)
Detector type	Temperature stabilized digital GE DXR detector array, 200 μm pixel size, 2000 × 2000 pixels, 400 × 400 mm, extremely high dynamic range >10000:1
Measurement accuracy	4+L/100 μm referring to VDI 2630-1.3 guideline
Manipulation	Granite based precision 5-axes manipulator (6-axes with detector shift)
Software	Phoenix datos|x 3D computed tomography acquisition and reconstruction software

2.3 The Devices/Equipment Used

The 3D micro-CT system used is *GE phoenix v |tome| x m.* The specification of the CT system components can be found in Table 1. The setting parameters were varied during the measurements (see in Sect. 2.4 Experimental Design).

Measurements of the CT data were carried out using software VG Studio Max 2.2 from Volume Graphics. Measurements were performed on voxel data.

Previous studies show [1] that the optimum setting for the measurement uncertainties and errors for the reconstruction phase is as follows: the Al threshold is manual, the Ruby threshold is manual, ROI setting is off. During the dimensional measurements these parameters were not varied, these values were fixed.

2.4 Experimental Design

During the examination of the CT measurement process several parameters were changed, such as type of magnification, number of views and the scanning parameters of the CT equipment. CT measurements are considered expensive, so it is advisable to reduce the number of experiments as much as possible. The method for this is DOE. In the course multi-factor experimental design the main and the interaction effects of the given parameters can be examined. If the quadratic effects of the given parameters are also taken into consideration, the RSM method is recommended.

Response surface methodology (RSM) method with face centred CCD (central composite design), where all parameters could be examined at 3 levels. The basis of the experimental design was a design consisting of 22 experimental runs, in which three parameters (such as magnification; number of views and scanning parameters) were changed systematically, including a measurement in the central point of the design and its repeat measurement:

- magnification (continuous variable, SDD: source detector distance, SOD: source object distance)

  - Magnification = 2.72 (SDD = 817 mm; SOD = 300 mm)
  - Magnification = 2.33 (SDD = 817 mm; SOD = 350 mm)
  - Magnification = 2 (SDD = 817 mm; SOD = 410.3 mm).
- number of views (three levels: 720, 1080, 1440, continuous variable)
- scanning parameters (the histogram shape for the setting is almost the same in both cases)
  - level 1: U = 250 kV; I = 360 μA; Timing = 330 ms; filter 0,5 mm Cu + 0,5 mm Sn
  - level 2: U = 280 kV; I = 250 μA; Timing = 500 ms; filter 1 mm Cu + 0,5 mm Sn.

The values set in the experimental runs are in Table 2. The experimental runs were prepared in random order (see Run order column) to avoid the optional effect of time. The factors of the CCD were set in a way that the effects of the main factors, interactions and quadratic effects could be estimated orthogonally. Therefore 22 scans by CT were performed, and the distance between the two ruby spheres (small ball bar) were measured in each case.

Table 2. Levels of the experimental runs

Std order	Run order	Magnification (M)	Number of views (NoV)	Scanning parameters (SP)
1	2	2.72	720	Level 1
2	8	2.00	720	Level 1
3	1	2.72	1440	Level 1
4	7	2.00	1440	Level 1
5	3	2.72	1080	Level 1
6	9	2.00	1080	Level 1
7	13	2.33	720	Level 1
8	18	2.33	1440	Level 1
9	19	2.33	1080	Level 1
10	12	2.33	1080	Level 1
11	20	2.33	1080	Level 1
12	21	2.72	720	Level 2
13	5	2.00	720	Level 2
14	11	2.72	1440	Level 2
15	10	2.00	1440	Level 2
16	15	2.72	1080	Level 2
17	4	2.00	1080	Level 2
18	14	2.33	720	Level 2
19	6	2.33	1440	Level 2
20	16	2.33	1080	Level 2
21	22	2.33	1080	Level 2
22	17	2.33	1080	Level 2

3 Results and Evaluation

The measurement values for the ball bar classified by the setting parameters are in Fig. 2. The measurement errors are between – 12 μm and 9 μm. In case of smaller magnification (2.00) the measured distances are larger than in case of higher magnification (2.72).

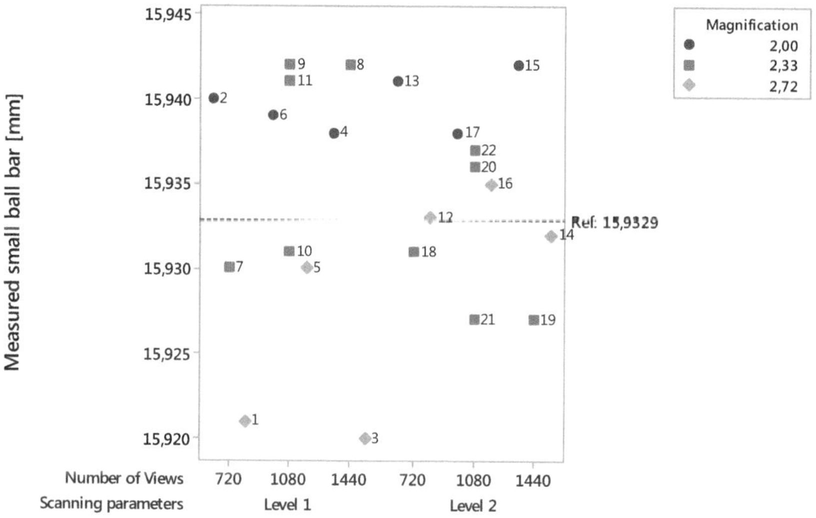

Fig. 2. The measured values of small ball bar (the numbers refer to the Std Order in Table 2)

The measurement errors (measured ball bar distance minus the calibrated value of the ball bar) were calculated and the following phenomenological model was developed:

$$Meas.\ error = \beta_0 + \beta_1 \cdot M + \beta_2 \cdot NoV + \beta_3 \cdot SP + \beta_1 \cdot \beta_2 \cdot M \cdot NoV \\ + \beta_1 \cdot \beta_3 \cdot M \cdot SP + \beta_2 \cdot \beta_3 \cdot NoV \cdot SP + \beta_{11} \cdot M^2 + \beta_{22} \cdot NoV^2 + \varepsilon$$

where M is the magnification, NoV is the number of views and SP is the setting of scanning parameters.

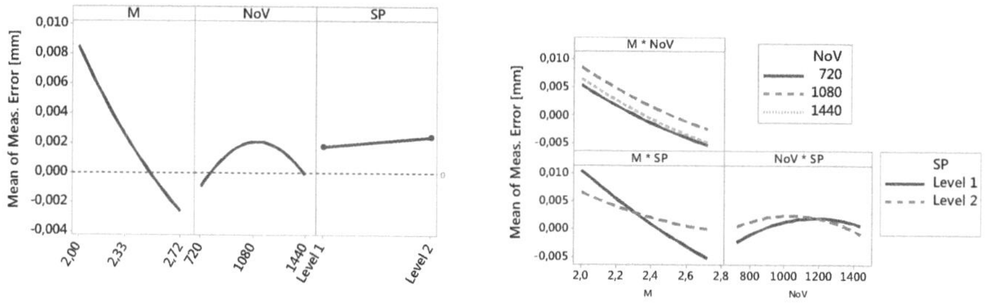

Fig. 3. The main effect and interaction plot for the measurement error

The main effect and interaction plots (Fig. 3) show how the parameter set change the measurement error on average. It is seen that the magnification has large effect on the measurement error. The smaller the magnification is the larger the value of the measurement error.

The absolute value of the measurement error (i.e. accurate measurement) has minimum at about 2.44 magnification. The difference between the two levels of magnification (2 and 2.72) is 18 μm. The change in NoV (from 720 to 1440) the curves related to the magnification show similar relationship (Fig. 3, right side, upper graph). Therefor the increase in measurement error values regarding the magnification factor is independent from the number of views parameter setting.

ANOVA analysis were performed to determine the significant factors at 95% significance level. The results show (Table 3) that the magnification factor (M) has only significant effect on the value of measurement error.

The setting of number of view (NoV) parameter does not affect the measurement error, i.e. the accurate dimensional measurements.

The residuals are normally and randomly distributed. The variance related to the repeated measurements is 0.000034 mm^2, the standard uncertainty is 0.0058 mm. This uncertainty comes from the repeated scans and related only to the size of the standard (ball bar).

Table 3. ANOVA table for measurement error

Source	DF	Adj SS	Adj MS	F-Value	P-Value
Model	8	0,000479	0,000060	1,79	0,169
Linear	3	0,000377	0,000126	3,76	0,038
Magnification (M)	**1**	**0,000374**	**0,000374**	**11,18**	**0,005**
Number of Views (NoV)	1	0,000002	0,000002	0,06	0,807
Scanning Parameters (SP)	1	0,000001	0,000001	0,03	0,857
Square	2	0,000035	0,000018	0,53	0,602
M*M	1	0,000001	0,000001	0,02	0,884
NoV*NoV	1	0,000035	0,000035	1,04	0,327
2-Way interaction	3	0,000066	0,000022	0,66	0,591
M*NoV	1	0,000000	0,000000	0,00	0,952
M*SP	1	0,000052	0,000052	1,56	0,234
NoV*SP	1	0,000014	0,000014	0,42	0,528
Error	13	0,000435	0,000033		
Lack-of-Fit	9	0,000300	0,000033	0,99	0,548
Pure error	4	0,000135	0,000034		
Total	21	0,000914			

The surface plots for the measurement error in case of two type scanning parameter setting are in Fig. 4. It is seen that in case of SP Level 1, the measurement errors are in wider interval than in case of SP Level 2.

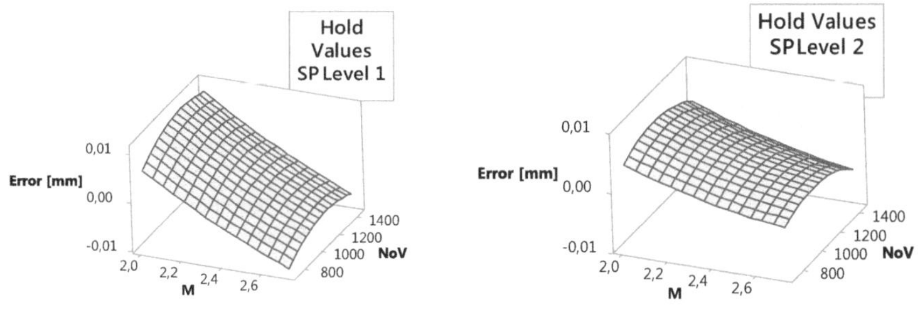

Fig. 4. Surface plots for the measurement error (in cases of scanning parameter two levels)

4 Conclusion

The manufacturer of the investigated CT equipment declared that the measurement accuracy is 4+L/100 μm which equals to 4.16 μm in the examined case. The results show that the standard uncertainty of measuring a calibrated ball bar in real environment (with Al test piece together) is 5.8 μm. These values are comparable with each other and the first one is related to the measured size of a manufactured object, the second refers to a measurement of a standard. Based on these results it can be stated that it is important to make adjustment on voxel data based on a calibrated size to get accurate data for the dimensional sizes of a manufactured part.

The mechanical magnification has large effect and the number of 2D X-ray images has no effect on the measurement accuracy in case of determination of the distance between two ruby spheres of a calibrated ball bar.

In future work the geometrical dimensions and shape parameters of the test piece will be determined after adjustment on voxel data. The effect of the CT parameters on these values will be investigated.

References

1. Bartscher, M., Sato, O., Härtig, F., Neuschaefer-Rube, U.: Current state of standardization in the field of dimensional computed tomography. Meas. Sci. Technol. **25**(6), 064013 (2014)
2. Drégelyi-Kiss, Á., Durakbasa, N.M.: Measurement error on the reconstruction step in case of industrial computed tomograph. In: The International Symposium for Production Research, pp. 309–323. Springer, Cham (2019)
3. Bartscher, M., Neuschaefer-Rube, U., Wäldele F.: Computed tomography - a highly potential tool for industrial quality control and production near measurements. NUMB, pp. 477–48 (2004)
4. Carmignato, S., Dewulf, W., Leach, R.: Industrial X-ray Computed Tomography. Springer, Berlin (2018)
5. Kruth, J.P., Bartscher, M., Carmignato, S., Schmitt, R., De Chiffre, L., Weckenmann, A.: Computed tomography for dimensional metrology. CIRP Ann. **60**(2), 821–842 (2011)
6. Lifton, J.J., Malcolm, A.A., McBride, J.W.: On the uncertainty of surface determination in x-ray computed tomography for dimensional metrology. Meas. Sci. Technol. **26**(3), 035003 (2015)

7. Arenhart, F.A., Baldo, C.R., Fernandes, T.L., Donatelli, G.D.: Experimental investigation of the influencing factors on the structural resolution for dimensional measurements with CT systems. In: 6th Conference on Industrial Computed Tomography, p. 12 (2016)
8. Hiller, J., Hornberger, P.: Measurement accuracy in X-ray computed tomography metrology: toward a systematic analysis of interference effects in tomographic imaging. Precis. Eng. **45**, 18–32 (2016)
9. Drégelyi-Kiss, Á., Marczis, A.: Distance measurements on compound workpiece with industrial computed tomography. In: IOP Conference Series: Materials Science and Engineering, vol. 426, no. 1, p. 012009. IOP Publishing (2018)

Constrained Least-Squares Fitting for Tolerancing and Metrology

Craig M. Shakarji and Vijay Srinivasan(✉)

National Institute of Standards and Technology,
Gaithersburg, MD 20899, USA
vijay.srinivasan@nist.gov

Abstract. Recent years have seen a remarkable emergence of a particular type of least-squares fitting, called *constrained* least-squares fitting, in national and international standards on tolerancing and metrology. Fitting, which is called an *association* operation in the international standards on tolerancing and related metrological practices, is an optimization process that associates an ideal-form feature to a non-ideal-form feature. Unconstrained least-squares fitting has been a very well-known practice in general metrology for a long time, but it has not been the standardized operation in tolerancing and related metrology standards till recently. Instead, other fitting methods, such as Chebyshev fitting, have dominated the standards thus far. Now constrained least-squares is emerging as an attractive alternative in these standards, especially for the establishment of datums. This paper describes these recent developments, with particular attention to the mathematical and computational aspects of the optimization problem, and their impact on the digitization of industrial metrology.

Keywords: Metrology · Tolerancing · Fitting · Optimization · Least-squares · Constraints

1 Introduction

Least-squares fitting has a long and colorful history in science and engineering. It gained popularity in the early 19th century for "the combination of observations" to solve problems in astronomy and geodesy [1]. Carl Friedrich Gauss and Adrien Marie Legendre were the founders of the least-squares fitting, which was first published by Legendre as a method of "moindres quarrés" in 1805 but its priority of discovery was claimed by Gauss.

In that dispute Gauss prevailed, and his name is celebrated in tolerancing and related metrology standards even to this day – with the letter G and the phrase 'Gaussian association' to refer to the least-squares method [2]. In the later part of the 19th century, least-squares fitting became the primary tool for regression analysis and thus became an integral part of statistics [3, 4]. Starting with the work of Sir Francis Galton around 1885, regression was used to create a new way to think about multivariate data in all experimental sciences – and least-squares fitting was at the heart of this new revolution in statistical studies.

V. D. Majstorovic and N. Durakbasa (Eds.): IMEKOTC14 2019, LNME, pp. 255–275, 2019.
https://doi.org/10.1007/978-3-030-18177-2_24

As a statistical tool, the least-squares fitting has been used since its inception in general metrology for discovering the underlying trend in measured data and for suppression of unwanted noise. So it is natural to expect that in the industrial measurement of geometrical quantities, least-squares fitting should enable the following three goals of statistics outlined by Ronald Fisher [3]: (1) the study of *populations* of part geometries, (2) the study of *variation* of geometry within a part and among the parts, and (3) the study of methods of the *reduction of data* to extract a small number of geometrical parameters from a large amount of point coordinate data. However, this potential is being recognized only recently in national and international standards on tolerancing and metrology, and this welcome change is primarily due to the emergence of a digital transformation of manufacturing.

With these recent developments in mind, this paper focuses on dimensional and geometrical tolerancing, and related metrological practices. The industrial use of tolerancing and metrology is dominated by national and international standards, such as those issued by the American Society of Mechanical Engineers (ASME) and the International Organization for Standardization (ISO). ASME still refers to its flagship standard in tolerancing as 'dimensioning and tolerancing' [5]. In the ISO parlance, tolerancing and related metrological practices are collectively referred to as 'geometrical product specifications and verification,' and are standardized under the 'GPS' brand (for example, as in [2]). Sometimes they are also called ISO GPS standards to avoid any confusion with the more popular use of the acronym GPS for the Global Positioning System.

In these ASME and ISO GPS standards, various types of fittings (curiously, dominated by methods other than the least-squares fitting due to the erstwhile prevalence of physical datums and physical gages) have been defined and used for a long time to characterize the populations of geometrical parts for interchangeability, to accommodate and control their inevitable geometrical variations, and to reduce the measured data to a few important parameters. These are, of course, in line with the traditional goals of statistics as outlined by Fisher and mentioned above.

However, there is an additional and distinct role for fitting in tolerancing and metrology. This additional role is about the digital simulation of inspection fixtures and gages in the form of datums and mating envelopes. The need for such digital simulation has grown considerably in recent years because of the digital transformation of manufacturing and metrology, which is variously referred to as smart manufacturing, Industrie 4.0, and cyber physical metrology.

Here the emphasis is more on simulating in the digital world what has been practiced in the physical world, thereby enabling the use of such concepts as digital thread and digital twins. This will also reduce the industry's dependence on expensive inspection fixtures (such as surface plates, expanding mandrels, and collet chucks) and functional gages (such as go and no-go gages).

It is in the context of this digital transformation that *constrained* least-squares fitting has emerged as an attractive alternative to more traditional (which are not based on least-squares) fittings in metrology. There is already an intense interest in the ASME and ISO GPS standards to use constrained least-squares fitting as the default method for simulating datums in the digital world (as well as in the physical world); there is also an interest in using it for simulating mating envelopes that arise in digital assembly operations (as well as in physical assembly operations). In addition, it has been recognized that *unconstrained* least-squares fitting can eminently serve Fisher's three goals of statistics in the digital world (as well as in the physical world), and this is now getting a greater attention in the ISO GPS standards. These recent developments are described in this paper.

The major technical contribution of this paper is the exposition of *constrained* least-squares fitting as an attractive alternative to the more traditional fittings in tolerancing and metrology. As a minor contribution, *unconstrained* least-squares fitting gets an excellent rating for statistical characterization (*a la* Fisher) of measured geometrical quantities. The paper accomplishes these goals by a detailed study of different types of fittings, which are illustrated with carefully chosen examples. All of these fitting problems are formulated as optimization problems amenable to numerical solutions. The mathematical structure of the optimization problems and the behavior of the computational solution schemes are described with equations and illustrated with examples for comparison.

The paper is organized as follows. Section 2 poses fitting as an optimization problem. Unconstrained least-squares fitting is described in Sect. 3. This is followed by the somewhat specialized, but important, problems of maximum inscribing and minimum circumscribing fittings in Sect. 4. The topic of minimum zone fitting is addressed is Sect. 5. Then Sect. 6 describes and illustrates the constrained least-squares fitting, which is the major theme of this paper. Finally, Sect. 7 provides a summary and some concluding remarks.

2 Fitting as an Optimization Problem

Consider a surface feature F on a manufactured part as indicated in Fig. 1(a) with a two-dimensional illustration. This feature is continuous with an (uncountably) infinite number of points. It is deliberately shown with wrinkles to highlight the fact that the manufactured feature does not have an ideal geometric form. The ISO GPS standards and some research literature refer to such non-ideal form features as portions of a 'skin model' of the manufactured part, thus evoking an analogy to the skin of an organic entity with wrinkles, bumps, and blemishes.

The purpose of fitting is to associate an ideal-form surface feature S indicated in Fig. 1(a) to the non-ideal surface feature F. Here the feature S is a mathematical surface of a specified type (e.g., plane, cylinder, sphere) and has a finite number of undetermined parameters. Some of these parameters are intrinsic to S, such as its radius. Other parameters are used to position S relative to F by translation and rotation. So, the fitting can be posed an optimization problem of finding the undetermined parameters of S that shapes and moves S as close to F as possible.

To quantify the notion of the closeness of S to F, consider the shortest distance $d(\boldsymbol{q}, S)$ of a point $\boldsymbol{q} \in F$ to S as shown in Fig. 1(a). Then, taking an elemental area dA in F around the point $\boldsymbol{q}$, an optimization problem can be posed using the L_p-norm as

$$\min_{\text{parm}(S)} \left[\frac{1}{A} \int_F |d(\boldsymbol{q}, S)|^p dA \right]^{1/p} \tag{1}$$

subject to applicable constraints. Here A is the area of F and parm(S) are the parameters (both intrinsic and positional) of S. The power p in Eq. (1) can take any integer value; but the most popular values of p are 1, 2, and ∞. When $p = 2$, the objective function uses the L_2-norm, and the fitting is called the least-squares fitting or the Gaussian fitting. When p is chosen to be ∞, the objective function is said to employ the L_∞-norm, and the fitting is called the minimax fitting or the Chebyshev fitting. In any case, Eq. (1) shows how a fitting (which is called an 'association operation' in the ISO GPS standards) problem is defined when a geometrical product is specified, that is, when it is toleranced.

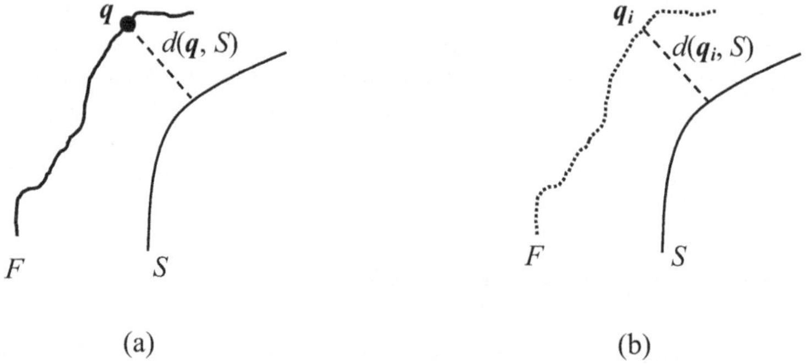

Fig. 1. Notations for defining the fitting problem.

When a manufactured part is verified using metrological techniques for conformance to specifications, only a finite number of points are sampled on the surface feature F, as illustrated in Fig. 1(b).

If the sampling is uniform on F, which is a big assumption that will be revisited later, then the optimization problem can be approximated by

$$\min_{\text{parm}(S)} \left[\frac{1}{n} \sum_{i=1}^{n} |d(\boldsymbol{q}_i, S)|^p \right]^{1/p} \tag{2}$$

subject to applicable constraints. Here n is the number of points sampled on F. The objective function in Eq. (2) uses the l_p-norm (to distinguish it from the integration formulation in Eq. (1) using the L_p-norm for the continuous feature F), with the l_2-norm leading to the least-squares fitting (also known as the Gaussian fitting) and the l_∞-norm leading to the minimax fitting (also known as the Chebyshev fitting).

An important observation related to Fig. 1 is that the distance $d(\boldsymbol{q}, S)$ is taken to be 'perpendicular' to S from q; because of this, the least-squares fitting is also known as orthogonal least-squares fitting, an instance of 'total' least-squares fitting. In the context of ASME and ISO GPS standards, any mention of least-squares fitting means (orthogonal) total least-squares fitting. Another important observation is related to Eqs. (1) and (2), and it pertains to the mention of 'applicable constraints.' If no constraints are specified, then it is an *unconstrained* optimization problem. In a *constrained* optimization problem, some constrains should be specified.

To enable the specification of such constraints, it is useful to assign signs to the distances $d(\boldsymbol{q}, S)$ and $d(\boldsymbol{q}_i, S)$ in Fig. 1. It is often necessary to keep the fitted surface feature S outside of (or external to) the material side of the manufactured feature F. This can be accomplished by first demanding that the fitted surface S is orientable (that is, no Mobius strips!), and then assigning a unique positive side to S so that $d(\boldsymbol{q}, S)$ is

- positive if $\boldsymbol{q}$ is on this positive side of S,
- equal to 0 if it is on S, and
- negative if $\boldsymbol{q}$ is on the negative side of S.

A similar convention applies to $d(\boldsymbol{q}_i, S)$ as well. Then the constraints of keeping the fitted surface S outside or inside of (equivalently, external or internal to) the material side of F can be mathematically posed as

$$d(\boldsymbol{q}, S) \geq 0, \boldsymbol{q} \in F \text{ in Eq. (1), and } d(\boldsymbol{q}_i, S) \geq 0, \forall i \text{ in Eq. (2).} \tag{3}$$

With these preliminaries for posing fitting as an optimization problem, the mathematical structure of the problem and the behavior of computational schemes to solve the problem will be explored with two concrete examples involving circles. Table 1 gives the x- and y-coordinates of nine input points as a 9×2 matrix $\boldsymbol{P}$ for Example 1. The actual units are not important – they can be centimeters or inches. These points represent a (non-uniform) sampling all around a circle, as can be seen in Fig. 2(a).

Table 1. Point coordinates for Example 1.

$$\boldsymbol{P} = \begin{bmatrix} 0.904085 & -0.013351 \\ 0.571958 & 0.567369 \\ 0.379183 & 1.006086 \\ -0.428696 & 0.850569 \\ -0.926511 & 0.365703 \\ -0.964090 & -0.326073 \\ -0.402405 & -1.030703 \\ 0.162633 & -1.090304 \\ 0.735787 & -0.751396 \end{bmatrix}$$

Table 2 provides the x- and y-coordinates of seven input points as a 7×2 matrix $\boldsymbol{P}$ for Example 2. These points represent a (non-uniform) sampling on an arc of a circle, as can be seen in Fig. 2(b); in fact, this arc is deliberately chosen to be less than a semi-circle. These two examples will be used for exploring various fitting problems in some detail in the following sections.

Table 2. Point coordinates for Example 2.

$$\boldsymbol{P} = \begin{bmatrix} -0.82705 & -0.64550 \\ -0.69887 & -0.94646 \\ -0.82899 & 0.94778 \\ -0.80932 & -0.69565 \\ -0.97495 & 0.60506 \\ -1.15281 & -0.15196 \\ -1.00753 & 0.43532 \end{bmatrix}$$

In many cases, the optimization problem will be solved iteratively, starting from a good initial solution. Fortunately, a good starting solution for circles can be found algorithmically [6, 7]. Figure 2 shows these initial circles for Examples 1 and 2. Table 3 presents the initial values for the center coordinates (x_0, y_0) and the radius r_0 that will be used in the iterative solutions.

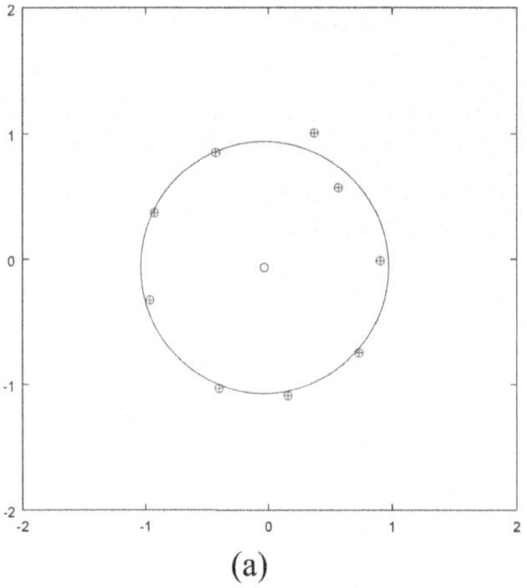

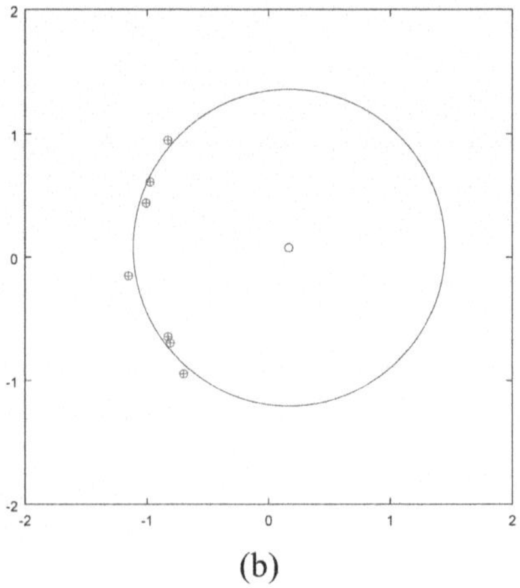

Fig. 2. Approximate fitting of circles: (a) for Example 1, and (b) for Example 2.

Table 3. Approximate solutions for starting circles.

	x_0	y_0	r_0
Example 1	−0.034197	−0.066199	1.0055
Example 2	0.167758	0.075117	1.2818

3 Unconstrained Least-Squares Fitting

Consider the unconstrained least-squares fitting of a circle to a set $\{\boldsymbol{q}_i\}$ of n points, each with coordinates (x_i, y_i), in a plane. The circle can be parameterized by its center coordinates (x_c, y_c) and its radius r_c. The distance between the center $\boldsymbol{c}$ of the circle and the point $\boldsymbol{q}_i$ is denoted by r_i. Figure 3 shows these notations, which will be used throughout this paper for circle fitting problems. The unconstrained optimization problem can then be posed as

$$\min_{x_c, y_c, r_c} \left[\frac{1}{n} \sum_{i=1}^{n} \left\{ r_c - \sqrt{(x_i - x_c)^2 + (y_i - y_c)^2} \right\}^2 \right]^{1/2}. \tag{4}$$

The objective function in Eq. (4) can be recognized as the RMS (Root Mean Square) value of the distances (i.e., deviations) of the points from the circle; so, the optimization problem minimizes the RMS value of the deviations. The objective function to be minimized is non-linear in the free parameters, but it is a smooth and continuous function of these three free parameters x_c, y_c, and r_c.

So, it is possible to find a solution to this problem by gradient descent methods [6, 8, 9], starting from a good initial solution.

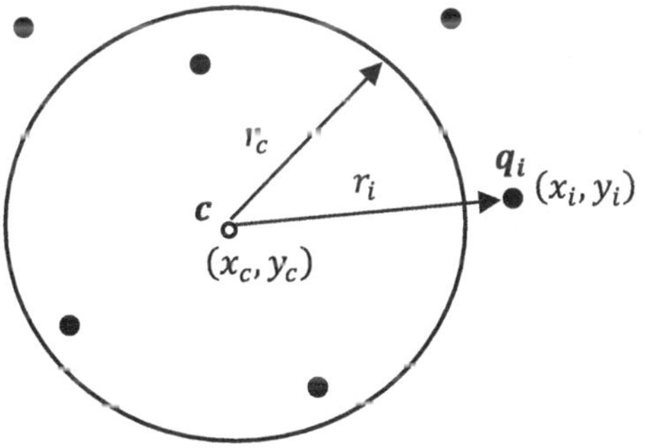

Fig. 3. Notations for circle fitting problems.

The structure of the optimization problem posed in Eq. (4) can be better visualized by splitting it into two problems. First, by fixing the center coordinates (x_c, y_c) and varying only the radius r_c, it can be seen that the minimum of the objective function can be obtained when $r_c = \frac{1}{n}\sum_{i=1}^{n} r_i$. That is, when the radius of the circle is the mean

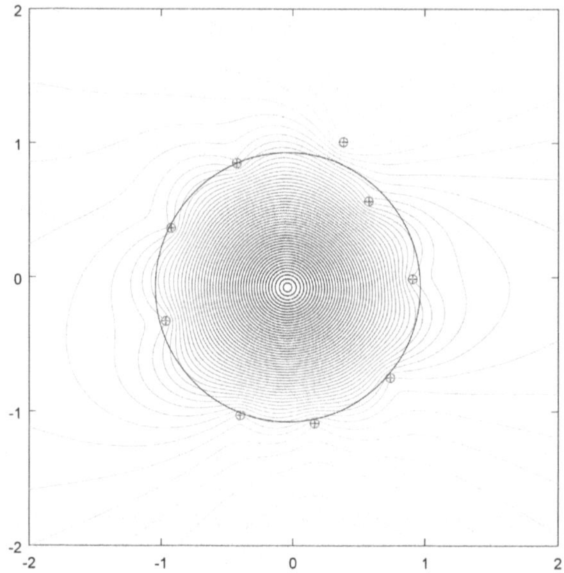

Fig. 4. Unconstrained least-squares fitting for Example 2, with contour plot. $\boldsymbol{c} = (-0.041347, -0.075330)$, $r_c = 1.0035$, $f = 0.071073$ is the minimized objective function value. f is also the RMS value of the deviations of the input points from the circle.

(average) of the distances of the points from the center. This then reduces the second part of the optimization problem of Eq. (4) to one whose objective function involves only two free variables, namely the center coordinates x_c and y_c.

A contour plot of this bivariate objective function for Example 1 is shown in Fig. 4, along with the unconstrained least-squares fitting circle. The smoothness and continuity of the objective function can be inferred from the contour plot. It can also be seen that the center of the fitted circle lies at the 'bottom of valley' created by the bivariate objective function. The optimization problem was solved by applying a gradient search method, starting with the solution of Table 3.

As indicated in the caption of Fig. 4, the minimized objective function is the RMS value of the deviations of the input points (of Example 1) from the fitted circle; it can be observed that it is also the standard deviation, σ, of these deviations. A similar plot and explanations for Example 2 are shown in Fig. 5. It is interesting to note that the bivariate objective function depicted by the contour plot in Fig. 5 is still smooth and continuous, and the unconstrained least-squares fitting gives a satisfactory solution even when the input points are only on a small arc (less than a semi-circle) of a circle.

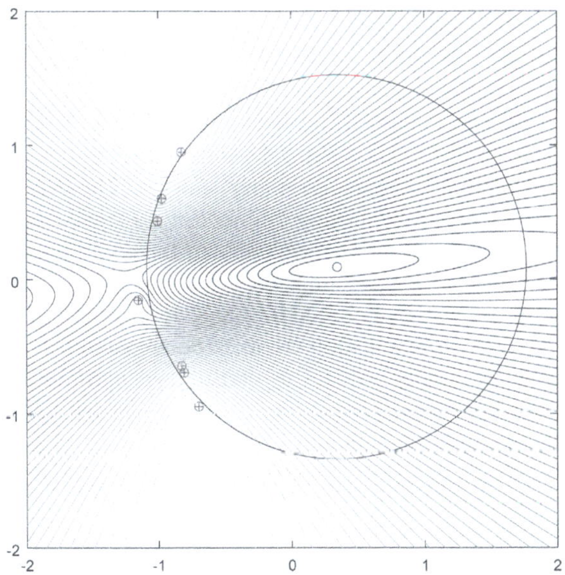

Fig. 5. Unconstrained least-squares fitting for Example 2, with contour plot. $\boldsymbol{c} = (0.341840, 0.092506)$, $r_c = 1.4311$, $f = 0.045317$ is the minimized objective function value. f is also the RMS value of the deviations of the input points from the circle.

The computed values of r_c and σ have special significance for specification and verification of a manufactured circular feature represented by the sampled input points. In recent ISO GPS standards, the diameter $2r_c$ of the unconstrained least-squares fitting circle is the Gaussian 'size' of the circular feature, and 6σ is a measure of the 'roundness' of the circular feature according to the Gaussian criterion.

The unconstrained least-squares fitting has been extended to several two- and three-dimensional features, such as straight lines, planes, cylinders, spheres, cones, and tori [6, 9].

Their computations have been implemented in commercial software systems, especially those that support coordinate measuring machines (CMMs), and have been widely tested for correctness.

However, a word of caution is in order regarding these computations. If the points are sampled uniformly by a CMM, then the discrete approximation of Eq. (1) by Eq. (2) can be justified, and the convergence of the discrete solutions to the continuous case as $n \to \infty$ can be established. But non-uniform sampling, such as those seen in Figs. 4 and 5 do not provide such comfort of convergence. This issue will be discussed further in Sect. 6 on constrained least-squares fitting.

Since the input points lie both inside and outside of the unconstrained least-squares fitting circle, it is not suitable for establishing datums or estimating mating sizes. Even if the least-squares circle is offset (by shrinking or expanding) internally or externally to keep the input points outside or inside, respectively, such an offset circle is not acceptable to the engineering community to serve as a datum or a mating envelope. These needs can be met by other types of fitting described in the following sections.

4 Maximum Inscribing and Minimum Circumscribing Fittings

There is a class of fitting problems – involving geometrical features such as circles, parallel lines, spheres, cylinders, and parallel planes – that admits the notion of maximum inscribing and minimum circumscribing features. These geometrical features belong to 'features of size' in tolerancing and metrology standards. Such fittings can serve as datum simulators and mating envelopes; in fact, all the national and international standards have relied on these fittings for these very purposes. However, it is also known that such fittings have some limitations and drawbacks.

This section poses the computation of maximum inscribing circles (MIC) and minimum circumscribing circles (MCC) as constrained optimization problems, and illustrates them using Examples 1 and 2. For this, consider the following constrained optimization problems using the notations of Fig. 3.

$$\text{For MIC: } \max_{x_c, y_c}(r_c) \text{ subject to } r_i \geq r_c, \forall\, i. \tag{5}$$

$$\text{For MCC: } \min_{x_c, y_c}(r_c) \text{ subject to } r_i \leq r_c, \forall\, i.$$

As will be explained later, such problems may not always be well defined, especially for MIC. To gain an understanding of the mathematical structure of these optimization problems and to compute their solutions, notice that Eq. (5) can be reformulated as the following equivalent problems.

$$\text{For MIC : } \max_{x_c, y_c}(r_{\min}) \text{ where } r_{\min} = \min(r_i), i = 1, \ldots, n. \tag{6}$$

$$\text{For MCC : } \min_{x_c, y_c}(r_{\max}) \text{ where } r_{\max} = \max(r_i), i = 1, \ldots, n.$$

In this reformulation, MIC becomes a maximin problem and MCC becomes a minimax problem. In practice, the MIC problem is posed as $\min_{x_c, y_c}(-r_{\min})$ for computational convenience.

Figure 6 shows the MIC and the related contour plot of the objective function for Example 1. It can be inferred from the contour plot that the objective function is not smooth – the contour curves seem to have some sharp corners. In fact, it can be shown that the objective function has first-derivative discontinuities, and the points at which such discontinuities occur are aligned along the nearest-neighbor Voronoi diagram of the input points as shown in Fig. 7. Also, the center of the MIC lies on a Voronoi edge or a Voronoi vertex. Figure 8 shows the MCC and the related contour plot of the objective function for MCC; here, it can be shown that the center of the MCC lies on the furthest-neighbor Voronoi diagram. These facts have been proved and exploited in finding algorithmic solutions to the MIC and MCC problems [10, 11]. The solution for the MIC may not be unique [12] – this is one of the drawbacks of this maximin problem.

It is also possible to find iterative solutions to the MIC and MCC problems, starting with a good initial solution. The presence of derivative discontinuities, and the fact that the solutions lie at these derivative discontinuities, require that a different search (i.e., not the gradient search) method be employed to find the optimum solution. A derivative-free Nelder-Mead (also known as 'downhill simplex') method can be used in these cases [8].

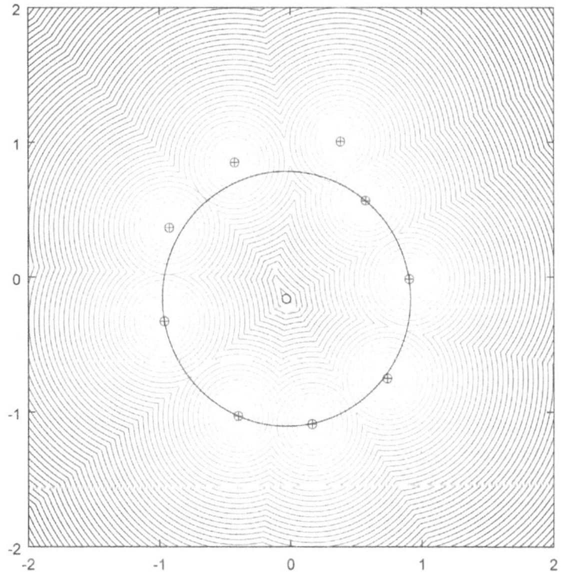

Fig. 6. MIC for Example 1, with contour plot. $\boldsymbol{c} = (-0.032819, -0.159977)$, $r_c = 0.94592$, $f = -0.94592$ is the minimized objective function value.

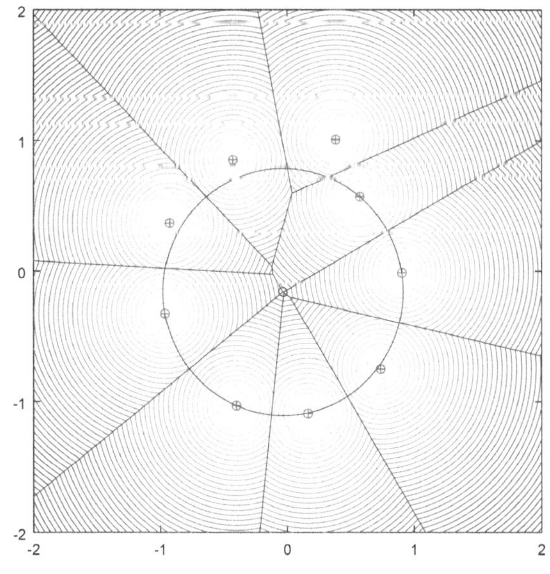

Fig. 7. MIC for Example 1, with contour plot and nearest-neighbor Voronoi diagram.

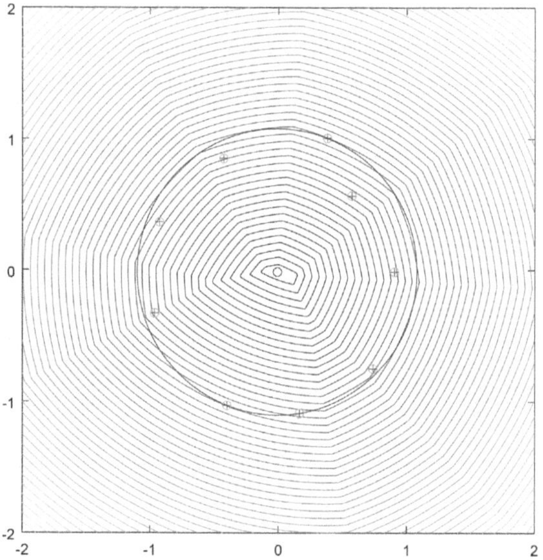

Fig. 8. MCC for Example 1, with contour plot. $\boldsymbol{c} = (-0.0093801, -0.0131645), r_c = 1.0908$, $f = 1.0908$ is the minimized objective function value.

The non-uniqueness of MIC may not be a show-stopper in practice. A more serious problem arises in cases like Example 2, where the MIC problem itself is not well defined and it does not have any finite solution at all. Figure 9 shows the contour plot of the objective function of MIC for Example 2. The solution is unbounded in this case.

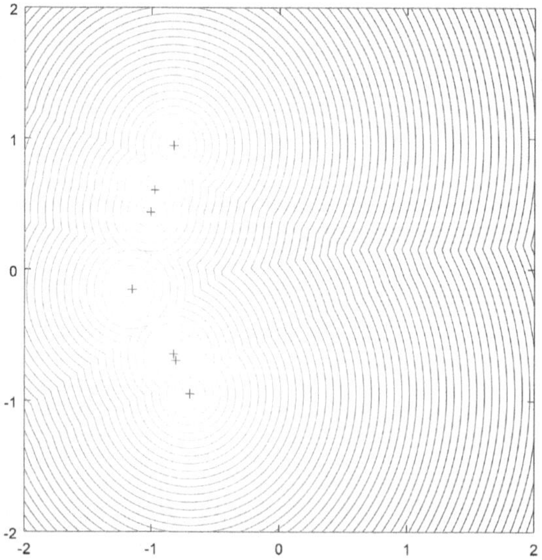

Fig. 9. MIC contour plot for Example 2. The solution is unbounded.

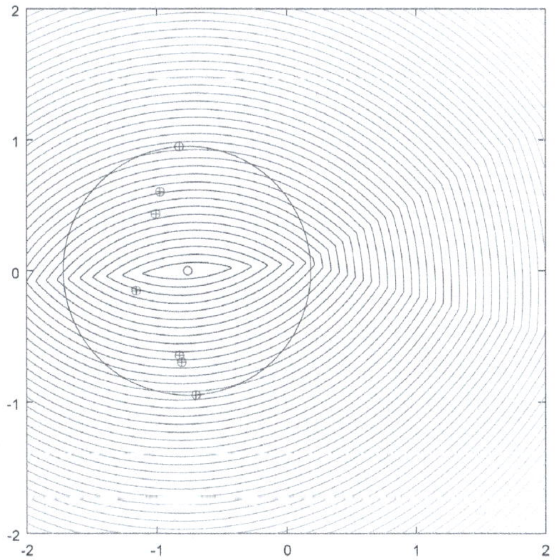

Fig. 10. MCC for Example 2, with contour plot. $\boldsymbol{c} = (-0.76382, 0.0006.6955), r_c = 0.94935$, $f = 0.94935$ is the minimized objective function value.

For cases like Example 2, MCC poses another practical problem. As shown in Fig. 10, the minimum circumscribing circle is not what the engineers want, either as a datum simulator or as a mating envelope – just contrast Fig. 10 with Fig. 5. Because of such fitting problems, arcs that are less than a semi-circle are excluded from considerations for MIC or MCC, and hence for datum simulators and mating envelopes. Similar limitations can be found for partial spheres and partial cylinders.

Despite these limitations, maximum inscribing and minimum circumscribing fittings are used quite extensively in standards and industry for datum establishment and mating envelopes. As mentioned earlier, their uses are restricted to features of size – that is, for features such as circles, spheres, cylinders, and slabs/slots. Software implementations of such fittings exist, but they are not as well tested as the unconstrained least-squares fitting.

An important observation for MIC and MCC is that only extreme points in the input set participate in the final solution; the interior points play no role. This means that uniform sampling is not a requirement for the discrete approximation to converge to the continuous case as the number of points $n \to \infty$. Such a munificent behavior is also exhibited by the minimum zone fitting, which will be described next.

5 Minimum Zone Fitting

Minimum zone fittings can be posed as optimization problems with the l_∞-norm of Eq. (2). Under this norm, the minimum zone circle (MZC) problem can be posed as the following minimax problem using the notations of Fig. 3.

$$\min_{x_c, y_c, r_c} (d_{\max}) \text{ where } d_{\max} = \max(|r_i - r_c|), i = 1, \ldots, n. \tag{7}$$

Equivalently, the optimization problem can also be posed as

$$\min_{x_c, y_c}(r_{\max} - r_{\min}) \text{ where } r_{\max} = \max(r_i), r_{\min} = \min(r_i), i = 1, .., n, \tag{8}$$

which illustrates the minimum zone nature of the fitting. Figure 11 shows the MZC for Example 1, along with the contour plot for the objective function of Eq. (8). It also shows the circles with radii $r_{\min}$ and $r_{\max}$. All the input points are contained within the zone (in this case, the minimum zone) bounded by the inner and outer circles. Figure 12 shows similar information for Example 2.

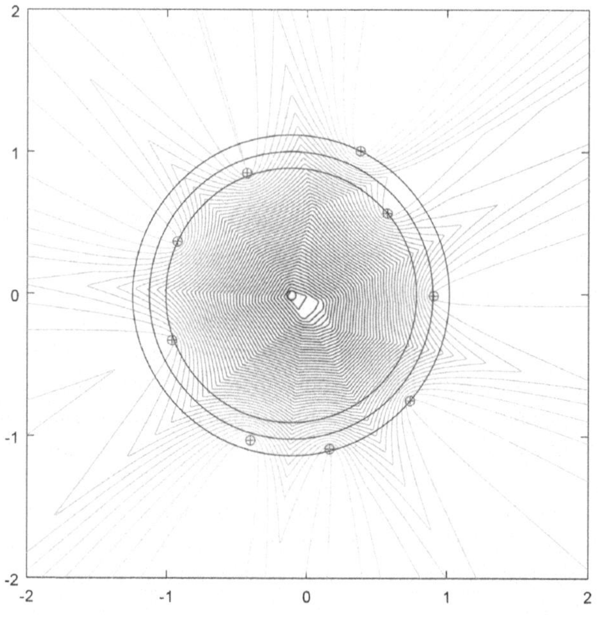

Fig. 11. MZC for Example 1, with contour plot. $\boldsymbol{c} = (-0.1132008, -0.0093477), r_c = 1.0120$, $f = 0.23295$ is the minimized objective function value; $r_{\min} = 0.89557, r_{\max} = 1.1285$, $f = r_{\max} - r_{\min}$ = minimum radial separation.

The contour plots in Figs. 11 and 12 indicate that the objective function is not smooth. It can be shown that the objective function has first-derivative discontinuities, and these occur along the nearest and furthest neighbor Voronoi diagrams of the input points. The minimum is achieved when the center of the circle is at the intersection of these Voronoi diagrams. A derivative-free search method, such as Nelder-Mead technique, can be employed to solve these optimization problems, starting with good initial solutions such as those in Table 3.

It can be seen from Figs. 11 and 12 that only extreme points participate in the solution; interior points don't matter. Therefore, uniform sampling of points is not necessary for the discrete approximations to converge to the continuous case as $n \to \infty$. It is also clear from these figures that MZC provides appropriate fits for both circles (as in Example 1) and arcs (as in Example 2) which were problematic in the MIC and MCC cases.

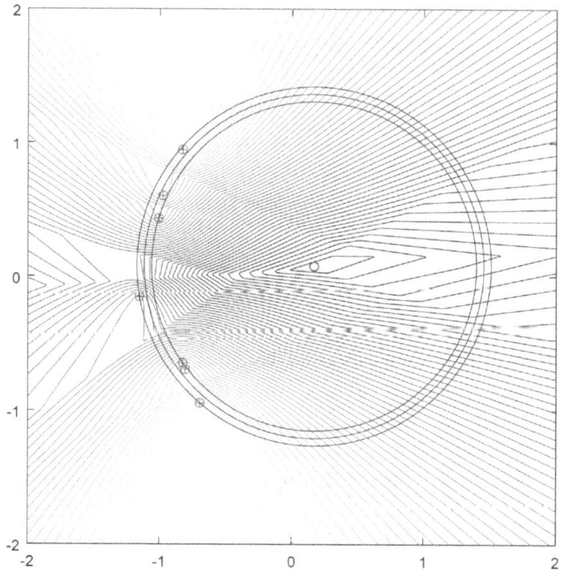

Fig. 12. Minimum zone fitting for Example 2, with contour plot. $\boldsymbol{c} = (0.168871, 0.076265)$, $r_c = 1.2856$, $f = 0.11128$ is the minimized objective function value; $r_{\min} = 1.2300$, $r_{\max} = 1.3412$, $f = r_{\max} - r_{\min}$ = minimum radial separation.

The MZC itself may be of some use to assess the size of the circular feature. However, it is the radial separation value related to the MZC that gets the maximum attention in standards and industrial metrology for an assessment of the roundness of circular features. Generalizing this idea, the minimum zone fitting (under the l_∞-norm) can be used to assess the form deviation of any geometrical feature. Because of this popular interest, minimum zone fitting has been implemented in commercial software systems. But its testing has not been as rigorous as the unconstrained least-squares fitting.

The inner and outer curves and surfaces that bound the minimum zones could also be considered for datum simulators and mating envelopes. For example, the inner and outer circles in Figs. 11 and 12 may serve this purpose. But in recent years constrained least-squares fitting has emerged as a more attractive alternative for the same purpose, and it will be described next.

6 Constrained Least-Squares Fitting

Consider the following constrained optimization problems under the l_2-norm (with the designation CL2IC for constrained least-squares inscribing circle and CL2CC for constrained least-squares circumscribing circle) and the notations of Fig. 3.

$$\min_{x_c,y_c,r_c}\left[\frac{1}{n}\sum_{i=1}^{n}\left\{r_c-\sqrt{(x_i-x_c)^2+(y_i-y_c)^2}\right\}^2\right]^{1/2}$$

$$\begin{array}{l}\text{subject to } r_i \geq r_c, \forall\, i \text{ for CL2IC,}\\ \text{subject to } r_i \leq r_c, \forall\, i \text{ for CL2CC.}\end{array} \tag{9}$$

It is interesting to compare Eq. (9) with Eqs. (4) and (5). The objective function of constrained least-squares fitting in Eq. (9) is the same as the objective function of the unconstrained least-squares fitting in Eq. (4). The constraints for the constrained least-squares fitting in Eq. (9) are the same as the constraints in the maximum inscribing and minimum circumscribing fittings in Eq. (5).

The tri-variate objective function in Eq. (9) is smooth, with a continuous gradient. This fact can be exploited in designing algorithms and heuristics to find the optimum solutions that satisfy the inequality constraints in Eq. (9). It is also possible to reduce the optimization problem to an equivalent form involving only a bivariate objective function by following two steps – reminiscent of the two-step process used in Sect. 3 to solve the unconstrained least-squares fitting problem.

In the first step, the circle center coordinates (x_c, y_c) are held fixed and the radius r_c is set to satisfy the constraints of Eq. (9). That is, $r_c = r_{\min} = \min_{i=1,\ldots,n} (r_i)$ for CL2IC and $r_c = r_{\max} = \max_{i=1,\ldots,n} (r_i)$ for CL2CC. Then, in the next step the following unconstrained optimization problems are posed with free variables x_c and y_c.

$$\min_{x_c,y_c}\left[\frac{1}{n}\sum\nolimits_{i=1}^{n}\left(r_{\min}-\sqrt{(x_i-x_c)^2+(y_i-y_c)^2}\right)^2\right]^{1/2} \quad \text{for CL2IC.}$$

$$\min_{x_c,y_c}\left[\frac{1}{n}\sum\nolimits_{i=1}^{n}\left(r_{\max}-\sqrt{(x_i-x_c)^2+(y_i-y_c)^2}\right)^2\right]^{1/2} \quad \text{for CL2CC.} \tag{10}$$

Reducing the objective function from a tri-variate function in Eq. (9) to a bivariate function in Eq. (10) allows an analysis of the mathematical structure of the optimization problem and computational methods to solve the problem. But it comes at a price – the bivariate objective function will not be smooth, as can be inferred from the contour plot involving CL2IC in Fig. 13 for Example 1.

The gradient discontinuities in the bivariate objective function in Eq. (10) can be explained using Voronoi diagrams. Figure 14 shows the nearest neighbor Voronoi diagram of the input points along with the CL2IC for Example 1. The derivate discontinuities occur along the nearest neighbor Voronoi diagram, and the minimum is achieved at a Voronoi edge. Recently, these facts have been thoroughly studied and exploited for computational purposes [7, 13]. Figure 15 shows the CL2IC for Example 2 along with the contour plot.

The presence of gradient discontinuities in the formulation of Eq. (10) indicates that a derivative-free approach should be sought to solve these problems iteratively. Again, a downhill simplex method like the Nelder-Mead technique can be used, starting with a good initial solution such as those in Table 3 or a solution to the unconstrained least-squares problem that has been solved quite satisfactorily in practice.

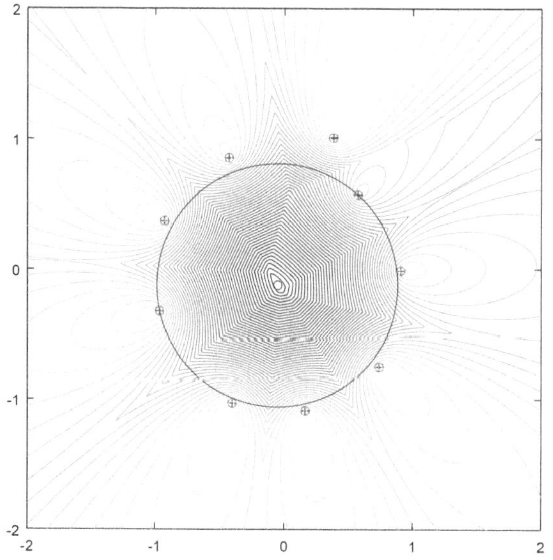

Fig. 13. Constrained least-squares fitting for inscribing circle for Example 1, with contour plot. $\boldsymbol{c} = (-0.052032, -0.126981)$, $r_c = 0.93353, f = 0.10843$ is the minimized objective function value.

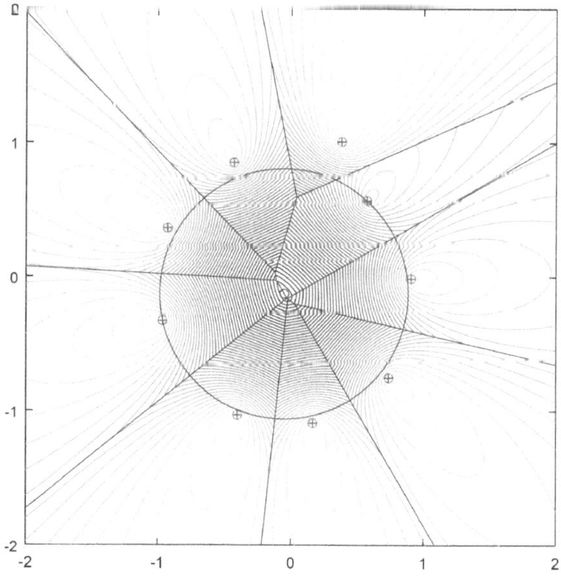

Fig. 14. Constrained least-squares fitting for inscribing circle for Example 1, with contour plot and nearest neighbor Voronoi diagram. $\boldsymbol{c} = (-0.052032, -0.126981)$, $r_c = 0.93353$, $f = 0.10843$ is the minimized objective function value.

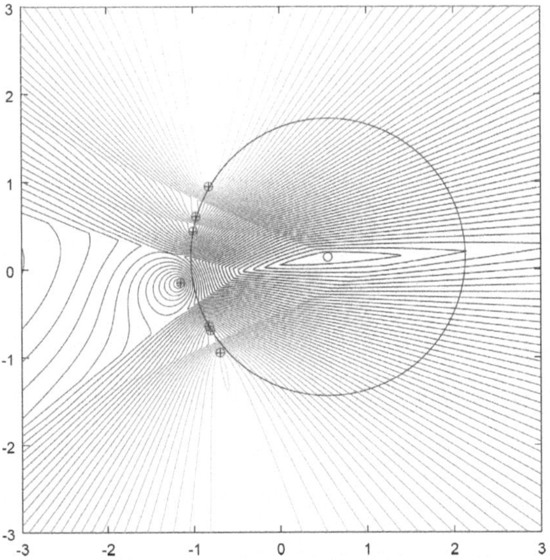

Fig. 15. Constrained least-squares fitting for inscribing circle for Example 2, with contour plot. $\boldsymbol{c} = (0.54809, 0.13960)$, $r_c = 1.5835, f = 0.060375$ is the minimized objective function value.

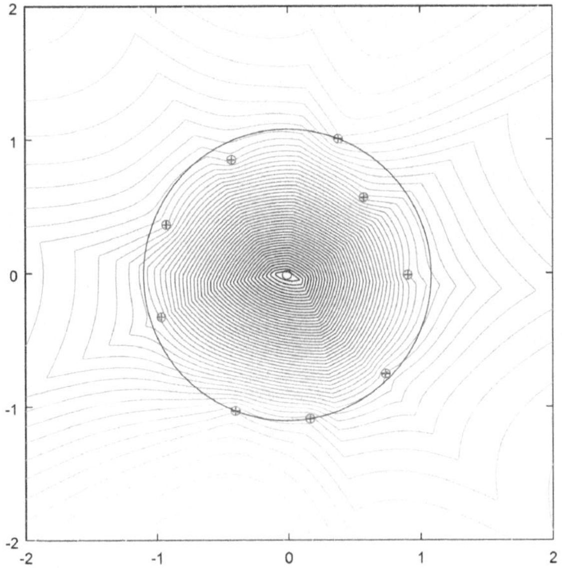

Fig. 16. Constrained least-squares fitting for circumscribing circle for Example 1, with contour plot. $\boldsymbol{c} = (-0.0094270, -0.0131575)$, $r_c = 1.0908$, $f = 0.12467$ is the minimized function value.

Figure 16 shows the CL2CC along with the contour plot for Example 1. Here the gradient discontinuities occur along the furthest neighbor Voronoi diagram of the input points. Figure 17 shows the same for Example 2.

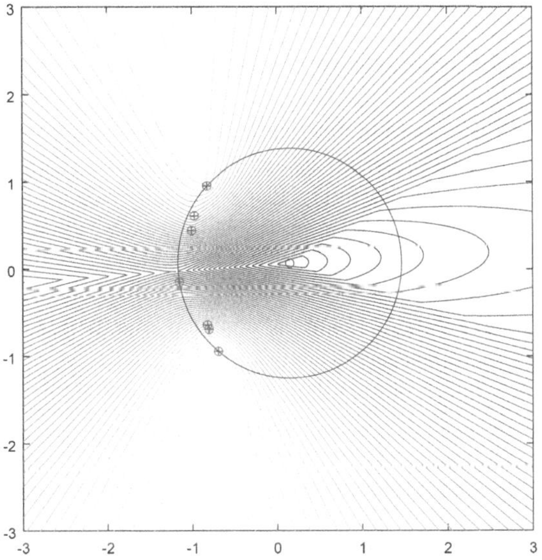

Fig. 17. Constrained least-squares fitting for circumscribing circle for Example 2, with contour plot. $c = (0.145994, 0.063188)$, $r_c = 1.3163$, $f = 0.073713$ is the minimized objective function value.

A close examination of all the fitting figures presented thus far reveals that the constrained least-squares fitting performs very well, both in terms of satisfying engineering requirements and computational efficacy. It shares the nice features of unconstrained least-squares fitting in taking all the points into account in arriving at a solution; that is, all points contribute to the final solution. As was seen earlier, this was not the case with maximum inscribing, minimum circumscribing, and minimum zone fittings; these depend only on the extreme points. Also, the constrained least-squares fitting accommodates both full and partial features (that is, both Examples 1 and 2), thereby avoiding the pitfalls of maximum inscribing and minimum circumscribing fittings.

While the contribution of all the input points to the final solution is a nice feature of the least-squares fitting (both constrained and unconstrained), it comes at a price. It depends on uniform sampling of points to ensure that the discrete solution based on n points will converge to the continuous case as $n \to \infty$. Alternatively, a more generalized objective function that weighs points appropriately could be used. This issue is currently being investigated.

Even though the examples in this paper were restricted to circles, the notion of constrained least-squares fitting is general and it can be applied to all geometric features. It is being considered for ASME and ISO standardization of datums [12], and it has the potential to address the needs of mating envelopes. Initial industrial implementations have been successful, but more implementations and testing are necessary.

7 Summary and Conclusions

This paper presented various fitting methods used in industry and codified in standards, to enable measurement of geometrical quantities. It provided an illustrative description of existing and new methods of fitting using carefully chosen examples involving circles. The structure of the optimization problems was explored using mathematical equations and contour plots, and computational methods to solve these problems were discussed. The main objective of this paper was to introduce a new fitting method – the constrained least-squares fitting – and to compare this with other existing fitting methods to show its usefulness and viability.

Constrained least-squares fitting has emerged as an attractive solution to the problem of establishing digital datums and digital simulation of mating features. National and international standards organizations are actively working on incorporating this fitting method as the default method for defining datums. Industrial implementation and testing of the algorithms and heuristics for constrained least-squares fitting has begun, and it is showing very good promise. There are still some research and development efforts that need to be undertaken to guide the industrial practice. This includes some of the sampling and convergence issues, as discussed in the body of the paper.

References

1. Stigler, S.M.: The History of Statistics. Harvard University Press, Cambridge (1986)
2. ISO 1101:2017: Geometrical product specifications (GPS) – Geometrical tolerancing – Tolerances of form, orientation and run-out. International Organization for Standardization, Geneva (2017)
3. Fisher, R.A.: Statistical Methods for Research Workers, 11th edn. Hafner, New York (1950)
4. Box, G.E.P., Hunter, W.G., Hunter, J.S.: Statistics for Experimenters. Wiley, New York (1978)
5. ASME Y14.5-2019: Dimensioning and tolerancing. American Society of Mechanical Engineers, New York (2019)
6. Forbes, A.B.: Least-squares best-fit geometric elements. NPL Report DITC 140/89, Teddington, England (1991)
7. Shakarji, C.M., Srinivasan, V.: On algorithms and heuristics for constrained least-squares fitting of circles and spheres to support standards. ASME J. Comput. Inf. Sci. Eng. (2019, to appear)
8. Gill, P.E., Murray, W., Wright, M.H.: Practical Optimization. Emerald Group Publishing, Bingley (1982)

9. Shakarji, C.M.: Least-squares fitting algorithms of the NIST algorithm testing system. NIST J. Res. **103**(6), 633–641 (1998)
10. O'Rourke, J.: Computational Geometry in C, 2nd edn. Cambridge University Press, Cambridge (1998)
11. Anthony, G.T., et al.: Chebyshev best-fit geometric elements. NPL Report DITC 221/93, Teddington, England (1993)
12. Shakarji, C.M., Srinivasan, V.: Toward a new mathematical definition of datums in standards to support advanced manufacturing. In: ASME MSEC2018-6305, Proceedings of the ASME 2018 Manufacturing Science and Engineering Conference, College Station, Texas (2018)
13. Shakarji, C.M., Srinivasan, V.: Optimality conditions for constrained least-squares fitting of circles, cylinders, and spheres to establish datums. ASME J. Comput. Inf. Sci. Eng. **18**(3), 031008 (2018)

Additive Miniaturized-Manufactured Gear Parts Validated by Various Measuring Methods

N. M. Durakbasa[1], P. Demircioglu[2], J. Bauer[1], I. Bogrekci[2], G. Bas[1], O. Bodur[1(✉)], and G. Poszvek[1]

[1] Institute for Production Engineering and Photonic Technologies, Department of Industrial Metrology and Adaptronic Systems, TU Wien (Vienna University of Technology), BA 09, 1060 Vienna, Austria
numan.durakbasa@tuwien.ac.at, doktorjbauer@yahoo.com.ar, goekcen.bas@ift.at, bodur.osman@hotmail.com, guenther@poszvek.net

[2] Faculty of Engineering, Department of Mechanical Engineering, Aydın Adnan Menderes University, Aydın, Turkey
{pinar.demircioglu,ibogrekci}@adu.edu.tr

Abstract. Recently, miniaturization has become an important topic to both scientists and engineers. The manufacturing trends are following the compact size manufactured-components. Miniaturization challenges engineers to obtain smaller size of the components, to reduce their weights & power consumption and to take less space utilization. The functional specifications of the parts must be clarified beforehand, so that the functions are not lost in miniaturization. With consideration of the geometrical product specification (GPS), it is possible to better limit the functional properties and thus succeed in miniaturization. In this work, the problems with the reduction of the gear components are explained and the subsequent assessment are presented with different methods such as contact and noncontact metrology methods. Tactile and optical methods are used to determine the surface structure. Coordinate measuring machines (CMM) are one of the geometry based tactile methods. The optical methods give more information about the geometry and microstructure of technical surfaces by using computed tomography (CT) and confocal laser scanning microscopy (CLSM). According to the measurement results, the measurement data belonging to CMM measurements for the partial circles showed the same results with CT measurements. The surface roughness values were varied from the existed geometry to the miniaturized geometry using CLSM. In porosity measurement with CT, porosity decreased in the micro geometry of the gear components. The miniaturized geometry had less porosity related to the gaps volume, than the normal geometry. The minimum gap volumes depend on the scanning resolution in CT. The gaps in macro geometry were logically more frequent than the micro geometry.

Keywords: Additive manufacturing · Computed tomography · Confocal Laser Scanning Microscopy · Gear part · Geometrical product specification (GPS) · Miniaturization

V. D. Majstorovic and N. Durakbasa (Eds.): IMEKOTC14 2019, LNME, pp. 276–290, 2019.
https://doi.org/10.1007/978-3-030-18177-2_25

1 Introduction

Miniaturizing the component size meets the need to manufacture a component in compact size. Without losing the functions of the components, miniaturizing and evaluation steps are important in order to verify the geometrical product specifications (GPS). In the practical applications of the tools provided in the relevant international (ISO-GPS) standards, a large number of questions usually arise. These begin with an understanding of the content of the standard and go beyond the correct interpretation of drawing entries to the solution of constructive tasks. Further questions arise when drawing the new geometric tools in CAD programs. Standard-compliant design drawings are based on the new ISO-GPS standards. These have below features:

- reduce manufacturing and testing costs,
- increase product value,
- reduce product liability risks to customers and suppliers,
- reduce the internal and external voting requirements,
- make the conformity to ISO 9001,
- secure the long-term competitiveness of the company [1].

The statements are repeatedly referred to the companies demanding for miniaturization. Under consideration of these conditions, the gearbox was analyzed accordingly. The aim of this study is to miniaturize any component without losing its functional properties and to validate the functional and geometrical properties with various measuring methods such as contact (CMM) and non-contact (CT, CLSM) measuring methods.

2 Methods and Theoretical Parts

The methodology of this paper was carried out as follows: A CAD model is constructed, and then converted to STL format. The STL format was used in order to produce miniaturize the gear box. Important subsequent of the RP4 gearbox was created. No backlash was observed. Tactile and optical measurement methods were utilized with the quality assurance.

2.1 Geometrical Product Specification and Verification

The designer has the difficult task of translating his/her idea of the function of a component into a form and position tolerance. These ensure this function is correct, complete and unambiguous. It is necessary to set the fundamental tolerancing principle. ISO 8015 [2] is to be chosen if the principle of independency is applied. ISO 14405 [3] is chosen if the principle of envelope is used. This is represented by the circled letter E in or on the title block. Then the general tolerance is set. Based on the manufacturing process for the cutting production, ISO 2768 [4] is normally used. Here the question arises which geometric elements are essential for the function of the component. Every geometric element can be designed and manufactured with complete implementation of GPS into metrology methods and analysis by achieving moreover lower costs and higher quality in production [5]. Only for the functionally importance of a form and position tolerance is provided, which restricts the general tolerances. Geometric features can be defined in three areas:

- the area of nominal definition
- the area of specification
- the area of verification.

The components in the automotive industry, with its high volumes, has benefited from Geometric Dimensioning and Tolerancing (GDT). The geometric product specifications serve to implement function-dependent requirements in parts, especially for the workpieces created. This is based on the following requirements:

- mathematical rules and methods,
- consideration of macro and micro geometry,
- possibilities for measuring of quantities and especially tolerance quantities and
- evaluation of uncertainty [6].

In a general approach, a smart design system defined by the technical drawing describes the shape (geometry), dimensions and surface properties of a workpiece. To ensure that the optimal function of a part is determined, certain manufacturing tolerances are used. In order to compare the workpieces, the specified specifications, as required by the geometric product specification system and verification, are measured. First step in an assembly is the determination of the permissible deviations and the properties of the components.

The geometric deviation is used to customize the manufacturing processes in order to adapt. The experienced metrologists interpret the specification under consideration of the non-ideal surface model in order to know the registered properties (ISO 17450-1:2011) [7]. Based on the real surface, the individual steps of the measuring schedule of the workpiece are decided depending on the measuring equipment. In comparison, the conformity of the stated properties with the measurement result is determined.

Before reducing the size of the parts, the functional specifications must be clarified. The functions should not be lost in miniaturization. With the consideration of the geometric product specification system, it is possible to have better functional properties with succeeding in miniaturization. This is only possible with a high-precision measurement technologies accompanying the entire process.

It is inquired whether it is satisfactory only comparing nominal or target values with measured values to confirm that the functional requirements are fulfilled according to ISO 17450-2:2013 [8]. Here the correlation uncertainty is crucial [9, 10]. The importance of parallelism and complementary of the actions that shall be distinguished in the specification procedure and the verification procedure in the process of a product development (Fig. 1) is highlighted in the ISO 17450-1:2011 [7, 11].

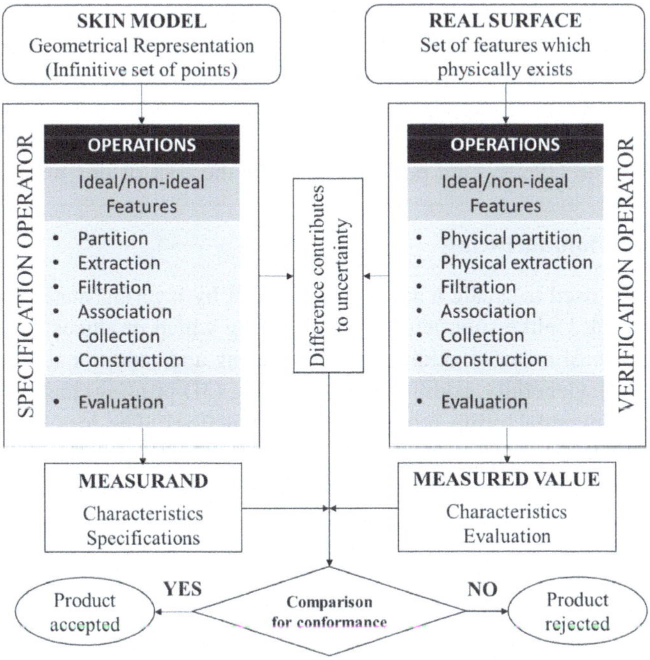

Fig. 1. The GPS duality principle of the specification and verification procedures [11].

The specification operator is considered as the virtual measuring procedure. The sequence of operations used during the measurement processes is determined by the verification operator. The theoretically exact verification operator is the implementation of the specification operator.

The processes of partitioning, extraction, filtration, association, collection and construction, which were introduced during the measurement as a metrology activity without the designer's instructions, are the most common cases [11].

2.2 Functional Specifications and Problem of Miniaturization

During the design phase, the geometric specifications are determined according to the functional specification of a workpiece, which are derived from the functional requirement (Fig. 2). The geometric requirements can be communicated to the manufacturing personnel and the machines. The ideal shape must be translated into the language of the ISO-GPS, which is based on standard geometric elements.

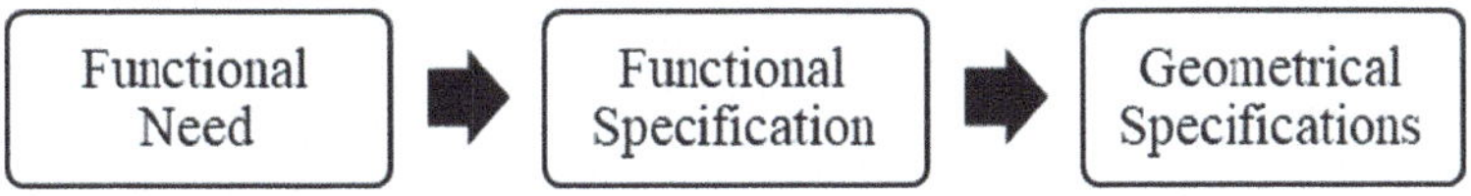

Fig. 2. Functional need, functional specification, geometrical specifications [7].

It is important to observe the normative regulations when determining technical drawings. Increasing or decreasing the size of an object is not a challenge for today's CAD programs. The limits of production methods are clearly shown in production environments. During miniaturization, it should be paid attention to the functionality. The reassessment of functional relationships takes place with the change in size. The extent to which the GPS system performs better is the task of this research.

2.3 Additive Manufacturing

This is a process used to create a 3D physical object by layering materials one by one on a digital model. Unlike subtractive manufacturing which produces the final product by cutting a block of material; additive manufacturing adds additional parts to form its final product [12]. Generally, additive manufacturing (3D printing) has more advantage than subtractive manufacturing (conventional production). The loss of material does not occur in additive manufacturing.

Production of a part with additive manufacturing technologies starts with 3D modeling. Then it is necessary to organize the data preparation and include the definition of the part direction, the location of the support structures and the slicing of the model. Post-processing operations are required after the part has been produced.

There are different printing processes like material jetting, binder jetting, material extrusion, sheet lamination, powder bed fusion, directed energy deposition, laser sintering or photo polymerization. Some of these procedures are very similar in principle and differ only in a few (partly patented by companies) modifications. Simplified, similar to the 3DP[1] process in selective laser sintering method, the print material is in powder form. However, the individual layers are not connected to a liquid adhesive, but fused together using a high-power laser (CO2 laser) under a protective atmosphere. In addition to plastics, it can also be used to process materials such as metals, ceramics and sand.

2.4 Measuring Methods

Measurement methods are used to understand the process, to remove variation sources to balance the process, and to apply statistical analysis to monitor the ongoing production process. Thus, geometric tolerance changes can be easily detected in additive manufacturing [13]. Before measurement methods, an important subsequent of the RP4 gearbox, which provides the elliptical movement, has been manufactured using additive manufacturing methods.

Tactile method (CMM[2]). CMM allows to measure the deviations of dimensions, form and position accurately [6]. When workpiece tolerances are more accurate than tolerance grade IT5[3], it is necessary to make use of coordinate metrology [6]. The optimized process in quality assurance and measurement technology enables high precision manufacturing. Figure 3 shows geometrical assessment of the gears by CMM.

[1] 3DP: 3D Printing with powder.

[2] CMM: Coordinate Measuring Machine.

[3] IT: International Tolerance.

Fig. 3. Geometrical assessment of the gears by CMM [14]

Non-contact measurement methods (CT[4] and CLSM[5]). As an non-contact measurement method, two measurement systems were used. One of the measurement method is Computed Tomography (CT), which uses the coordinate metrology for 3D measurement [15]. A high-resolution industrial computed tomography system provides accurate measurement results when determining geometric characteristics and evaluating deviations from geometric tolerances. Figure 4 demonstrates the computed tomography (CT) measurement of the gear component according to ISO 1101:2017 [16].

Fig. 4. Measurement of 3D printed component with CT [17]

[4] CT: Computed Tomography.

[5] CLSM: Confocal Laser Scanning Microscopy.

CLSMs are one of the most important devices used to monitor the components in 3D measurements. The confocal microscope is a microscope that works similarly to the fluorescence microscope, and laser beams are used as lighting sources. CLSM works with using a point laser light source, an illuminating aperture and a confocal aperture. CLSM, with three-dimensional imaging capabilities, can provide surface topography image of sample using a series of adjacent 2D images attained from different depths.

In this study, surface quality of miniaturized-manufactured gear parts was evaluated in micro/nano-scale (Figs. 5, 6). The surface investigations were measured with a rotating (Nipkow disc) confocal microscope (NanoFocus – µsurf) [18].

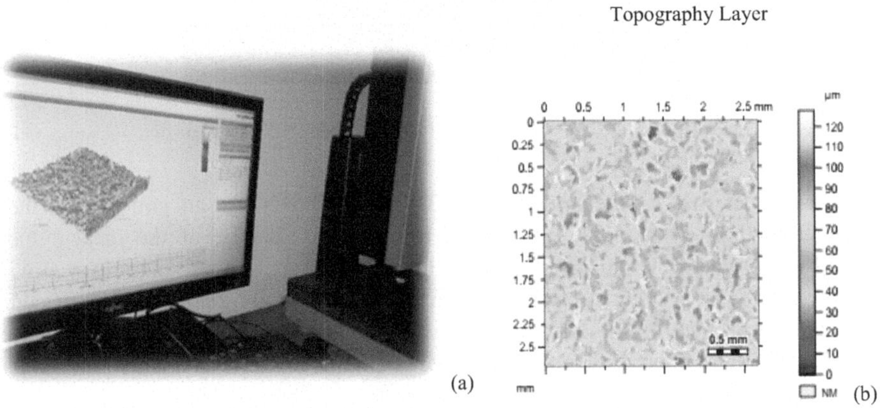

Fig. 5. Confocal Laser Scanning Microscope [18] (NanoFocus - µsurf) during the roughness measurements of the miniaturized gear part (a) 2D surface profile (b)

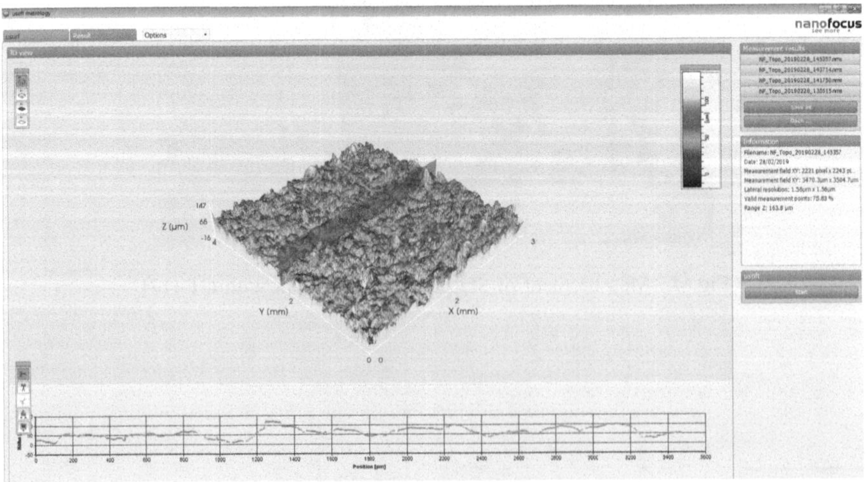

Fig. 6. 3D surface profile [18]

3 Results

Coordinate metrology (CMM and x-CT) and optical metrology are primary precision measurement techniques used for quality control of the additive manufactured parts [23]. The industrial CT equipment is used not only for non-destructive analysis but to perform geometrical evaluations [19]. The surface structure of the existing geometry and miniaturized geometry of the gear components determined and the geometrical dimensions of the macro geometry are analyzed through statistical process control in various measurement methods which are tactile and non-contact methods. First of all, the important geometrical structure and tolerances are defined according to functional specifications of the gear components. Important dimensions and geometrical tolerances [16] are given with red color on the technical drawings in Fig. 7.

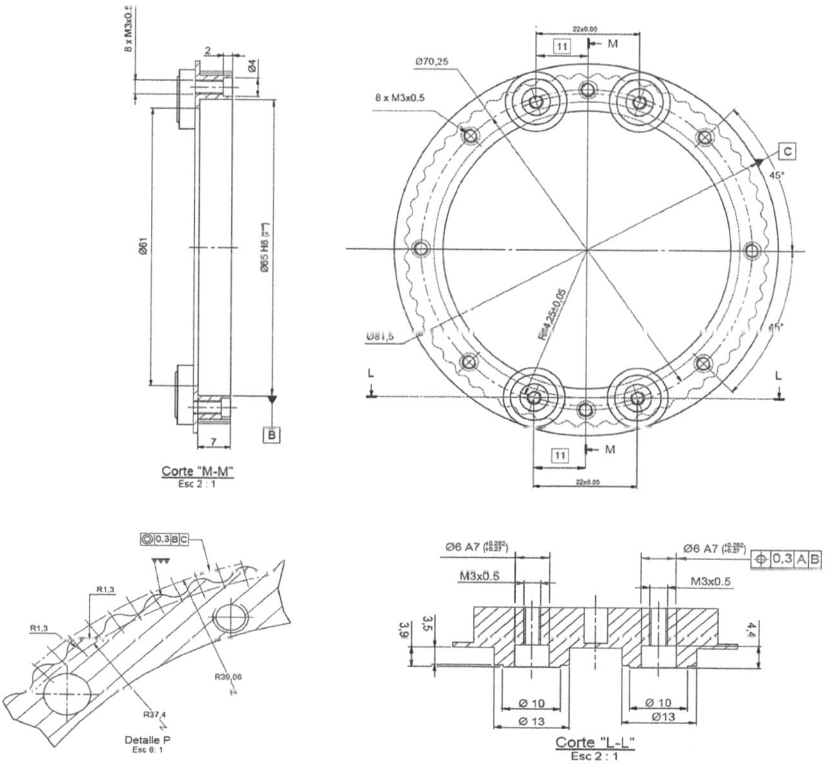

Fig. 7. Front and section view's technical drawing of the gear component

There are 5 geometry components and 5 miniaturized with 1:5 scaled geometry components manufactured 3D gear components according to STL model of technical drawing in Fig. 8. PA[6] 2200 Balance 1.0 was used as printing material in selective laser sintering (SLS) printer [14]. The characteristics and properties are listed in Table 1.

Table 1. The characteristics of PA 2200 Balance 1.0

Characteristic	Explanation
Application method	Laser sintering, rapid prototyping
Delivery forms	Powder
Chemical resistance	General chemical resistance
Certificates	FDA approval according to USP Biological test (classification VI / 121 ° C)
Properties	
120 μm layer thickness	
930 kg / m³ density	
White powder color	
176 ° C Melting temperature (20 ° C / min)	
0,5 mm flammability	

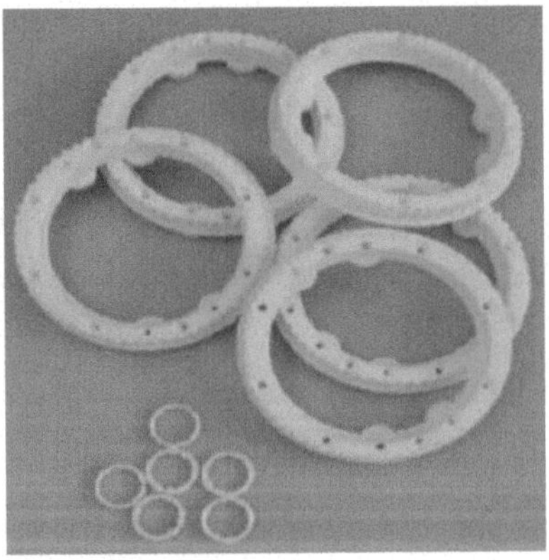

Fig. 8. 3D manufactured gear parts

After manufacturing gearbox, the geometrical deviations and important dimensions of the parts were measured through computed tomography (Fig. 9). The measurements are carried out for assessment of tolerance ranges from the ISO 286-1:2010 and ISO 1101:2017 [16, 20] (Table 2).

[6] PA: Polyamide.

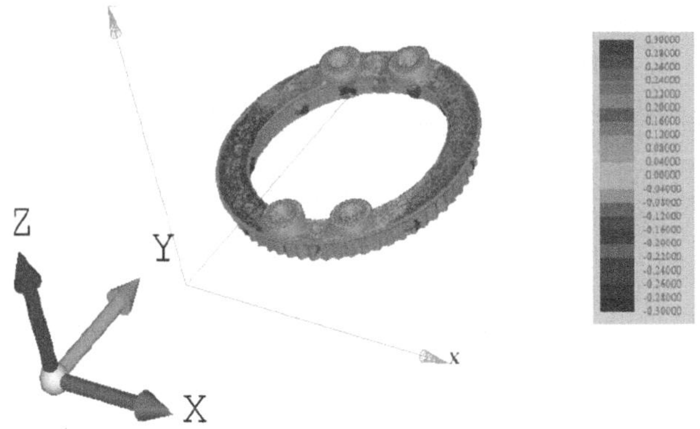

Fig. 9. Assessment of form deviation of miniaturized gear component with 1:5 scale using CT

Table 2. The CMM measurement results

CMM values		
Dimensions of gear component	Mean	Std Dev
Θ65H8	65.1081	0.0004
R34.25	34.2284	0.0002
Θ6A7	6.1026	0.0028
11 mm	10.9844	0.0003

Table 3. The CT measurement results

CT values		
Dimensions of gear component	Mean	Std Dev
Θ65H8	65.1562	0.0525
R34.25	34.2634	0.0517
Θ6A7	6.1245	0.0400
11 mm	10.9821	0.0020
R39.06	39.0590	0.0073
R37.4	37.2811	0.0175

The measurement data belonging to CMM measurements for the partial circles presented comparable results with the CT measurements (Table 3).

Table 4. The CLSM roughness measurement results (R_a)

	M1	M2	M3	M4	M5	M6	M7	M8	M9	M10	Mean	Std Dev
Sn1	12.1	11.1	12.5	10.9	9.8	10.1	11.5	12.6	12.5	11.9	11.5	1.01
Sn2	10.8	10.2	9.5	10.3	9.3	9.7	12.3	12.7	11.3	11.7	10.78	1.19
Sn3	13.1	12.9	12.7	11.9	10.8	12.1	12.4	12.6	13.3	13.9	12.57	0.85
Sn4	10.1	11.6	11.5	10.4	10.6	9.1	12.9	11.6	11.5	10.1	10.94	1.08
Sn5	12.4	12.1	11.9	11.8	10.2	10.3	10.5	10.5	11.9	10.7	11.23	0.86
Sm1	8.7	9.3	7.9	10.7	9.5	8.2	9.4	10.4	9.1	9.3	9.25	0.87
Sm2	8.1	10.1	12.5	10.9	9.8	10.1	11.5	12.6	12.5	11.9	11	1.48
Sm3	9.2	9.4	9.5	9.9	10.2	10.3	9.7	10.6	8.5	9.9	9.72	0.61

Investigating surface roughness of existed and miniaturized geometry of gear samples, there were five samples of existed geometry and three samples of miniaturized geometries, besides those 10 measurements were obtained from each sample [21, 22]. Mean of the average roughness values of measured sample (R_a) was calculated for each sample. Means of R_a were calculated as 11.5 μm, 10.78 μm, 12.57 μm, 10.94 μm, 11.23 μm, respectively for each normal sample (*Sn1, Sn2...Sn5*) of gear components. Means of R_a were calculated as 9.25 μm, 11.5 μm, 9.72 μm, respectively, for each miniaturized sample (*Sm1, Sm2...Sm5*) of gear components. It was observed that the surface roughness values were varied from existed geometry to miniaturized geometry (Table 4).

Another measurement method is to observe the porosity of additive-manufactured gear components through CT. Since the selective laser sintering method to manufacture the workpiece is used, solidity ratio is set as 100%. There seems to be always some gaps in micro/nano scale in the manufactured parts. These gaps exist in 3D axis and they are not possible to observe them by bare eyes. Porosity reflects the condition of the manufactured parts. In characterization of these gaps, porosity is a percent result of the volume ratio of the gaps to the workpiece volume. There are effects of porosity on the mechanic specifications of the material and geometric specifications. Figure 10 depicts the material and gaps for whole volumes of the normal and miniaturized parts. In Fig. 11, it is also possible to observe the calculated volumes of the material and gaps at different colors, respectively, blue and green.

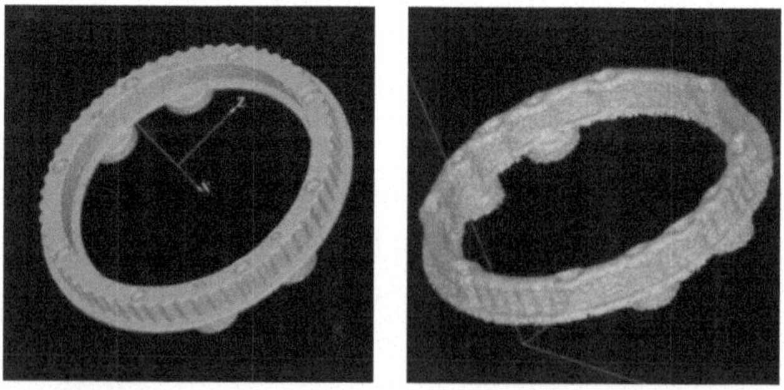

Fig. 10. STL geometries of the whole volumes for macro and micro components

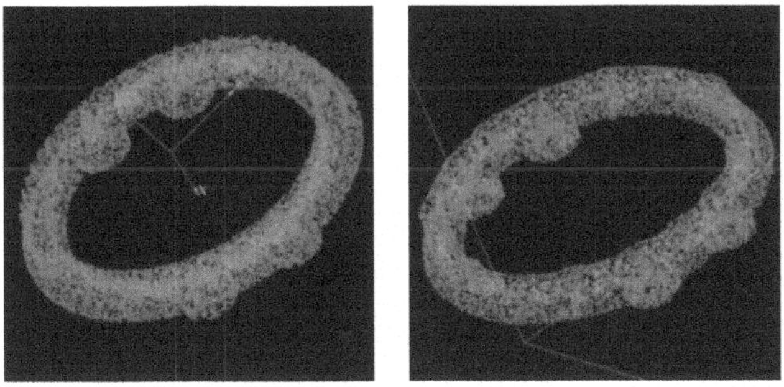

Fig. 11. Calculated material and gaps volumes of the gear components

The porosity of the manufactured normal geometry and micro geometry with 1:5 scale of the gear components in 3D axis is shown (Fig. 12). According to the calculation in software, the volumes of the material and gaps are presented at the Table 5. The miniaturized geometry has less porosity, which is related to the gaps volume, than the normal geometry. Furthermore; the maximal, minimal and average volumes of the gaps are calculated and presented in Table 6 by multi-distance analysis for each gap volumes in CT. An approximate value is defined in μm^3. Also, the gaps in macro geometry are logically more frequent than the micro geometry (Fig. 13).

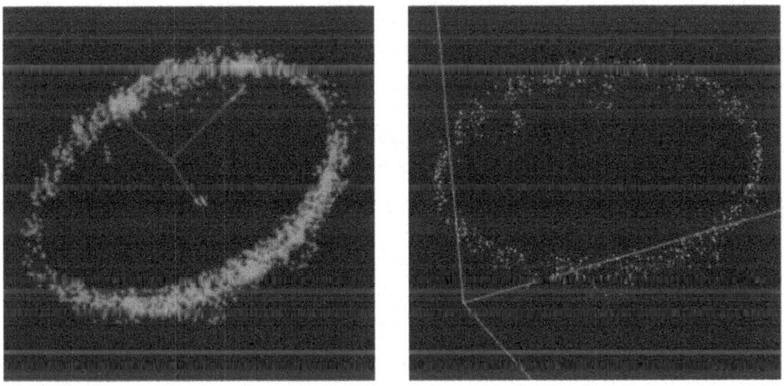

Fig. 12. The gaps volumes in the three-dimensional axis of the gear components

Table 5. The Porosity of the macro and micro gear components

Components	Macro geometry	Micro geometry
Total volume (mm^3)	11,644.5454	84.6695
Gaps volume (mm^3)	6.1789	0.3190
Porosity (%)	0.053	0.037

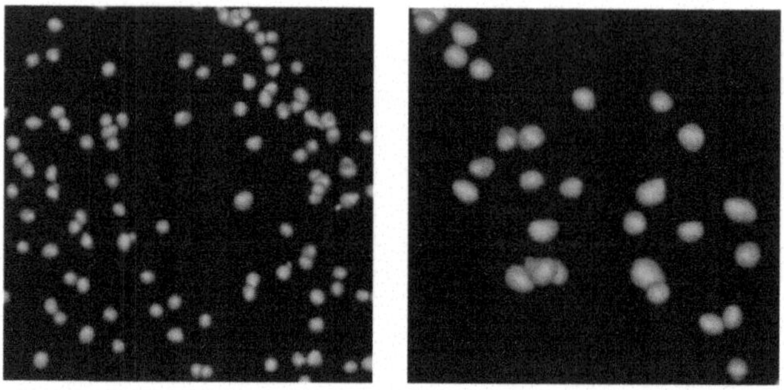

Fig. 13. The gaps volumes of the macro and micro geometry

Table 6. The Multi distance analysis results

Components	Macro geometry	Micro geometry
Min. gap volume (μm^3)	Up to 60	Up to 10
Max. gap volume (mm^3)	0.0067	0.0026
Average gap volume (mm^3)	0.0023	0.0004

4 Conclusions

Before reducing the size of the parts, the functional specifications must be clarified. The functions should not be lost in miniaturization. With the consideration of the geometric product specification system, this is only possible with a high-precision measurement techniques accompanying the whole process. The quality assessment of the additive manufacturing parts is carried out by non-contact (CLSM, CT) and tactile metrology methods (CMM). By using CT, geometrical specifications are determined and evaluated to control of the deviations from geometrical tolerances in the desired range. In particular, mechanical parts in the micro/nano-scale are better resolved by CT to detect smaller tolerances.

The aim of this application was to recognize the differences between the surfaces of the gear parts and those of the miniaturized-manufactured gear parts with measurements.

This study highlighted the initial conditions when measuring roughness of the gear part's surfaces using Confocal Laser Scanning Microscopy (CLSM), which is used for surface characterization. The results from the surface roughness measurement ascertained that the direct sliding motion was affected by the surface finish quality.

The technical drawing of gear component was created according to functional and geometrical specifications. The surface structure of gear component was determined by CMM and CT. Using various measurement methods, geometrical specifications and tolerances were evaluated statistically. As partial circular, some specific geometrical tolerances are not possible to measure through a tactile method as CMM for small parts such as gear component. Computed Tomography is great solution to detect smaller tolerances which are not touchable by probes in CMM and requires high precision measurement technology. Furthermore, it was also feasible to observe the form deviations of each macro and miniaturized with 1:5 scale gear geometrical components. After miniaturizing the gear component by additive manufacturing, because of increasing uncertainty and accuracy; desired functional and geometrical specifications were lost considering the same conditions as material, ambient etc.

In porosity measurement with CT, porosity decreased in the micro geometry of the gear components. The miniaturized geometry had less porosity, which was related to the gaps volume, than the normal geometry. The minimum gap volumes depend on the scanning resolution in CT. The gaps in macro geometry were more frequent than the micro geometry as expected.

References

1. Steinbeis-Beratungszentrum (2019). https://www.toleranzen-beratung.de/en/downlads/informations broschucrcn/
2. ISO 8015:2011, Geometrical product specifications (GPS) - Fundamentals - Concepts, principles and rules (2011)
3. ISO 14405-3:2016, Geometrical product specifications (GPS) - Dimensional tolerancing - Part 3: Angular sizes (2016
4. ISO 2768-2: 1989, General tolerances - Part 2: Geometrical tolerances for features without individual tolerance indications (1989)
5. Durakbasa, M.N., Osanna, P.H., Afjehi-Sadat, A.: A general approach to workpiece characterization in the frame of GPS (Geometrical Product Specification and Verification). Int. J. Mach. Tools Manuf. **41**, 2147–2151 (2001)
6. Durakbasa, M.N.: Geometrical product specifications and verification for the analytical description of technical and non-technical structures. Printed in Austria (TU Wien: Abteilung Austauschbau und Messtechnik) (2003). ISBN 3-901888-26-8
7. ISO 17450-1:2011, Geometrical product specifications (GPS) - General concepts - Part 1: Model for geometrical specification and verification (2011)
8. ISO 17450-2:2013, Geometrical product specifications (GPS) - General concepts - Part 2: Basic tenets, specifications, operators and uncertainties (2013)
9. Weckenmann, A., Hartmann, W.: Function-oriented method for the definition and verification of micro structured surfaces. Precis. Eng. **37**, 684–693 (2013)
10. Sommer, K.D., Weckenmann, A., Siebert, B.: Qualität für Messergebnisse-Messunsicherheit nach GUM. Tech. Mess. **73**(4), 189–199 (2006)
11. Humienny, Z.: State of art in standardization in GPS area. CIRP J. Manuf. Sci. Technol. **2**, 1–7 (2009)
12. The future of making things, 3D printing additive manufacturing (2019). https://www.autodesk.com/solutions/additive-manufacturing
13. Durakbasa, N.M., Bauer, J.M., Bodur, O., Poszvek, G.: Challenges of miniaturizing a precision gear. In: Proceedings of the International Symposium for Production Research 2018, pp. 239–253. Springer Nature Switzerland AG (2019)
14. FORMIGA P 110 - laser sintering 3D printer - EOS (2019). https://www.eos.info/systems_solutions/plastic/systems_equipment/formiga_p_110
15. Aberlink innovative metrology (2019). https://www.aberlink.com/products/cmm/axiom-toohs/
16. ISO 1101:2017, Geometrical product specifications (GPS) - Geometrical Tolerancing - Tolerances of form, orientation, location and run-out (2017)

17. TomoScope®XS, Werth, Inc. (2018). http://werthinc.com/products/tomoscope-xs/
18. NanoFocus - μsurf (2018). http://www.nanofocus.com/products/usurf/usurf-explorer/
19. Drégelyi-Kiss, Á., Durakbasa, N.M.: Measurement error on the reconstruction step in case of industrial computed tomograph. In: Durakbasa, N., Gencyilmaz, M. (eds.) Proceedings of the International Symposium for Production Research 2018, ISPR 2018. Springer, Cham (2019)
20. ISO 286-1:2010, Geometrical product specifications (GPS) - ISO code system for tolerances on linear sizes - Part 1: Basis of tolerances, deviations and fits (2010)
21. DIN EN ISO 4287:2010, Geometrical Product Specifications (GPS) - Surface Texture: Profile Method - Terms. Definitions and surface texture parameters (2010)
22. ISO 25178-1:2016, Geometrical product specifications (GPS) - Surface texture: Areal - Part 1: Indication of surface texture (2016)
23. Sagbas, B., Boyacı, T.H., Durakbasa, N.M.: Precision metrology for additive manufacturing. In: Durakbasa, N., Gencyilmaz, M. (eds.) Proceedings of the International Symposium for Production Research 2018, ISPR 2018. Springer, Cham (2019)

Measuring Quality Orientation in Organisations: A Cognitive-Behavioural Approach

Ina Heine(✉) and Robert Schmitt

Laboratory for Machine Tools and Production Engineering,
RWTH Aachen University, Campus-Boulevard 30, 52074 Aachen, Germany
i.heine@wzl.rwth-aachen.de

Abstract. This paper explores the application of a cognitive-behavioural approach for measuring quality orientation in organisations. It aims at providing a method for improving the alignment of an organisation's strategy with its culture. The explored and presented approach is called Situational Judgment Test (SJT) and is based on the Critical Incident Technique (CIT), a method to gather information about reoccurring human-system problems and their causes. Critical incidents allow the systematic identification of behaviour that contribute to success or failure in specific situations. Thereby, they form a solid basis for performance appraisals as well as personnel development and selection procedures. We followed a deductive test development approach and conducted two studies with Subject Matter Experts (SMEs) to validate the developed items. The SMEs evaluated 40 situational descriptions with regard to their degree of realism as well as relevance and assigned each situation to one predefined variable related to quality orientation. Analysis of the data led to the exclusion of 14 items. The interrater agreement for the remaining 26 items is considered as fair and reflects the difficulty associated with construct-based SJT development. For future research, the combination of critical incidents with Virtual Reality (VR) as simulation method appears to be promising for increasing fidelity. The presented cognitive-behavioural approach provides a novel method for measuring and strengthening organisation-wide quality initiatives, going beyond the implementation of established quality tools and techniques.

Keywords: Critical incidents · Cognitive-behavioural approach · Organisational culture · Situational Judgment Test · Quality orientation

1 Introduction

Quality has been a major focus for organisations' strategic positioning in the last century. Starting with the application of statistical methods to quality control in the nineteen-thirties, the quality understanding evolved from a strong product perspective to a more holistic and organisation-wide approach. Nowadays, quality is considered on the product, process, as well as system level (Schmitt and Pfeifer 2015). The importance of quality for organisational performance appears self-evident, but still empirical

V. D. Majstorovic and N. Durakbasa (Eds.): IMEKOTC14 2019, LNME, pp. 291–306, 2019.
https://doi.org/10.1007/978-3-030-18177-2_26

evidence is scarce and quality as well as quality management remain concepts with manifold and fuzzy definitions (Dahlgaard-Park et al. 2013).

One major challenge lies in the measurement of the different relevant concepts like quality, quality performance, and quality orientation. Approaches to measure organisational quality orientation for empirical purposes include for instance the assessment of implemented quality management tools and techniques like cause and effect diagrams, also considered as 'hard' elements (e.g. Fotopoulos and Psomas 2009). However, in our understanding the determination of *how* employees deal with quality-related is more significant than *what* tools and techniques are implemented. We refer to the idea of acting in accordance with a predefined understanding of quality as *quality orientation* and propose a cognitive-behavioural measurement approach.

In the following, we provide a summary of the paper's theoretical background including the person-environment fit theory and psychological testing in organisations, focusing on Situational Judgment Tests (SJTs). Next, the methodological approach is outlined and followed by the results section. The paper closes with a discussion of the results and a conclusion.

2 Theoretical Background

This section provides a brief overview of the primary theoretical background for the proposed measurement approach. The first subsection refers to the concept of Person-Organisation (P-O) Fit as belonging to the more general Person-Environment Fit Theory. It gives a theoretical basis for the assumption that depending on the agreement between a person's knowledge, skills, abilities, and other characteristics (KSAOs) and an organisation's profile, there might be a good fit or not.

With the objective of selecting a suitable measurement approach for assessing quality orientation, the second subsection provides insights about psychological testing and assessment in organisations. In addition, we present a general process for constructing measures and combine it with the Situational Judgment Tests (SJTs) method, which is based on critical incidents and possible reactions to these incidents.

2.1 Person-Organisation Fit

While Person-Organisation (P-O) Fit is part of the Person-Environment Fit Theory, in Organisational Behaviour research, the focus is on the relationships between (1) person-job, (2) person-team, and (3) person-organisation (Anderson et al. 2004; Lauver and Kristof-Brown 2001). P-O Fit might be briefly summarised as the degree to which individuals and organisations share the same values (Lauver and Kristof-Brown 2001).

Values in organisations are often associated with the concept of organisational culture, which is defined by Martin and Fellenz (2010) as *"set of shared, often implicit assumptions, beliefs, values, and sensemaking procedures that influences and guides the behaviour and thinking of organisational members, and is continuously enacted and in turn reinforced – or changed – by the behaviour of organisational members"* (p. 481). Thus, there appears to be a reciprocal relationship between personal values and behaviour by individuals and an organisation's culture. In this context, we understand quality culture as the shared agreement on organisational level (in terms of implicit assumptions) with regard to quality-related issues.

If the fit between a person and organisation is not given, but there is a strong organisational culture (i.e. core values are widely shared), then this person will probably either assimilate or leave the organisation. If there is no strong organisational culture and the individual's value system is contradicting the organisation's strategic positioning, the missing fit will further add to this misalignment. It is important to note that we consider heterogeneity beneficial for organisational performance, but that homogeneity regarding a specific set of core values (*here:* values related to quality orientation) is necessary to strengthen an organisation's quality culture. Since there is no established approach for measuring quality orientation (neither on individual nor on organisational level) that goes beyond assessing the implementation of quality tools and techniques, we propose an approach that focuses on the behavioural and cognitive dimensions. The next subsection provides a brief overview about psychological assessment methods in organisational settings.

2.2 Psychological Testing and Assessment in Organisations

In the early 20th century, the first Binet-Simon Intelligence Scale was published and Cattell, Thorndike, and Woodworth founded the first major company for publishing tests (Aiken 1985). Standardised psychological testing has the objective to assess individual attributes that are typically not directly observable, but referred to as constructs, and which are therefore difficult to measure. In comparison to, for instance, a kitchen table, which is a solid object with certain attributes like the top's thickness that can be determined more or less precisely by a ruler or sliding caliper, constructs like quality orientation are not directly accessible. This inaccessibility leads to a complex development and validation process of new assessment instruments, a discipline called psychometrics. Aguinis et al. (2002) provide an eight steps general process for constructing psychological measures. These are: (1) determining the purpose, (2) defining the attribute, (3) developing a measure plan, (4) writing items, (5) conducting a pilot study and item analysis, (6) selecting items, (7) establishing norms, and (8) determining the reliability and validity.

In the field of Industrial Psychology, psychometrics is especially relevant for personnel selection (Kline 2000). The applied tests vary between intelligence tests, ability tests, tests of motor skills, aptitude tests, personality tests, questionnaires, projective and objective tests, motivation and interest tests, and attitude scales. According to Urbina (2004), the term 'test' should only be used if the test taker's responses are evaluated or scored as right or wrong. Instruments that are not evaluative (e.g., no personality is incorrect), should be preferably called inventory or questionnaire. However, many use these terms interchangeably (Hubley and Zumbo 2013).

Schmidt and Hunter's (1998) meta-analysis about the validity and utility of selection methods provides an overview of 19 different personnel measures and their test statistical performance. According to the authors, only work samples are slightly more valid than measures of general mental ability (GMA). Work samples belong to the class of simulation tests with high-fidelity and tend to be costly in their application. A less cost-intensive alternative are low-fidelity simulations like Situational Judgment

Tests (SJTs), which consist of written or spoken descriptions of situations and written or spoken responses to these situations (Motowidlo and Dunnette 1990). Thus, the overall approach is comparable to situational interviews, but considered as more structured because respondents choose from predefined responses (Ployhart and MacKenzie 2011; Ployhart and Ward 2013).

The general process for SJT development includes the three steps: (1) situation generation, (2) response option generation, and (3) scoring (Ployhart and MacKenzie 2011). The first step comprises the collection of critical incidents (CIs) through Subject Matter Experts (SMEs), the selection of CIs (i.e. item stems), and their length editing. The critical incident technique (CIT) is a method developed by Flanagan (1954) to gather information about reoccurring human-system problems and their causes. CIs allow the systematic identification of behaviour that contribute to success or failure in specific situations. The second step involves the identification of one or more responses to each situation through another group of SMEs, a review of all responses, and the preparation of the potential responses list. Response options are possible effective and ineffective behavioural responses to the respective situations and require the cognitive process of making a choice. The effectiveness of responses depends on the outcome of the situation. If the problematic situation were solved by the behavioural response, this response would be considered effective. If the problematic situation were not solved by the behavioural response, this response would be considered ineffective. The third step refers to the development of a scoring key (McDaniel and Nguyen 2001; Ployhart and MacKenzie 2011). Figure 1 shows an exemplary SJT item for measuring employee integrity, illustrating its stem and possible response options as main components. In this case, response option D correlates highly with other established measures of integrity and is therefore considered the 'right' choice (Becker 2005).

The following section provides an overview of our methodological approach to develop and validate the item stems for measuring quality orientation.

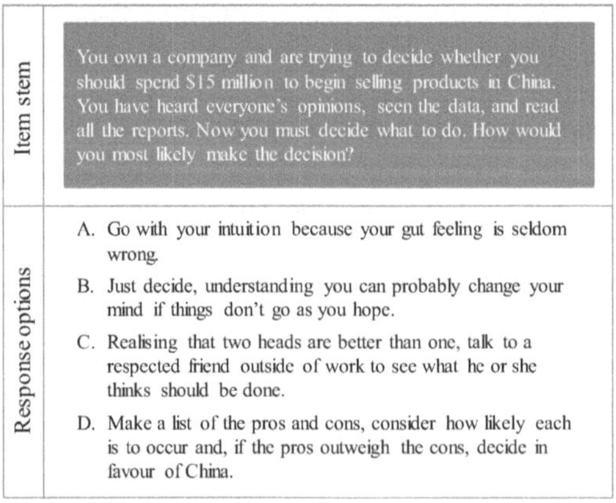

Fig. 1. Exemplary SJT item for measuring employee integrity (based on Becker 2005; Heine 2016).

3 Methodology

The paper's aim is to provide a measuring approach for the assessment of quality orientation in organisations. Due to the bidirectionality between cognition and behaviour in explaining organisational culture and the high performance of simulation tests for personnel selection, we chose the development of a SJT. For test development purposes, we conducted two studies with SMEs. This paper presents the results of the first study. The following subsection provides a description of the used materials and the second subsection gives an overview of the study's procedure.

3.1 Materials

In the first study, the SMEs received a brief written explanation with background information including the structure of CIs focusing on the item stem, a formal definition of quality orientation, and task instructions. In addition, the participants received a list of 40 critical incidents that we had developed based on a literature-based definition of quality orientation as set of the following five variables (1) internal customer focus, (2) external customer focus, (3) continuous improvement orientation, (4) systems-thinking perspective, and (5) data orientation (Heine et al. 2016). In a second study, the participants received again background information including the structure of CIs, but this time focusing on the items' responses, the formal definition of quality orientation, and task instructions. The second group received 27 item[1] stems with 5–9 response options for each item. Following sections deal primarily with outlining the first study.

3.2 Procedure

We decided for a deductive development of the critical incidents (i.e. item stems) because of two reasons. First, the empirical collection through querying job incumbents and/or supervisors might be biased due to their direct job and organisational involvement. Second, we required the construct-based reporting of CIs, which might be more difficult for persons that are not familiar with the idea of constructs. In total, we developed for each variable related to quality orientation, eight critical incidents that were supposed to present a representative sample for the respective variable and middle management as main target group, resulting in an initial list of 40 items. Each item stem was edited to a maximum of 100 words to keep the level of complexity reasonable while still including some situational details to increase fidelity. The item stems' formal design is in accordance with two of the three critical incidents criteria (i.e. antecedent information and description of the situation) formulated by Butterfield et al. (2005). The third criterion 'outcome of the incident' is expressed in the items' response options.

[1] In the second study, one new item was added to the remaining 26 items to have at least 5 items per variable.

After we had developed this first initial list of items, we conducted the first study with six SMEs, which were asked to participate if they had at least three years of theoretical and practical experiences in the field of quality management. Each SME rated all 40 items with regard to two aspects: (1) the situations' degree of realism and (2) whether the incidents reflect problematic situations for middle management or not. For this, the SMEs indicated to what extent they agreed or disagreed with the following two statements on a Likert-type scale from 1 to 7 (strongly disagree to strongly agree):

1. *'This incident reflects a realistic situation in organisations.'*
2. *'This incident reflects a problematic situation for middle management.'*

Furthermore, the SMEs were asked to answer the question *'In order to handle this situation, which of the values related to quality orientation becomes most relevant?'* for each item. If the SMEs considered important incidents to be missing, they could use an additional text field or add comments to each item. Based on the SMEs' ratings, 14 situations were excluded from further test development. For the remaining 26 item stems and one new situation description, we developed 5–9 possible response options per item and conducted a second study.

4 Results

This section is organised into two subsections. The first subsection provides a summary of the descriptive statistics for the 40 item stems with regard to their degree of realism and importance. The second subsection summarises the agreement between the experts' item allocation to the five variables related to quality orientation.

4.1 Item Stems' Degree of Realism and Relevance

The first group of experts rated each of the 40 item stems with regard to their degree of realism and relevance. Table 1 provides an overview of the descriptive statistics. To be included for further development, an item was expected to be rated with a mode of ≥ 6 and a mean value of ≥ 5 for both ratings. Items that clearly missed to fulfill these criteria were excluded from further development. Items that barely missed to fulfill the criteria were checked again before deciding for inclusion or exclusion as the mean is a comparatively sensitive parameter with small samples. In total, 25 items met the expected values. The items ICF05, ICF06, ICF08, ECF02, ECF07, and STP01 were also included after an additional review by the item developer as their ratings were just below the threshold but appeared to be useful for sampling situations, in which quality orientation can manifest. Thus, 9 of 40 items were excluded due to low realism and/or importance ratings (i.e. ICF02, ICF04, ECF06, ECF08, CIO01, CIO03, STP06, STP08, DO03).

Table 1. Descriptive statistics of item stems ratings.

Variable	N	Item's degree of realism				Item's degree of relevance			
		Min.	Max.	Mean	Mode	Min.	Max.	Mean	Mode
Internal Customer Focus (ICF)									
ICF01	6	6	7	6.17	6	6	7	6.33	6
ICF02[a]	6	4	6	5.17	6	3	7	4.83	4
ICF03	6	5	7	6.00	6	1	7	5.67	7
ICF04[a]	6	2	6	4.00	4	3	6	5.17	6
ICF05	6	4	7	5.67	6	3	6	4.83	6
ICF06	6	5	7	6.00	6	4	7	5.67	4[d]
ICF07	6	6	7	6.33	6	2	7	5.83	7
ICF08	6	2	7	4.67	5	2	6	5.00	6
External Customer Focus (ECF)									
ECF01	6	3	7	5.50	6	4	7	5.83	6
ECF02	6	2	7	5.17	5[c]	1	7	4.83	6[d]
ECF03[b]	6	3	7	5.83	7	4	7	5.67	6
ECF04	6	5	7	6.17	6	3	7	5.67	6
ECF05	6	6	7	6.50	6[c]	3	7	5.67	6
ECF06[a]	6	2	7	4.67	6	2	7	3.67	2
ECF07	6	2	6	4.83	6	5	7	6.17	6
ECF08[a]	6	2	7	4.50	2[c]	3	7	5.33	5[d]
Continuous Improvement Orientation (CIO)									
CIO01[a]	6	5	7	5.83	6	3	7	5.00	3[d]
CIO02	6	6	7	6.50	6[c]	5	7	6.17	6
CIO03[a]	6	5	6	5.50	5[c]	3	7	5.00	4[d]
CIO04	6	4	6	5.50	6	5	7	6.00	5[d]
CIO05	6	5	6	5.83	6	2	6	5.00	6
CIO06	6	5	6	5.83	6	6	6	6.00	6
CIO07[b]	6	3	6	5.17	6	3	7	5.17	6
CIO08	6	6	7	6.33	6	4	7	6.00	6
Systems-thinking Perspective (STP)									
STP01	6	6	6	6.00	6	1	6	4.33	6
STP02	6	5	7	6.00	6	2	7	5.50	6
STP03[b]	6	5	7	5.83	6	2	7	5.17	5[d]
STP04	6	6	7	6.50	6[c]	5	7	6.17	6
STP05	6	6	7	6.33	6	6	7	6.33	6
STP06[a]	6	3	7	4.83	6	2	6	4.67	6
STP07	6	5	7	6.00	6	2	7	5.17	6[d]
STP08[a]	6	3	6	4.67	4[c]	3	7	5.33	5[d]
Data Orientation (DO)									
DO01[b]	6	1	7	5.50	6	3	7	5.83	6
DO02	6	5	7	6.00	5[c]	4	7	6.17	7
DO03[a]	6	3	7	4.83	4	2	7	4.33	2[d]
DO04	6	4	7	5.67	6	5	6	5.83	6
DO05	6	5	7	6.33	7	5	7	6.00	5[d]

(*continued*)

Table 1. *(continued)*

Variable	N	Item's degree of realism				Item's degree of relevance			
		Min.	Max.	Mean	Mode	Min.	Max.	Mean	Mode
DO06	6	3	7	5.50	6	2	7	5.17	5[d]
DO07	6	5	7	6.00	6	3	7	5.67	6
DO08[b]	6	5	7	5.83	6	2	7	5.00	6

Note. [a]Item has been excluded for further development. [b]Item has been excluded due to low interrater agreement. [d]Indicates that there are several modes. Here, the smallest mode is shown.

4.2 Interrater Agreement of Item Assignment

Besides rating the items' degree of realism and importance, the experts were asked to assign each item to one of the five variables related to quality orientation. Table 2 provides an overview of the interrater agreement between the six SMEs across all 40 items. The algorithmic mean results in an overall agreement index of 0.31. According to Landis and Koch (1977) values between 0.21 and 0.40 are considered fair, 0.41–0.60 reflects moderate agreement, and values above 0.60 are regarded as good.

Table 2. Summary of interrater agreement (40 items).

Cohen's Kappa					
Rater	1	2	3	4	5
1	-	-	-	-	-
2	.46	-	-	-	-
3	.18	.19	-	-	-
4	.43	.27	.13	-	-
5	.27	.36	.30	.30	-
6	.50	.37	.29	.38	.37

Based on the diverging ratings, we excluded the items ECF03, CIO07, STP03, DO01, and DO08 from further development due to low agreement between the experts' ratings, resulting in a final set of 26 items. The item ICF03 was reassigned to the category STP and the items ICF07 and STP02 to CIO. Table 3 provides an overview of the interrater agreement between the six SMEs across the remaining 26 items. The algorithmic mean results in an overall agreement index of 0.40.

Table 3. Summary of interrater agreement (26 items).

Cohen's Kappa					
Rater	1	2	3	4	5
1	-	-	-	-	-
2	.50	-	-	-	-
3	.23	.27	-	-	-
4	.46	.40	.20	-	-
5	.27	.41	.30	.50	-
6	.66	.57	.37	.49	.51

5 Discussion

The results presented in the previous section have one immediate and two indirect implications for further research. First, on the one side the results provide evidence for the difficulty of developing construct-based SJTs. This is in accordance with the perspective that SJTs measure one specific construct that predicts job performance, like practical intelligence or judgmental ability, and cannot be specifically designed to assess any construct (Weekley and Ployhart 2006). On the other side, the results also show that there is at least fair to moderate agreement between the raters. This indicates that, in general, a construct-based design might be possible but requires careful item development. Since quality and quality orientation appear to be constructs that are not unequivocally defined and hard to grasp, the construct-based development seems especially challenging.

Secondly, after careful item development, the SJT should be tested with regard to criterion (concurrent and predictive) and construct (convergent and discriminant) validity (Christian et al. 2010). For examining the validity of data generated by the SJT, there needs to be a theoretical understanding of related and unrelated constructs to quality orientation represented in a nomological network. In another first empirical study, we considered conscientiousness as being positively and resilience as being negatively correlated to quality orientation as measured by the Heine Scale for Managerial Quality Orientation (HSMQ). However, statistical analysis of the empirical data did not provide evidence for the hypothesised relationships (Heine 2016). Future research should repeat the study with a revised version of the HSMQ, after higher interrater agreement has been achieved within the item development phase. The conceptualisation of quality orientation as set of five related variables and their formal definitions should be also revaluated as internal customer focus and systems thinking perspective might overlap.

Lastly, we continue to consider SJTs as promising alternative to work samples for identifying Person-Organisation (P-O) Fit. More precisely, we see high potential in new technologies like Virtual Reality (VR) to enhance fidelity and reduce possible confounding effects (e.g. cognitive load) associated with written SJTs. The representation format of SJTs has been an ongoing subject in recent literature and Multimedia Situational Judgment Tests (MMSJT) are expected to increase fidelity (Lievens et al. 2008; Ployhart and MacKenzie 2011). However, until now multimedia has been primarily associated with video-based representation of SJTs, which has a reduced potential for immersion in comparison to VR. The assessment of how employees react to quality-related situations represented in VR appears to be a promising new approach to determine quality orientation.

6 Conclusion

The aim of this paper was to explore the application of a cognitive-behavioural approach for measuring quality orientation in organisations. The presented approach is based on the Situational Judgment Test (SJT) method and Critical Incident Technique (CIT). Results of this study as well as the additional referenced research indicate that the conceptualisation of quality orientation and the developed items require further refinement to improve interrater agreement. Strong agreement in assigning the items to the variables associated with quality orientation is the prerequisite for a construct-based measurement instrument.

Appendix

Internal Customer Focus (ICF)	01	For some time, there seems to have been an uneasy atmosphere in the marketing department that you took over two months ago when you came freshly into the organisation. There is only very little cooperation and knowledge sharing among your employees. You do not know the underlying cause for this. What would you do?
	02	As manager of a department, it is very important for you that your employees are motivated and engaged in their work, not least because motivated employees show better work performance. Lately, you have noticed an increased turnover rate and demotivation in your department. You do not know whether the cause for this lies within your department or is an organisation-wide issue. What would you do?
	03	As manager of a quality management department, you have agreed to meet certain objectives regarding quality-related key performance indicators (KPI). These objectives conflict with targets of other managers and departments like purchasing. You notice that this causes a competitive situation with adjacent departments. The competitive behaviour spreads throughout all employees and the lack of collaboration causes unnecessary delays and risks. What would you do?
	04	You are manager of a sales department and you have noticed that recently customers treat your employees disrespectfully and unfairly, which causes demotivation and stress in your employees – not least because of your organisation's strong customer focus. What would you do?
	05	As manager of a purchase department, the output of your and your employees' work strongly influences the work of other departments because of its direct input for various other projects within the organisation. Lately, you have received many internal complaints about the length of time taken before a supplier is selected and assigned. What would you do?
	06	You are manager of a quality management department and you have noticed that an increasing number of projects are behind schedule. Since the results of most projects are input for other organisational processes, this delay negatively affects various adjoining departments or process owners, resulting in tensions. What would you do?

	07	You are head of a department. Your performance is assessed through certain Key Performance Indicators (KPI) on a yearly basis. Last year, your KPI values were unsatisfactory but you and your team have worked hard during the last year and therefore your KPIs are now on a satisfactory level. Still, there is room for improvement. What would you do?
	08	As head of the human resource department, you provide internal services like, for example, a selection procedure process or support during change management projects to the other departments. Repeatedly, you have noticed dissatisfaction from other departments about the services provided. What would you do?
External Customer Focus (ECF)	01	As part of the middle management, you participate in a new voluntary programme in which managers can meet their external customers personally. During the last meeting, you noticed that the general attitude of customers towards the organisation has gotten worse. However, the reason for this does not seem to be very clear. The last general customer satisfaction survey six months ago did not reveal anything particular. What would you do?
	02	As manager of a public relations department, you receive information about unsatisfied and influential or famous customers quickly. Because of their visibility, they heavily influence the public opinion of your organisation. Yesterday, you saw a newspaper article about a famous singer who burnt one of your products after a concert because he was so annoyed about it not working well. What would you do?
	03	You are head of a quality management department and you regularly assess the satisfaction of your customers. The organisation's goals regarding customer satisfaction is an index of 7.5 (maximum = 10). Similar to previous years, you record a value of 7.7 but the organisation still loses market share. What would you do?
	04	You are head of a development department. The last two products that have been developed do not seem to meet customer requirements to a good degree because the sales figures are far below the expected values. Although this can have many causes, a qualitative customer survey has revealed that customers miss certain product features. What would you do?
	05	You are head of a quality management department. One of your key organisational processes is managing customer complaints. One of the most important performance indicators is how fast customers receive a reaction to their complaint. Therefore, your department has implemented an automatic response mechanism. Incidentally, you have read customer comments on the internet that reveal dissatisfaction with the fast but automatic responses. They criticise the delay in personal reaction. What would you do?
	06	You are manager of a public relations department. In the community in

		which your organisation is located, many of the day-care facilities for children are in bad conditions. This issue has been frequently on the local news but the lack of financial and human resources makes it difficult for the community to overcome this problem. What would you do?
	07	You are manager of a sales department and during the last week, you have visited some of the sales personeel and participated in customer interactions to get a better impression about their working environment and key issues. You noticed that most interactions were more driven by what the sales person likes about a product and not so much by what the customer actually needs, not leading to a purchase. What would you do?
	08	You are head of a marketing department and because of recent market analyses, you are knowledgeable about current and expected customer needs. You frequently observe that employees from other departments act against these customer needs. What would you do?
Continuous Improvement Orientation (CIO)	01	As manager of the quality management department, you regularly interact with the development department. The designing engineers are expected to create innovative and good quality products that are better than previous generations. Therefore, they sometimes perform costly experiments. This causes conflicts and a lack of appreciation because not everyone sees the value in experimenting. What would you do?
	02	You are head of a quality management department and you have worked for your organisation for more than 15 years. With new generations of young professionals, you notice that new ideas and approaches reach the organisation. However, many of your long-term colleagues are rather conservative or dismissive about trying out new courses of action. What would you do?
	03	You have been head of a human resource department for two months. Previously, you had a group leader function in one of the production plants, where errors were frequently considered to be employees' personal fault and covered up. What would you do?
	04	You have been working as head of the quality management department for two months. You notice that the annual customer satisfaction survey is outdated or at least it is not the standard survey that is used in your branch of industry. This makes it difficult to compare to other organisations. What would you do?
	05	You are manager of a production department and the organisation has implemented a suggestion system for employees with the goal of using the employees' knowledge and increasing their involvement. However, it seems that this system is not receiving much attention from the employees and thus there are only a few internal ideas about how to improve organisational performance. What would you do?
	06	As head of a sales department, you are especially interested in the annual sales figures. Of course, the goal is to meet the predefined target,

		which is based on improvement over the previous year. The new target value is far beyond your current sales figures and, from your point of view, difficult to realise without drastic organisational changes. What would you do?
	07	You are a new manager within the human resource department of an organisation and your team works regularly on internal projects to improve different operations. For you, the actions' quantitative effects are often not visible. What would you do?
	08	You are head of a marketing department and there are various work processes within your department that seem to be outdated. It appears that most of your employees follow their own way of doing things, which causes high variance in performance. What would you do?
Systems-Thinking Perspective (STP)	01	You are manager in a production plant. In the last two months, there has been an accumulation of complaints about products that have been produced in this plant. In order to remedy this, vendor parts have to be exchanged because the material does not seem to be enduring enough. You know that these vendor parts are also used in other plants for similar products. What would you do?
	02	As head of a quality management department, your decisions affect the work of all other departments. Last month you received a new target value for the time [in days] in which customer problems must be solved from one of the managing directors. The new target value is significantly lower than the current mean value. What would you do?
	03	You are head of a human resource department and you frequently hear colleagues saying that improving processes is not helping to solve performance problems. According to them, the only thing that the organisation and its employees require is a 'mind change'. What would you do?
	04	You have been head of the quality management department in an international organisation for two months. Your first major task is to unify and standardise the different quality management practices and establish one common system. What would you do?
	05	You are supervisor of a development department and you have noticed that most of your employees have little contact with colleagues from other departments. This sometimes leads to misunderstandings and frustrations in your employees because important information for the development process is lacking. In fact, the last product development contained a bug, which was very costly and could have been prevented. What would you do?
	06	You are head of a quality management department and you have regular management meetings. During the last meeting, one of your colleagues suggested a new production process that is supposed to improve the throughput time but is questionable with regard to sustainability. Your colleague's forecast predicts a 4.6% bonus increase for all managers if

		the organisation changes the respective process. What would you do?
	07	You are head of the development department. Your work results are mainly used by the production department. During meetings with the head of the production department, you find out that your work results and the defined processes and specifications lead to very high efforts in the production department. To decrease efforts in the production department you would have to accept slightly higher efforts in your department. What would you do?
	08	As head of an organisational development department, you implement actions to improve performance. These interventions have been mainly targeted at the individual level, for example, leadership development. You receive the feedback that the interventions' positive effect is often lacking. What would you do?
	01	As head of a marketing department, you are responsible for presenting the product in such a way that customers want to buy it. The sales figures from last year are rather disappointing and you could not meet the target value determined by the managing directors. What would you do?
	02	You have been manager in a quality management department for two months. You notice that there is a lot of data collection inside your organisation but it seems that employees do not make much use of it and therefore it is difficult for them to provide you with data-based propositions. What would you do?
Data Orientation (DO)	03	You are manager in a human resource department and you get a call from the head of the department. She is worried because the number of applicants has been quite small for the last few vacancies and you have the feeling that she makes you responsible for that because you took over the responsibility for recruiting just four months ago. What would you do?
	04	As new manager within a purchase department, you are responsible for choosing suppliers that best fit the projects' requirements and contribute to improving the performance of your organisation. The current internal database with all supplier details contains only those projects that have been assigned to the respective supplier and whether the project was accomplished on schedule or not. For you, this information is not sufficient to make important decisions. What would you do?
	05	You are manager in a quality management department and in order to ensure high quality standards in your organisation, process owners and project managers have to provide you with a performance overview of current projects on a quarterly basis. Because of recent incidents, you have doubts regarding the quality of the provided data, making them useless for important decisions. What would you do?
	06	You are head of a marketing department and you are planning a big campaign to advertise a new product that was just released by the organisation. Since there has not been a comparable product on the

		market before, you do not have historical data for your estimations. Of course, the general manager expects you to deliver a sales forecast anyway. What would you do?
	07	You are the new head of a production department and your organisation is going through turbulent times. It is very important that you get a quick overview of the department's performance and resources in order to make an action plan. What would you do?
	08	As head of a development department, your annual budget contains financial resources for undertaking research activities. Repeatedly, there have been mismatches between your forecast and expenditures, which means the planning was imprecise, and your department has to return the surplus. This has negative effects on future budgets. What would you do?

* Items shaded in grey have been included in the further test development process.

References

Aguinis, H., Henle, C.A., Ostroff, C.: Measurement in work and organizational psychology. In: Anderson, N., Ones, D.S., Sinangil, H.K., Viswesvaran, C. (eds.) Handbook of Industrial, Work and Organizational Psychology, Volume 1: Personnel Psychology, pp. 27–50 (2002)

Aiken, L.R.: Psychological Testing and Assessment, 5th edn. Allyn and Bacon, Boston (1985)

Anderson, N., Lievens, F., van Dam, K., Ryan, A.M.: Future perspectives on employee selection: key directions for future research and practice. Appl. Psychol. Int. Rev. **53**(4), 487–501 (2004)

Becker, T.E.: Development and validation of a situational judgment test of employee integrity. Int. J. Sel. Assess. **13**(3), 225–232 (2005)

Butterfield, L.D., Borgen, W.A., Amundson, N.E., Maglio, A.T.: Fifty years of the critical incident technique: 1954–2004 and beyond. Qual. Res. **5**(4), 475–497 (2005)

Christian, M.S., Edwards, B.D., Bradley, J.C.: Situational judgment tests: constructs assessed and a meta-analysis of their criterion-related validities. Pers. Psychol. **63**(1), 83–117 (2010)

Dahlgaard-Park, S.M., Chen, C., Jang, J., Dahlgaard, J.J.: Diagnosing and prognosticating the quality movement – a review on the 25 years quality literature (1987–2011). Total. Qual Manag. Bus. Excel. **24**(1–2), 1–18 (2013)

Flanagan, J.C.: The critical incident technique. Psychol. Bull. **51**(4), 327–358 (1954)

Fotopoulos, C.B., Psomas, E.L.: The impact of “soft” and “hard” TQM elements on quality management results. Int. J. Qual. Reliab. Manag. **26**(2), 150–163 (2009)

Heine, I.: Quality orientation in middle management - development and validation of a situational judgment test, Dissertation RWTH Aachen University, Apprimus Verlag (2016)

Heine, I., Schmitt, R., Beaujean, P.: Critical incidents of quality orientation in lower and middle management. TQM J. **28**(5), 734–744 (2016)

Hubley, A.M., Zumbo, B.D.: Psychometric characteristics of assessment procedures: an overview. In: Geisinger, K.F., Bracken, B.A. (eds.) APA Handbooks in Psychology. APA Handbook of Testing and Assessment in Psychology, pp. 3–19. American Psychological Association, Washington, DC (2013)

Kline, P.: The Handbook of Psychological Testing, 2nd edn. Routledge, London, New York (2000)

Landis, J.R., Koch, G.G.: The measurement of observer agreement for categorical data. Biometrics **33**(1), 159–174 (1977)

Lauver, K.J., Kristof-Brown, A.: Distinguishing between employees' perceptions of person-job and person-organization fit. J. Vocat. Behav. **59**(3), 454–470 (2001)

Lievens, F., Peeters, H., Schollaert, E.: Situational judgment tests: a review of recent research. Pers. Rev. **37**(4), 426–441 (2008)

Martin, J., Fellenz, M.R.: Organizational Behaviour and Management, 4th edn. Cengage Learning, Andover (2010)

McDaniel, M.A., Nguyen, N.T.: Situational judgment tests: a review of practice and constructs assessed. Int. J. Sel. Assess. **9**(1–2), 103–113 (2001)

Motowidlo, S.J., Dunnette, M.D.: An alternative selection procedure: the low-fidelity simulation. J. Appl. Psychol. **75**(6), 640–647 (1990)

Ployhart, R.E., MacKenzie, W.I.: Situational judgment tests: a critical review and agenda for the future. In: Zedeck, S. (ed.) APA Handbook of Industrial and Organizational Psychology, Volume 2: Selecting and Developing Members for the Organization, pp. 237–252. American Psychological Association, Washington, DC (2011)

Ployhart, R.E., Ward, A.: Situational judgment measures. In: Geisinger, K.F., Bracken, B.A. (eds.) APA Handbooks in Psychology. APA Handbook of Testing and Assessment in Psychology, pp. 551–564. American Psychological Association, Washington, DC (2013)

Schmidt, F.L., Hunter, J.E.: The validity and utility of selection methods in personnel psychology: practical and theoretical implications of 85 years of research findings. Psychol. Bull. **124**(2), 262–274 (1998)

Schmitt, R., Pfeifer, T.: Quality Management: Strategies, Methods, Techniques (Qualitätsmanagement: Strategien - Methoden – Techniken), 5th edn. Carl Hanser Verlag, München (2015)

Urbina, S.: Essentials of Psychological Testing. Essentials of behavioral science series. Wiley, Hoboken (2004)

Weekley, J.A., Ployhart, R.E.: Situational Judgment Tests: Theory, Measurement, and Application. SIOP Organizational Frontiers Series. Lawrence Erlbaum Associates, Mahwah (2006)

Cyber-Physical Approach to Coordinate Measurement of Flexible Parts

Samir Lemeš[1(✉)], Nermina Zaimović-Uzunović[1], Josip Kačmarčik[1], Almira Softić[2], and Hazim Bašić[2]

[1] University of Zenica, Fakultetska 1, 72000 Zenica, Bosnia and Herzegovina
slemes@unze.ba, {nzaimovic,kjosip}@mf.unze.ba
[2] University of Sarajevo, Vilsonovošetalište 9, 71000 Sarajevo, Bosnia and Herzegovina
{softic,basic}@mef.unsa.ba

Abstract. This research implements the Finite Element Analysis of the contact between the flexible part and the CMM touch-probe in order to support the Cyber-Physical Manufacturing Metrology Model and its metrology integration into coordinate measuring machine (CMM) inspection of flexible parts. Although optical metrology offers a solution for flexible part measurements, because there is no contact between the measurand and the probing system, not all geometrical features are accessible by the optical CMM probe. Therefore, it is important to quantify and to validate the deformation introduced by the contact measuring process. The Cyber-Physical Systems connect the virtual and physical worlds in a way that intelligent objects communicate and interact with each other. This research will connect the physical measurement system consisting of CMM and flexible plastic part with their digital representation, where contact is simulated by means of Finite Element Method.

Keywords: Finite Element Analysis · Coordinate measuring machine · Flexible parts

1 Introduction

1.1 Cyber-Physical Manufacturing Metrology Model

Wide use of Information and Communications Technologies (ICT) in industrial practice leveraged the research and development activities in Cyber-Physical Systems (CPS) applications, where real and virtual worlds are integrated in internet environment in industry [1]. CPS is a system that combine a physical system with an embedded information processing system such that the resulting system has novel capabilities that could not be achieved by either the physical or the computational entity alone [2]. In CPS, smart sensors are used to collect vast amount of information, and one can conclude that coordinate measuring machine (CMM) is already a CPS enabled device, because it uses force and displacement sensors to collect data about the geometry. However, a true CPS could use the advantages of computer simulation to compensate the material deformation occurring due to measuring process [3].

V. D. Majstorovic and N. Durakbasa (Eds.): IMEKOTC14 2019, LNME, pp. 307–317, 2019.
https://doi.org/10.1007/978-3-030-18177-2_27

The integration of CPS into coordinate measurements was performed by Majstorović et al. in [4], where an algorithm was developed for collision avoidance in order to integrate the Cyber-Physical Manufacturing Metrology Model into CMM inspection process planning for prismatic parts. The Intelligent Model for Inspection Planning was developed for use in case of measuring path planning for geometrically complex prismatic parts with large numbers of tolerances.

1.2 Measuring Flexible Parts on CMM

The flexible parts require special attention, not only in measurements, but also in manufacturing. Deng et al. in [5] derived an algorithm to extract the intrinsic attributes for deformation compensation in manufacturing process. Measurement of nonrigid parts is an issue often solved by using non-contact methods, but sometimes these cannot be applied due to obscured or somehow inaccessible geometry features. When injection moulded parts are measured by means of touch-probe CMM, the contact between the touch-probe and the part introduces the deformation which could be as big as 25% of the real geometry deviation [6].

One approach to measure nonrigid part geometry, is to use a complex fixturing, as shown in [7]. The fixtures do reduce the deformation caused by measuring force, but the fixturing assembly can be extremely complex and time-consuming. Another approach is the application of computer simulations to approximate the deformation that appends when the part is installed in the fixturing device [8]. In computer simulation, virtual models of the CMM, the workpiece and the measurement process are created, estimates of the individual error sources are provided, and these data are then applied in repeated virtual measurements, in order to determine the measurement variability and uncertainty [9]. Radvar-Esfahlan and Tahan in [10] sought out geometric properties that are invariant to inelastic deformations, and presented a systematic comparison of some well-known dimensionality reduction techniques in order to evaluate their accuracy and potential for non-rigid metrology, by using a simulated state of use.

Even international standards such as ASME Y14.5 [11] and ISO 1101 [12] state that manufactured part requirements are to be evaluated in a free-state unless otherwise specified, but exemptions to this rule are given for nonrigid parts. One must have an understanding of the nonrigid part's compliant behavior and function requirements, as well as an understanding of the specification method allowed by the standards in order to select a proper specification method [13].

According to all above mentioned references, it is justifiable to examine the possibility and to evaluate the applicability of CPS approach by combining finite element computer simulation with physical geometry measurement on CMM.

2 Measurements on CMM

For the purpose of this research, two flexible parts were used, made of mild steel grades for cold forming, manufactured by Salzgitter Flachstahl GmbH, having properties according to DIN EN 10130-02:99, quality DC04 A (deep-drawing steel), which is equivalent to the steel SAE 1010 (Modulus of elasticity E = 211.669 MPa, Poisson's ratio ν = 0,291, Shear modulus G = 82.047,6 MPa).

The measurements were performed with 3D Coordinate measuring machine Zeiss Contura G2 700 Aktiv, measurement range: 700 × 1000 × 600 mm, measurement uncertainty according to ISO 10360-2: MPE_E = (1,8 + L/300 μm, MPE_P = 1,8 μm). VAST XT measuring head can vary the measuring force between 50 and 1.000 mN.

Two parts with different geometry were measured, (a) curved and (b) flat. Figure 1 shows the geometry properties, and Figs. 2 and 3 show measurement setups on CMM.

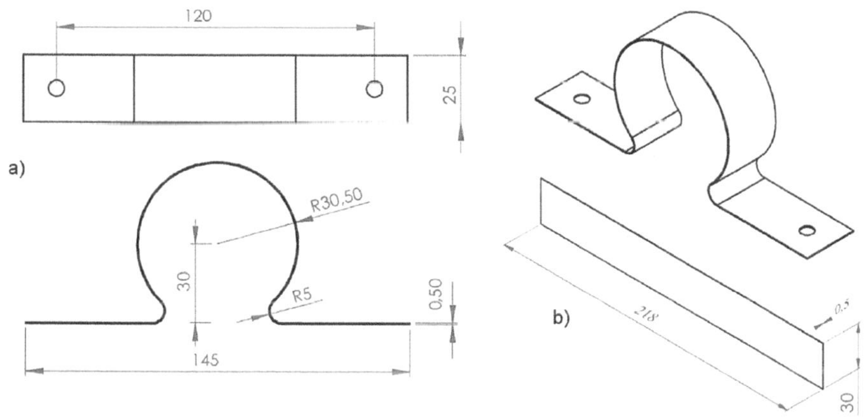

Fig. 1. The geometry of flexible parts measured on CMM.

Fig. 2. Measuring curved part on CMM.

Fig. 3. Measuring flat part on CMM.

The measurements were performed with two different forces (forces that trigger the registration of coordinates when touch-probe comes in contact with measured part), namely 100 mN and 200 mN. The curved part was measured in two points, and the flat part was measured in 6 points, at the distances of 5, 35, 65, 95, 125 and 155 mm from the fixture. There were 3 parts, and every sample was measured three times under the same conditions and using the same measurement strategy. In that way, one could compare the differences between the samples, showing the manufacturing quality, and the differences between the measurements of the same sample, showing the measurement repeatability.

Table 1 summarizes the measurement results of the curved part, and Table 2 gives a summary of measurement results of the flat part.

Table 1. Measurement results of the curved part.

Measuring force (mN)	100		200		Deformation
Coordinate (mm)	Average	St. dev.	Average	St. dev.	
Sample 1	59,9408	0,0001	59,9151	0,0001	−0,0257
Sample 2	59,9521	0,0002	59,9269	0,0003	−0,0252
Sample 3	59,9834	0,0001	59,9595	0,0003	−0,0239
All samples	59,9588	0,0191	59,9338	0,0198	−0,0250

It is obvious that larger force deforms a measured part, since measurement results with 200 mN force are in average smaller by 0,025 mm. Small standard deviation between the consequent measurements of 3 samples shows good repeatability of the measurement, and larger standard deviation of all samples shows that the parts were poorly manufactured.

Table 2. Measurement results of the flat part.

Measuring force (mN)	100				200			
Deformation (mm)	Sample 1	Sample 2	Sample 3	Average	Sample 1	Sample 2	Sample 3	Average
5	0,0027	0,0040	0,0058	0,0042	0,0059	0,0086	0,0124	0,0090
35	0,0343	0,0346	0,0441	0,0377	0,0687	0,0693	0,0883	0,0754
65	0,1707	0,1698	0,1872	0,1759	0,3415	0,3395	0,3744	0,3518
95	0,5091	0,5036	0,5209	0,5112	1,0182	1,0072	1,0418	1,0224
125	1,1362	1,1225	1,1225	1,1271	2,2723	2,2451	2,2451	2,2542
155	2,0945	2,0692	2,0473	2,0703	4,1891	4,1385	4,0946	4,1407

The 3 samples of the flat part were measured in 6 evenly distributed points. All standard deviations were smaller than the CMM measurement uncertainty (0,0018 mm). These results will later be compared with the analytical and numerical results, in order to estimate the influence of the measuring force on deformation of non rigid parts.

3 FEM Simulation

The contact between the measured non rigid part and the stylus was simulated by means of Finite Element Method (FEM), using software Solid Works 2017. The CAD model of the part and the stylus was used to define the geometry.

The boundary conditions were defined in a way that the top surface of the stylus has limited motion only in one direction – the direction of the contact force. The force acts on the same top surface, and the force intensity was 100 mN and 200 mN. The circular holes on the flat section of the part are clamped (no motion and no rotation in any direction), and the top flat surface is defined as slider, without translation perpendicular to the surface, as seen in Fig. 4.

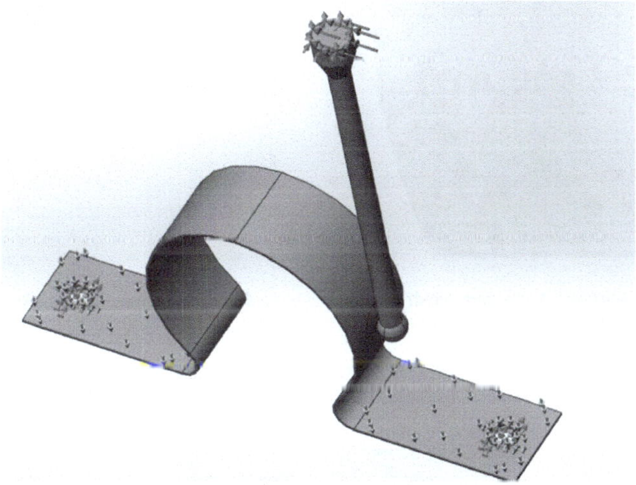

Fig. 4. The boundary conditions defined on the curved part.

The curved part was meshed with solid and shell finite elements in different variants, with same mesh size control parameters: mesh size around the contact is defined as 0,1-0,25-0,5 mm in concentric circles with diameters 0,2-5-12,5 mm (Fig. 5).

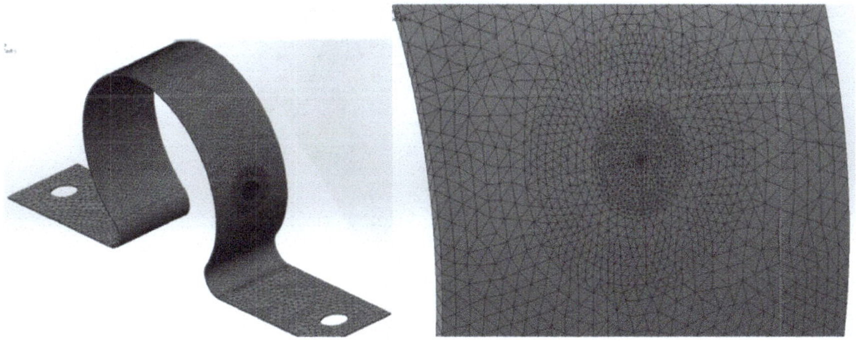

Fig. 5. The finite element mesh of the curved part.

The stylus was modelled with solid finite elements in two variants, as deformable (linear elastic) and as rigid body. The mesh in both cases had the same size parameters. Figure 6 shows the simulation where solid mesh was used in both stylus and the curved part, and Fig. 7 shows the contact simulated between the solid stylus and the shell part.

Figure 8 shows how the surface to surface contact (no penetration) was defined in the simulation.

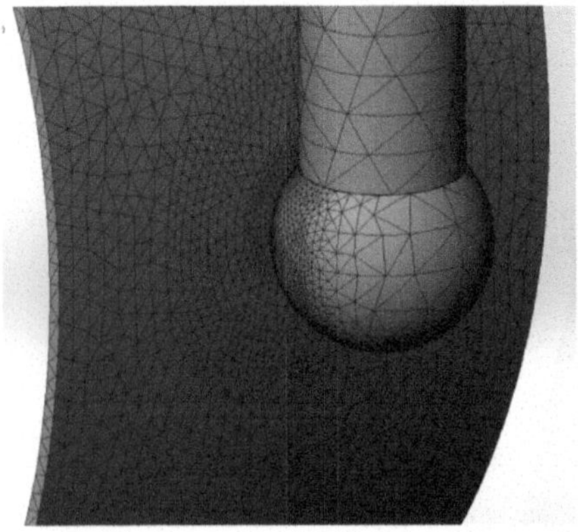

Fig. 6. Solid-solid contact mesh.

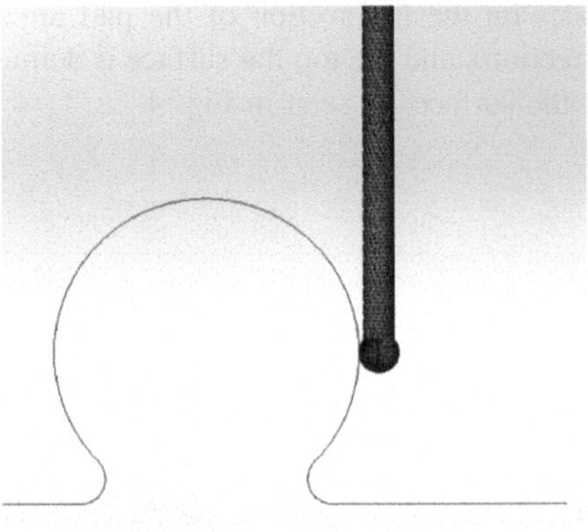

Fig. 7. Solid-shell contact mesh.

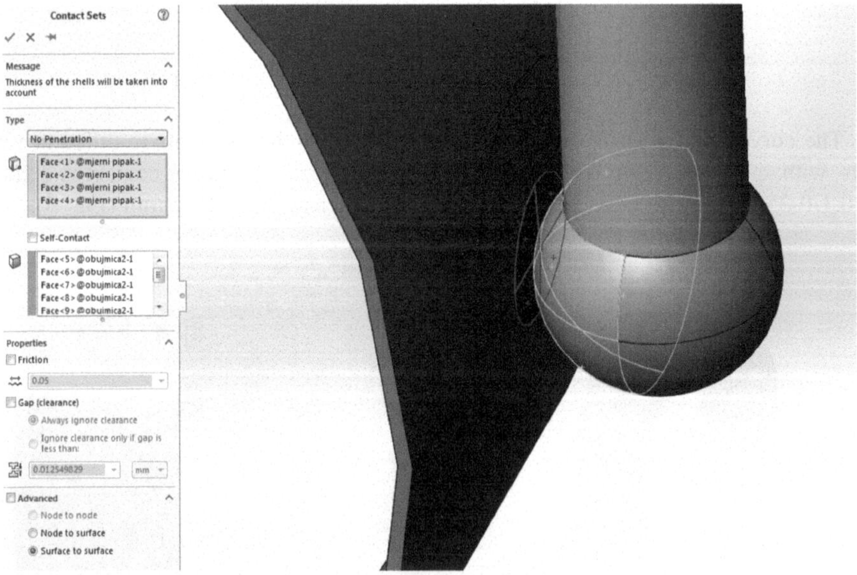

Fig. 8. The definition of the contact set.

The third model was made without a stylus, where stylus was considered rigid and thus replaced only with the contact force acting instead of the stylus, as seen in Fig. 9. Simulation results are presented in Figs. 10 and 11 and in Table 3.

Table 3 shows that the combination shell-solid gives slightly smaller results than the solid-solid and the solid-force variants. As expected, larger contact force (200 mN) induced approximately double deformation than the smaller force (100 mN).

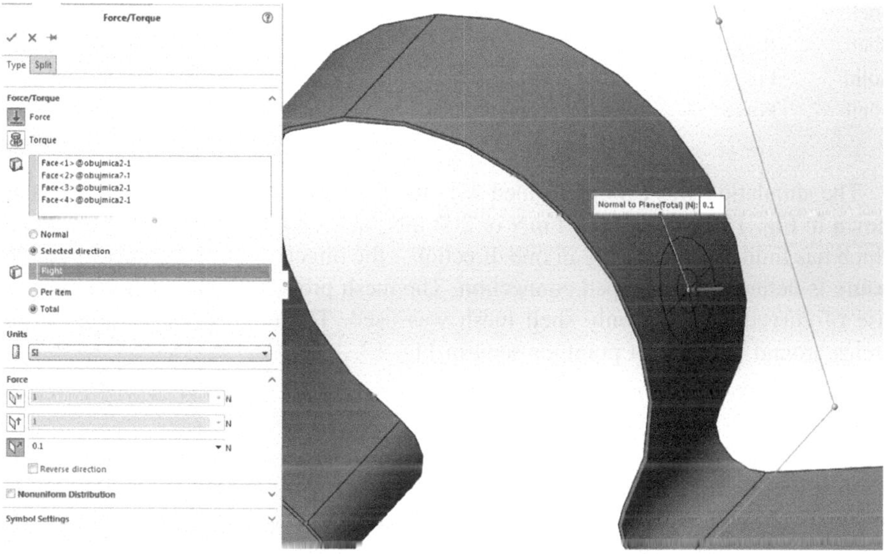

Fig. 9. The stylus replaced with a single force.

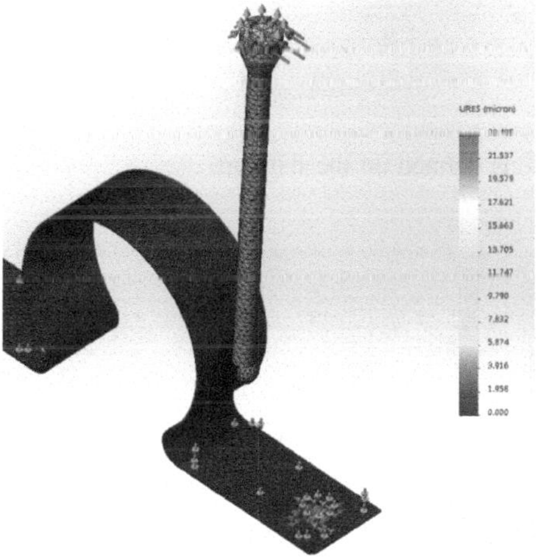

Fig. 10. Static displacement of solid-solid mesh.

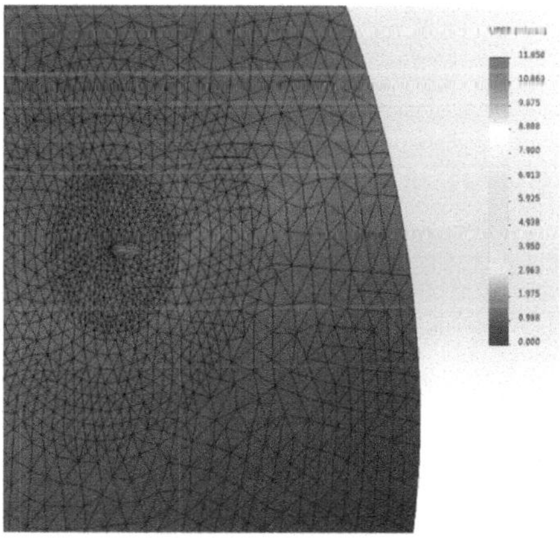

Fig. 11. Static displacement of shell mesh when stylus is represented only as a force.

Table 3. Simulation results of the curved part – displacement (mm).

Contact force (mN)		100		200	
Curved part	Stylus	Stylus tip displacement	Contact point displacement	Stylus tip displacement	Contact point displacement
Solid	Solid	0,0086	0,0087	0,0173	0,0174
Solid	Rigid	0,0086	0,0087	0,0173	0,0174
Shell	Solid	0,0085	0,0085	0,0170	0,0170
Shell	Rigid	0,0084	0,0085	0,0169	0,0170
Solid	Force	-	0,0087	-	0,0171
Shell	Force	-	0,0085	-	0,0175

The simulation was also performed with the flat part. The boundary conditions are shown in Fig. 12 The force (100 mN or 200 mN) acts on the top surface of the stylus, which has limited motion only in one direction – the direction of the contact force. The fixture is defined as a clamped connection. The mesh properties were defined as in the case of curved part, but only shell mesh was used. The mesh density varied in the circles around the contact point, as seen in Fig. 13.

Fig. 12. The boundary conditions defined on the flat part.

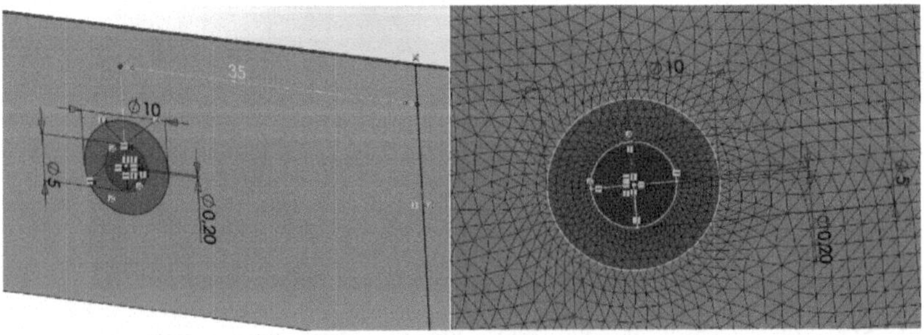

Fig. 13. The finite element mesh on the flat part.

The simulation was performed as a linear static design study, where variable was the distance between the contact point and the clamp. Figure 14 illustrates the simulation results for one case. Table 4 summarizes the simulation results, including the analytical calculation results (simple static calculation of cantilever beam) and the results of CMM measurements.

For calculation of cantilever beam, the following equation was used:

$$d = \frac{Fl^3}{3EI} \tag{1}$$

where d is the deflection (mm), F is the contact force (N), l is the distance between the clamp and the force (mm), E is Young's modulus of the material (MPa) and I is the moment of inertia (mm^4).

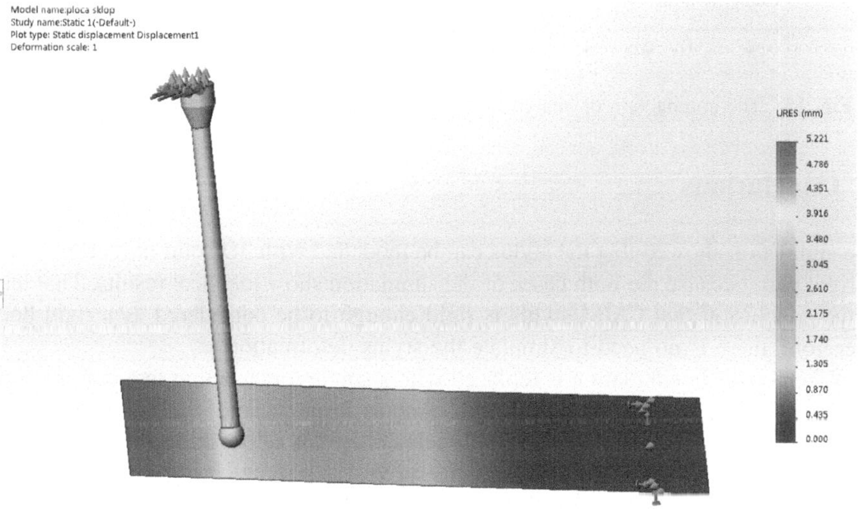

Fig. 14. The simulation results - static displacements of the flat part.

Table 4. Simulation results of the flat part – displacement (mm).

Contact force (mN)	100			200		
Deformation (mm)	Simulation	Calculation	CMM	Simulation	Calculation	CMM
5	0,001	0,000	0,004	0,001	0,000	0,009
35	0,021	0,022	0,038	0,042	0,043	0,075
65	0,134	0,138	0,176	0,268	0,277	0,352
95	0,420	0,432	0,511	0,840	0,864	1,022
125	0,979	0,984	1,127	1,957	1,968	2,254
155	1,840	1,877	2,070	3,679	3,753	4,141

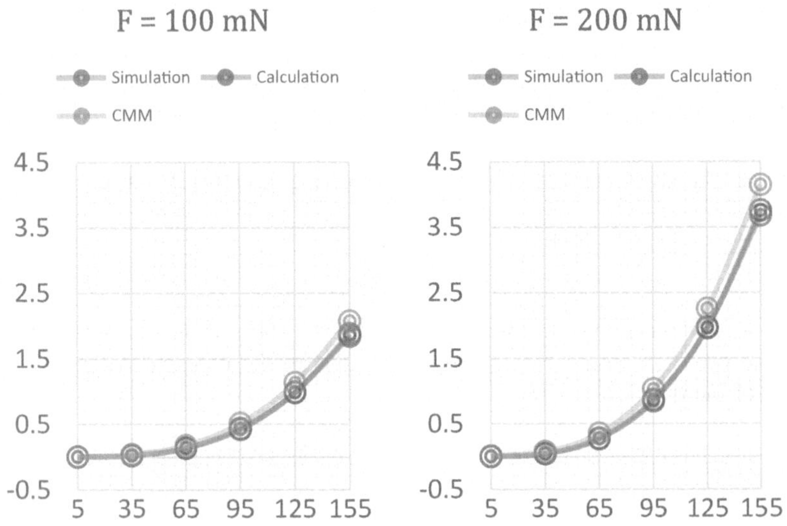

Fig. 15. The comparison of analytical calculation, simulation and CMM measurements.

4 Conclusions

The simulation showed that the stylus can be modelled with 3D solid finite elements or as rigid part, because the both cases of the simulation show identical results. That leads to the conclusion that CMM stylus is rigid enough to be considered as a rigid body. Therefore, there is no need to simulate the stylus deformation.

When stylus in simulation is replaced with a single force, acting in a contact point, the simulation results are practically the same as in the case of full 3D model of the stylus. That contributes to the previous conclusion, pointing out that stylus deformation is neglectable when nonrigid parts are measured.

Since simulation with shell finite element mesh is less time consuming than 3D solid mesh, nonrigid parts should be simulated with shell elements. These simulation results showed that the same displacement was obtained in both solid and shell elements.

When results from Table 4 are graphically represented, as seen in Fig. 15, one can conclude that the simulation corresponds very well with the analytical solution, but the experimental results differ slightly, most probably because the real material properties (modulus of elasticity) differ from the values used in simulation.

Finally, this case shows that the deformation of the nonrigid parts is not neglectable and it should be taken into account when these parts are measured by the touch-probe CMMs. The results also show that the numerical simulations can be used to compensate the part deformation in coordinate measurements because they provide relevant information about the part deformation, but only when real material properties are known.

References

1. Jakovljević, Ž., et al.: Cyber-Physical Manufacturing Systems (CPMS). In: Proceedings of 5th International Conference on Advanced Manufacturing Engineering and Technologies, pp. 199–214. Springer (2017). https://doi.org/10.1007/978-3-319-56430-2
2. Karsai, G., Sztipanovits, J.: Model-integrated development of cyber-physical systems. In: IFIP International Workshop on Software Technologies for Embedded and Ubiquitous Systems, pp. 46–54. Springer (2008). https://doi.org/10.1007/978-3-540-87785-1_5
3. Lemeš, S.: Validation of numerical simulations by digital scanning of 3D sheet metal objects. Ph.D. thesis, University of Ljubljana (2010)
4. Majstorović, V., et al.: Path planning for inspection prismatic parts on CMM as a part of cyber-physical manufacturing metrology model. Proc. Manuf. Syst. **11**(1), 3–8 (2016)
5. Deng, Y., Lu, Q., Chen, J., Chen, S., Wu, L., Tang, L.: Study on the extraction method of deformation influence factors of flexible material processing based on information entropy. Adv. Mech. Eng. (2014). https://doi.org/10.1155/2014/547947
6. Lemeš, S.: Comparison of similar injection moulded parts by a coordinate measuring machine. SN Appl. Sci. **1**, 193 (2019). https://doi.org/10.1007/s42452-019-0191-3
7. Ascione, R., Polini, W.: Measurement of nonrigid freeform surfaces by coordinate measuring machine. Int. J. Adv. Manuf. Technol. **51**, 1055–1067 (2010). https://doi.org/10.1007/s00170-010-2684-5
8. Jaramillo, A.E., Boulanger, P., Prieto, F.: On-line 3-D system for the inspection of deformable parts. Int. J. Adv. Manuf. Technol. **57**, 1053–1063 (2011). https://doi.org/10.1007/s00170-011-3332-4
9. Baldwin, J.M., Summerhays, K.D., Campbell, D.A., Henke, R.P.: Application of simulation software to coordinate measurement uncertainty evaluations. NCSLI Meas. **2**(4), 40–52 (2007). https://doi.org/10.1080/19315775.2007.11721398
10. Radvar-Esfahlan, H., Tahan, S.A.: Performance study of dimensionality reduction methods for metrology of nonrigid mechanical parts. Int. J. Metrol. Qual. Eng. **4**(3), 193–200 (2013). https://doi.org/10.1051/ijmqe/2013051
11. ASME Y14.5 Dimensioning and tolerancing (ASME Y14.5 2018). The American Society of Mechanical Engineers National Standard. The American Society of Mechanical Engineers, New York (2018)
12. ISO 1101 Geometrical product specifications (GPS) - Geometrical tolerancing - Tolerances of form, orientation, location and run-out (ISO 1101:2017). International Organization for Standardization/TC 213 Dimensional and geometrical product specifications and verification (2017)
13. Abenhaim, G.N., Desrochers, A., Tahan, A.: Nonrigid parts' specification and inspection methods: notions, challenges, and recent advancements. Int. J. Adv. Manuf. Technol. **63**(5–8), 741–752 (2012). https://doi.org/10.1007/s00170-012-3929-2

In-Liquid Laser Nanomachining by Photonic Nanojet in Optical Tweezers Configuration

Reza Aulia Rahman(✉), Tsutomu Uenohara, Yasuhiro Mizutani, and Yasuhiro Takaya

Department of Mechanical Engineering, Osaka University, Suita, Osaka, Japan
rahman@optim.mech.eng.osaka-u.ac.jp

Abstract. Parallel direct laser machining in sub-micro scale patterning at a surface of material on a large scale remains a challenging task though the laser machining has been widely applied in various applications. A photonic nanojet comes up as a promising way to solve the problem by involving near-field focusing of light waves below the surface of an introduced microsphere to fabricate structures of micro- and nanometer size. By generating laser power, the workpiece is then modified only locally at the tip of photonic nanojet that has as small as 616 nm width which provide fine machining pattern. In order to control the position of microsphere in nanomachining process, optical tweezers is introduced to the optical system. An in-liquid processing nanomachining by generating photonic nanojet in optical tweezers configuration from a laser beam is then be a subject to study by controlling the parameters investigate the viability of machining process.

Keywords: Photonic nanojet · Optical tweezers · Laser trap · Nanomachining

1 Introduction

In the rapidly growing multidisciplinary field of nanotechnology recently, many challenges exist to develop three-dimensional (3D) nanoscale materials and devices [1]. In the recent period of manufacturing nanoscale devices, many industries have been emphasized the machining method to fabricate small device, which refers to removing material from a piece of raw material using tools to shape it into intended design. Hence, in the micro/nanomachining there is a high requirement to have a high resolution, flexible, and fast technique to accommodate high volume of production. However, fabrication of nanoscale device is still quite complex and expensive. In order to meet those three requirements, a laser beam, by virtue of its characteristics such as monochromatic, small size, high collimation, and high intensity, is become capable to do rapid replication of nanoscale pattern and structures with a high degree of robustness and high throughput in complex shapes product [1].

In the conventional laser machining, a focusing lens is normally used to focusing the laser beam into the raw material or workpiece. However, there is a constraint of working distance or depth of focus of this conventional laser machining, the depth of focus is relatively small and if we missed this focusing spot, it will reduce the resolution of the product. In order to do nanomachining, one important point is the ability

V. D. Majstorovic and N. Durakbasa (Eds.): IMEKOTC14 2019, LNME, pp. 318–328, 2019.
https://doi.org/10.1007/978-3-030-18177-2_28

to obtain a small beam diameter, it usually done by applying higher Numerical Aperture objective lens. There are also limitations of this objective lens so we need to invent another method to obtain nanoscale laser beam.

Photonic nanojet (PNJ) has emerged as a promising solution to tackle the problem. A PNJ involves near-field focusing of light waves, with waists in nanometer scale, in the below surface of a monolayer of a particle to fabricate structures of micro- and nanometer size by reaching the fluence, energy per area, above the machining threshold of the material to work with [2] as its schematic diagram is shown in Fig. 1. These photonic nanojet-based nanomachining method presents new opportunities for low-cost and high throughput fabrication of nanoscale device in the future. Due to its unique characteristics, PNJs are believed to have potential applications in nanoparticle sensing; sub-wavelength resolution direct-write nanopatterning and nanolithography; broadband low-loss waveguiding; ultramicroscopy; and ultrahigh-density optical data storage [3].

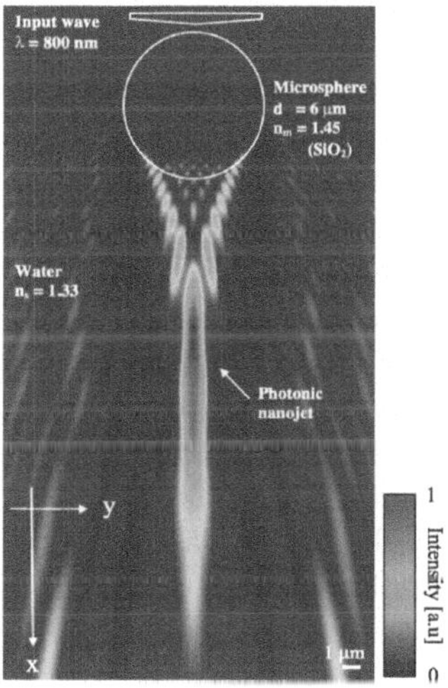

Fig. 1. PNJ intensity distribution along *x-y* plane

In this research, we introducing the use of the liquid as a medium in the machining process. In this process, work material is ablated by laser beam while it is submerged in water. Water is typically used since it is cheap, harmless, and recyclable. The under-water laser ablation method is by far the simplest approach since it require less transformation of the laser and optic system. During the ablation process, water or liquid can cool down the workpiece temperature and also carry away the cut debris, thus reducing the size of Heat Affected Zone (HAZ) and preventing the deposition of removed material [4]. By applying thin water layer on the surface of workpiece, pattern can be fabricated more efficiently, faster, and with a better quality than in air atmosphere. Moreover, the laser trapping, which will be introduced later in this contribution, becomes easier due to higher value of trapping force that affect on less complicated laser and optical system.

2 Photonic Nanojet Investigation

The PNJ is a fine high-intensity light beam that generated at the backside of a transparent dielectric microcylinder, microdisk, or microsphere of diameter greater than the illuminating wavelength, which in this research focused on dielectric microsphere, that irradiated by a laser light. A PNJ has a small localized beam diameter of several hundred nanometers. Due to its small beam diameter, the intensity is increased from the incident laser light. Furthermore, a PNJ could propagate at a long distance with low divergence, over several wavelengths, while retaining its high intensity. A PNJ then could be used for laser fine machining owing to high intensity and small diameter of its beam.

To evaluate the performance of the PNJ, we employed high resolution three dimensional finite-difference time-domain (FDTD) computational electrodynamics modelling technique to analyze and execute numerical solution by elaborating Maxwell equation in time domain from Mie scattering theory [2]. This FDTD gives broadband output from a single execution of the program. As the number of unknown variables in photonic nanojet study increases, FDTD quickly outpaces other method in terms of efficiency. A Perfectly matched layer (PML) absorbing boundary condition is applied in FDTD simulations to efficiently terminate the outer boundary of the computational region of interest. Then, we conducted several simulations based on real experimental parameter to explore PNJ intensity distribution and gain insight into physical mechanism of the PNJ and its influence on the final machining results.

In order to characterize the PNJ, five important parameters of PNJ for laser nanomachining are the beam diameter; the distance from microsphere surface to peak intensity position, namely working distance; depth of focus; the intensity enhancement ratio; and fluence as it shown in Fig. 2. The intensity distribution of PNJ is then could be controlled by changing the illuminating laser light parameters, the morphology of the investigated microsphere, surrounding medium of the microsphere. Complete parameters that need to be controlled for each part is showed in Table 1.

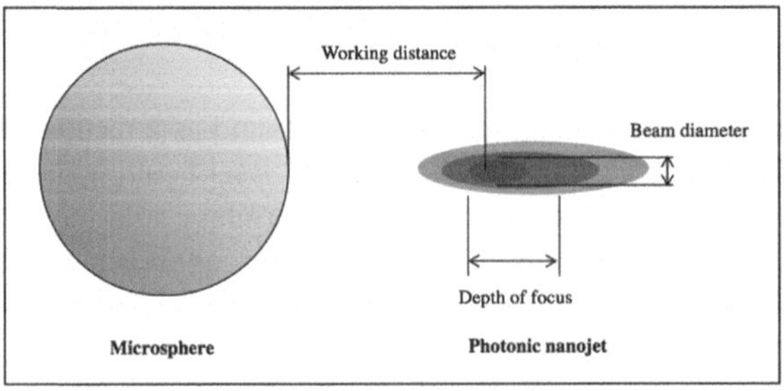

Fig. 2. Schematic picture of generated PNJ in FDTD simulation

In this contribution, we investigate the effect of different size of microsphere, different material of microsphere, and different liquid. As for all simulations, we used 800 nm fs laser for machining with 1 W power is going to be utilized just before hit the

Table 1. Parameter of each component in PNJ

Component	Parameter
Laser	Wavelength (λ_{inp})
	Beam profile
	Polarization direction
	Incident angle
Microsphere	Shape
	Size
	Refractive index (n_m)
Surrounding	Refractive index (n_s)

microsphere. This 800-nm fs laser has Gaussian beam profile with 16 μm of Full Width at Half Maximum (FWHM) beam waist with linear polarization. At first, we tried to investigate PNJ with mostly available in the market5, 8, and 10 μm of diameter (d_m) of silicon dioxide (SiO_2) microsphere used that submerged in the water ambient. As the result compared side by side in Fig. 3. As shown in Fig. 3a, the PNJ generated by using 5 μm microsphere has 789 nm of beam diameter with 4.50 μm of working distance calculated from the backside surface of the microsphere. In the other hand, by using bigger diameter of microsphere, 10 μm, we could have 936 nm of beam diameter and 10.96 μm of working distance, while 8 μm diameter of microsphere produced beam diameter and working distance in the middle of it. Those result is accumulated in graph shown in Fig. 4 for beam diameter and Fig. 5 for working distance and the tendency is adequate for bigger or smaller size of microsphere. Intensity distribution are different depends on the size of the microsphere.

In the case of different material of microsphere, it will be resulting a different value of refractive index (n_m). In this simulation, we employ 8 μm silicon dioxide and polystyrene microsphere with 1.45 and 1.55 of refractive index value respectively. As previous simulation, we used same parameter of laser which these microspheres are going to submerged in the water. The result shown in Fig. 6 indicated very distinguish result in terms of PNJ depth of focus. By using SiO_2 microsphere, shown in Fig. 6a, we could acquire 863 nm of beam diameter, while it is 616 nm by using polystyrene microsphere. In terms of working distance, the SiO_2 microsphere generated at 8.12 μm from the backside of microsphere while polystyrene only 4.30 μm as shown in Fig. 6b. SiO_2 microsphere producing 8.09 μm PNJ depth of focus in z direction, while the polystyrene is nearly half of it, 4.38 μm. From this simulation, one could refer that by applying bigger refractive index value of microsphere, small beam diameter will be obtained. In the case of applying those PNJ in silicon that has 200 mJ/cm^2, PNJ from polystyrene microsphere produced smaller ablation area and confirming that we could do the nanomachining by using this PNJ technique.

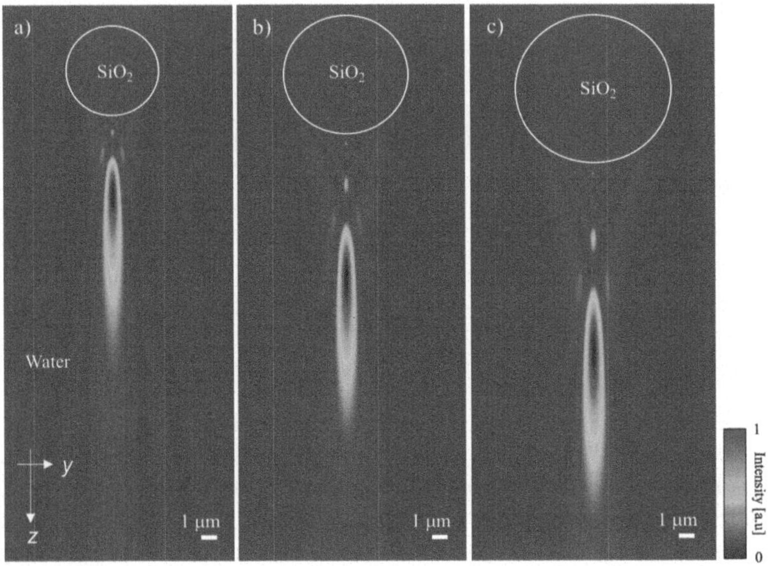

Fig. 3. PNJ intensity distribution along *y-z* plane calculate in 3D FDTD simulation with different diameter of microsphere (a) 5 μm, (b) 8 μm, (c) 10 μm

Regarding different surrounding medium, it also resulting a different value of refractive index (n_s). In this case, we generate PNJ from 8 μm Polystyrene as it is submerged in water, 1.33, and immersion oil, 1.51 with same parameter laser as previous are going to be used. From the 3D FDTD result shown in Fig. 8, interestingly, the result show much different rather than previous parameters simulated. In the case of water as shown in Fig. 7a, PNJ has 616 nm of beam diameter, 4.30 μm of working distance while in immersion oil, it has 2175 nm of beam diameter and 26.62 μm of working distance which are quite big and we could not acquire sub-micron machining. From here, surrounding medium play a big part in PNJ-based nanomachining technique. To acquire smaller beam diameter, one must apply smaller refractive index value surrounding medium. Furthermore, by immersed the sample and so-called microsphere in liquid, PNJ is known to have much longer propagating distance than in air. The implication of this generated PNJ for efficient surface patterning, the sample needs to be kept in near-field contact with particles in air surrounding medium. In the other hand, laser machining by PNJ in water-immersed sample might be easier to control in which does not need to place the sample very close with microsphere. By having longer working distance.

By utilizing three-dimensional (3D) FDTD, we could optimize the parameter of each component in PNJ to obtain nanometer scale beam diameter by optimizing each component in PNJ listed is Table 1 as follows:

1. Smaller diameter of microsphere, in case of sphere shape;
2. Higher refractive index value of microsphere;
3. Lower refractive index value of surrounding medium;
4. Higher ratio between the refractive index value of microsphere and surrounding medium

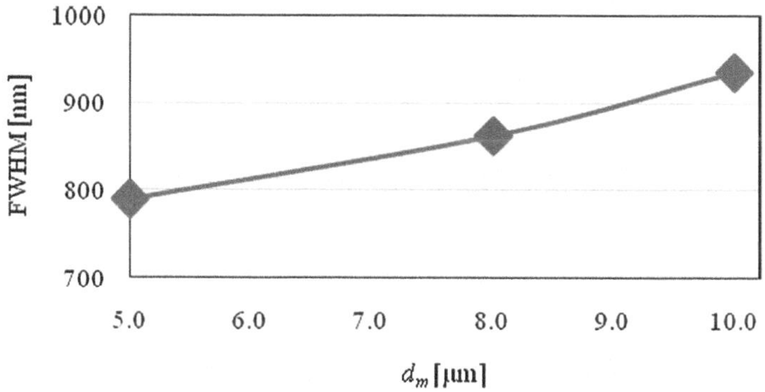

Fig. 4. Resulting PNJ beam diameter with different diameter of microsphere

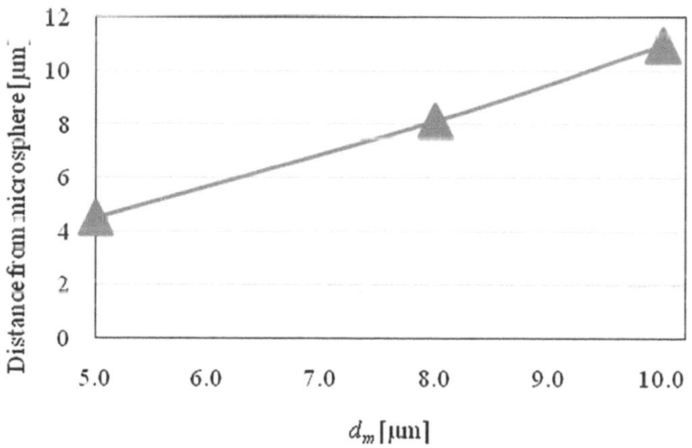

Fig. 5. Resulting PNJ working distance with different diameter of microsphere

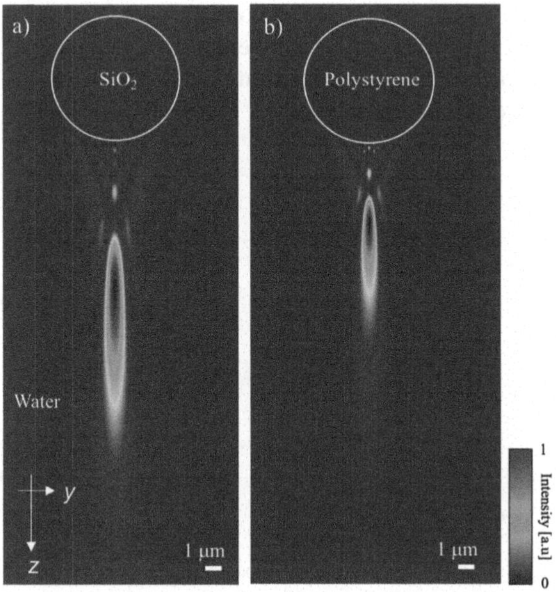

Fig. 6. PNJ intensity distribution along *y-z* plane calculate in 3D FDTD simulation with different material of microsphere (a) silicon dioxide, (b) polystyrene

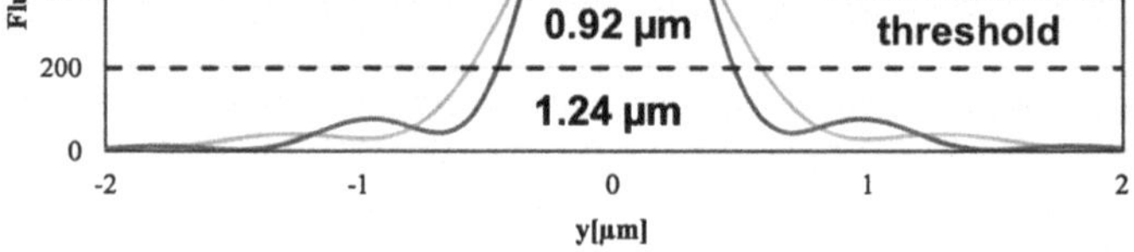

Fig. 7. Resulting ablation area in the silicon with different material of microsphere

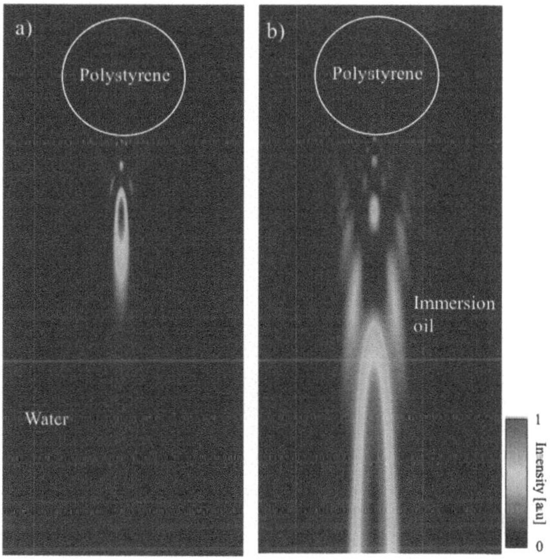

Fig. 8. PNJ intensity distribution along *y-z* plane calculate in 3D FDTD simulation with different surrounding medium (a) water, (b) immersion oil

3 Laser Trapping

In order to control the machining position to do the pattern we intended to fabricate, we introduced optical tweezers technique. Optical tweezers is a technique to trap and manipulate microscopic particle by a focused laser beam with high numerical aperture regime. Light or more specifically in this research laser beam, composed of particles called photons and carries energy and momentum with it. This laser beam also tends to apply a force, called radiation pressure, on a particle along the direction of beam propagation. When the light or photon impact an object, there is a momentum change due to reflection and refraction at the object boundary surface and generating force inside object, there is scattering force that moves away from the object toward the propagation direction of the light beam, and also gradient force which also called as trapping force which its schematic diagram of laser trapping mechanism is shown in Fig. 9. This trapping force essential to do the optical trapping and we could control this trapping force by adjusting the laser power. By immersing the sample in liquid condition, it is known that trapping becomes easier due to smaller value of gradient force compared in air medium. Furthermore, it will increase the surface quality of the product with microcrack and debris are efficiently carried away. And it is important to control the quality of trapping to have stiff and stable trapped microsphere.

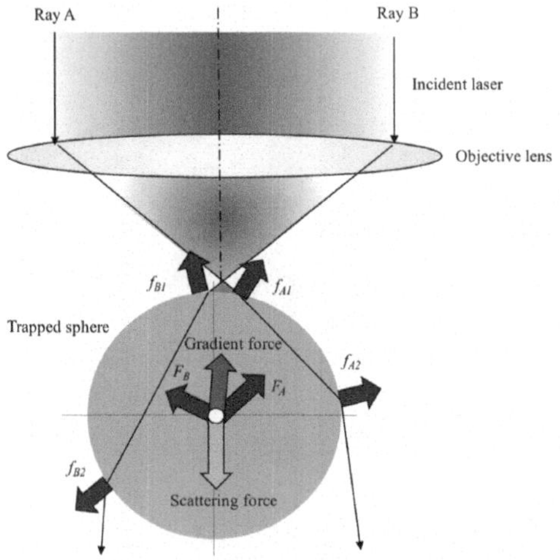

Fig. 9. Schematic diagram of laser trapping system

For most of the lasers used in practical, when the beam is passing through the objective lens, the focal spot of trapping beam is in the order of micron. In this intensity gradient near focal point, microsphere is subjected to a restoring force toward the focal point like a spring. Since the spring constant is small and in the order of 10^{-5} N/m [5], the trapped sphere becomes sensitive to external force. As an initial state, a trapped sphere retains a stable position under application of trapping force that is the resulting force of all radiation pressure by focused laser beam. If some external force, i.e. moving of input laser beam, applied to the trapped sphere, the balance is broken dynamically and the probe begins to shift. At this moment, the trapping force changes with the radiation pressure distribution depending on the illumination condition for the trapped sphere position. When an external force is released, the trapped sphere is accelerated in the direction of the original position and is finally recovered to the initial stable position precisely.

A novel experiment was carried out with such optical setup is shown in Fig. 10. With this experiment, 8 μm SiO_2 particle could be trapped easily by adjusting the power of YAG laser with 1064 nm of wavelength on 0.73 objective lens which is classified as a high Numerical Aperture (NA) lens. As it is explained before, the gradient or trapping force and the quality of trapping is important to do the optical trapping and by adjusting the laser power, we could have bigger trapping force which means it is easy to do the trapping.

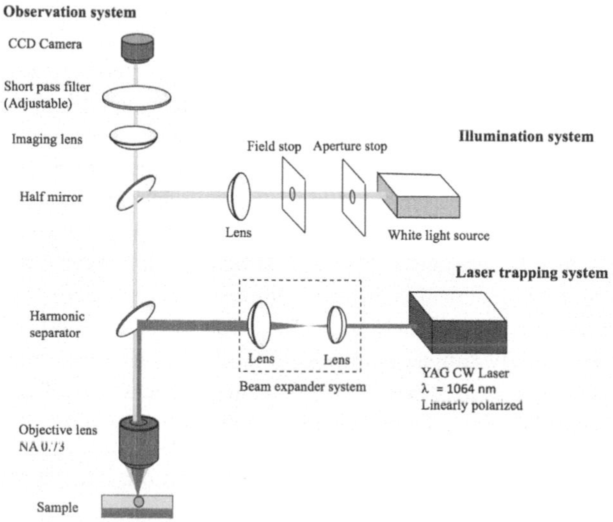

Fig. 10. Optical setup for laser trapping

In this experiment, we realized trapping of 8 μm SiO_2 microsphere in water medium. Firstly, we employ a very small amount of power, 10 mW, we could not do the trapping. As we higher up the laser power into 730 mW, the bead is then to more easily to manipulate that we could move it as we intended to while the other microsphere which not in the focal point of the laser is in idle state as it shown in Fig. 11. This simple experiment showed that we could realize the complicated and fast nanomachining technique by this configuration.

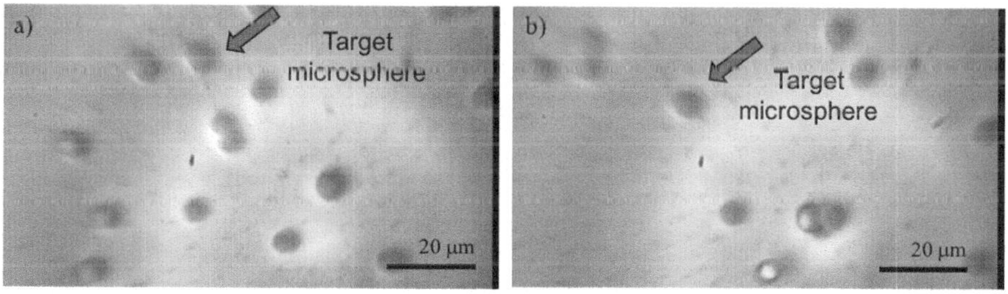

Fig. 11. Observation result by applying 730 mW laser power, (a) before, (b) after moving the microsphere

4 Summary

We already showed that a novel laser nanomachining method as small as 616 nm patterning size could be realized by combining photonic nanojet and laser trapping technique. We could also control the ablated region in the material by controlling the PNJ intensity distribution from their parameters such as microsphere and surrounding medium. For these novel research, the optical trapping sphere acts as an ideal framework to employ photonic nanojet. An understanding between the interaction of an optically trapped microsphere and its PNJ field structure is a fundamental importance in optical physics and has practical significance in another application such as imaging, nanolithography, detection, and metrology. By combining PNJs and optical tweezers, they appear to particularly simple, convenient, and highly useful to implement for the patterning of very large areas when micro- or nanoscale amounts of material must be processed in a controlled way. To enhance the research, we would like to measure the microsphere position in more detail and do the nanomachining on silicon wafer. Then we also introducing Electron-enhancement Raman Spectroscopy (EERS) system to observe machining process in atomically as it is a new concept in machining terms.

Acknowledgement. This work was supported by JICA Innovative Asia Program Grant Number D1707479.

References

1. Ali, M., Wagner, T., Shakoor, M., Molian, P.A.: Review of laser nanomachining. J. Laser Appl. **20**(3), 169–184 (2008)
2. Chen, Z., Taflove, A.: Photonic nanojet enhancement of backscattering of light by nanoparticles: a potential novel visible-light ultramicroscopy technique. Opt. Express **12**(7), 1214–1220 (2004)
3. Heifetz, A., Kong, S., Sahakian, A.V., Taflove, A., Backman, V.: Photonic nanojets. J. Comput. Theor. Nanosci. **6**(9), 1979–1992 (2009)
4. Charee, W., Tangwarodomnukun, V., Dumkum, C.: Ultrasonic-assisted underwater laser micromachining of silicon. J. Mater. Process. Technol. **231**, 209–220 (2016)
5. Michihata, M., Takaya, Y., Hayashi, T.: Development of the nano-probe system based on the laser trapping technique. CIRP Ann. **57**(1), 493–496 (2006)

Correction to: Practical Aspects in the Application of Geometrical Product Specifications and Verification (GPS) in the Micro and Nano-Scale Manufacturing

N. M. Durakbasa and G. Poszvek

Correction to:
Chapter "Practical Aspects in the Application of Geometrical Product Specifications and Verification (GPS) in the Micro and Nano-Scale Manufacturing" in: V. D. Majstorovic and N. Durakbasa (Eds.): *Proceedings of the 12th International Conference on Measurement and Quality Control - Cyber Physical Issue*, LNME,
https://doi.org/10.1007/978-3-030-18177-2_21

In the original version of this chapter, the authors had reproduced five verbatim excerpts from References [5], [7] and [9]. Though these references were properly mentioned in chapter-end references' list, they were reproduced without quotation marks and without proper citations.

Now, the chapter has been updated with proper acknowledgement of the missing citations.

The updated version of this chapter can be found at
https://doi.org/10.1007/978-3-030-18177-2_21

V. D. Majstorovic and N. Durakbasa (Eds.): IMEKOTC14 2019, LNME, p. C1, 2019.
https://doi.org/10.1007/978-3-030-18177-2_29

Author Index

V. D. Majstorovic and N. Durakbasa (Eds.): IMEKOTC14 2019, LNME, pp. 329–330, 2019.
https://doi.org/10.1007/978-3-030-18177-2

The manufacturer's authorised representative in the EU is Springer Nature Customer Service Centre GmbH, Europaplatz 3, 69115 Heidelberg, Germany. If you have any concerns regarding our products, please contact ProductSafety@springernature.com

Printed and bound by CPI Group (UK) Ltd, Croydon, CR0 4YY
05/10/2026
02228135-0015